Computational Fluid Dynamics

ABOUT THE AUTHORS

Imane Khalil (San Diego, CA) is an associate professor of mechanical engineering at the University of San Diego and the associate dean of graduate programs. She was born in Beirut, Lebanon and immigrated to the United States in 1989. Imane earned her PhD degree in mechanical engineering from the University of California at San Diego in 2003. She worked at Sandia National Laboratories between 2004 and 2014, first as a scientist and then as a manager where she managed different departments including the one that performed structural analysis for *Curiosity*, the rover that landed on Mars in 2012. Imane joined the University of San Diego in 2014 as an assistant professor of mechanical engineering. Her research interests are in numerical methods applied to solid and fluid mechanics, thermal hydraulics, reactor safety, and uncertainty quantification applications. Imane is a Fellow member of the American Society of Mechanical Engineers.

Issam Lakkis (Beirut, Lebanon) is professor and chair of the Mechanical Engineering Department at the American University of Beirut (AUB), Beirut, Lebanon. He graduated from AUB with a BE and ME in mechanical engineering in 1991 and 1993 respectively. He then joined the reacting gas dynamics lab at MIT in 1994 and earned his PhD degree in 2000. From 2000 until 2003, he worked at Coventor on computer-aided design of MEMS and RF circuits. In 2003, he joined AUB and has been a professor since 2017. His research interests span species transport in stochastic fields, with applications to pollution transport in the ocean, atmosphere, and urban environments; development of grid-free computational methods for continuum and non-continuum flows; and modeling, design, analysis and simulation of multi-scale/multi-physics micro-devices.

Computational Fluid Dynamics

An Introduction to Modeling and Applications

Imane Khalil and Issam Lakkis

With Contributions by Maram Ammar

New York Chicago San Francisco
Athens London Madrid
Mexico City Milan New Delhi
Singapore Sydney Toronto

Computational Fluid Dynamics: An Introduction to Modeling and Applications

1 2 3 4 5 6 7 8 9   LCR   28 27 26 25 24 23

Library of Congress Control Number is on file with the Library of Congress

ISBN 978-1-264-27494-9
MHID 1-264-27494-7

Sponsoring Editor
  Lara Zoble

Editorial Supervisor
  Janet Walden

Project Manager
  Alok Singh, MPS Limited

Acquisitions Coordinator
  Elizabeth M. Houde

Copy Editor
  Subashree Baskaran

Proofreader
  Raama

Indexer
  Ariel Tuplano

Production Supervisor
  Lynn M. Messina

Composition
  MPS Limited

Illustration
  MPS Limited

Art Director, Cover
  Jeff Weeks

Contents

Preface

This book is not a traditional computational fluid dynamics (CFD) textbook. While current CFD books explain and derive the numerical methods used to solve fluid dynamics problems, programming these methods on computers is a tedious process for undergraduate engineering students who are primarily interested in using computational modeling software to solve engineering problems without "re-inventing the wheel." Moreover, the engineering community is currently witnessing a rise in computing resources, including CFD softwares that are accessible by undergraduate students to solve problems of different levels of complexity. Hence, the purpose of this book is to provide a resource for undergraduate engineers who aim to learn how to use CFD software, and understand the basics of CFD at a level that is both technical and practical. It is intended to be used as a textbook when offering a CFD course for undergraduate engineering students.

This book informs students from all engineering disciplines to use Fluent as a tool to model fluid flow and heat transfer for different engineering applications. The Fluent tutorial in each chapter of the book contains extensive details such that students do not need to consult any additional resources. The tutorials were created and have been used for the past five years in the technical elective CFD course taught by Dr. Imane Khalil at the University of San Diego. Nevertheless the book also contains a concise yet complete overview of fluid dynamics laws and CFD theory, culminating from an extensive teaching experience at the American University of Beirut. It begins with an introduction to the physical laws governing the conservation of mass, momentum, and energy in thermal fluids. After explaining how to numerically solve conservation equations using the finite volume method, the book uses Fluent to demonstrate how these computational methods solve laminar and turbulent flows involving heat transfer.

Acknowledgments

The authors wish to acknowledge the following individuals for their contributions.

Maram Ammar, PhD student at the American University of Beirut working under the supervision of Dr. Lakkis. Her contribution, dedication, and commitment to writing the initial draft were invaluable.

Maher Salloum, principal member of technical staff at Sandia National Laboratories who reviewed, edited, and contributed to every chapter.

Terrie San Giorgio, technical editor who reviewed the full manuscript.

Quinn Pratt, former student who worked under the supervision of Dr. Khalil and is currently a PhD student at the University of California Los Angeles. Quinn reviewed the theoretical chapters and provided excellent feedback.

Natalie Khalil, Ahmad Arakji, Peter Soudah, and Elie Tannoury, engineering students at the University of Southern California and the American University of Beirut who tested the tutorials in this book and provided feedback on how to improve the instructions.

Gregory Wagner, professor of mechanical engineering at Northwestern University who provided guidance in writing the solution verification sections.

Dr. David Miller, Imane Khalil's graduate advisor at the University of California at San Diego who guided her through her undergraduate, master's, and doctorate degrees and has continued to mentor her in every step in her career.

All of Dr. Khalil's students at the University of San Diego who used the tutorials during the last few years and provided extensive feedback to improve them.

CHAPTER 1

Conservation Laws in Thermal-Fluid Sciences

List of Symbols	
Mass	m
Volume	$\mathcal{V}$
Infinitesimal area	$d\mathcal{S}$
Control surface	$\mathcal{S}$
Density	ρ
Time	t
Unit vector normal to the control surface	$\hat{n}$
Velocity vector	$\vec{u}$
Momentum	$\vec{\mathcal{M}}$
Characteristic length scale	L
Body force per unit mass	$\vec{f}_b$
Gravitational acceleration	g
Stress vector	$\vec{\sigma}$
Pressure	p
Temperature	T
Force	$\vec{F}$
Rate of heat transfer into the control mass or control volume	$\dot{Q}^{\leftarrow}$
Rate of work done on a control mass or control volume	$\dot{W}^{\leftarrow}$
Total energy	E
Internal energy	U
Kinetic energy	KE
Potential energy	PE
Specific total energy	e
Thermal conductivity	λ
Heat flux	$\vec{q}$
Coordinate in the normal direction	n
Rate of work	$\dot{W}$

Continued

List of Symbols	
Physical property, intensive macroscopic property carried by mass	ϕ
Stress tensor	σ
Identity matrix	I
Strain	ϵ
Coefficient of dynamic viscosity	μ
Constant volume specific heat	c_v
Reynolds number	Re
Knudsen number	Kn
Microscopic length scale	l_m
Speed of sound	a_s
Mach number	Ma
Diffusion coefficient	Γ
Source term	$\dot{S}$
Constant pressure specific heat	c_p
Flux	$\vec{j}$

In engineering applications, conservation laws are often used to study a physical process. The subject of study is also dictated by the engineering application itself. For example, when studying thermodynamic processes undergone by fluid in a sealed cylinder equipped with a movable piston, the subject of study is the trapped fluid as it undergoes processes that change the volume, pressure, and/or temperature. This is an example of a control mass, shown in Figure 1.1 (left). As opposed to a control mass, which is made up of matter of fixed identity, a control volume is a volume that commonly encases the subject of study, while allowing bulk flow to cross its boundary. An example

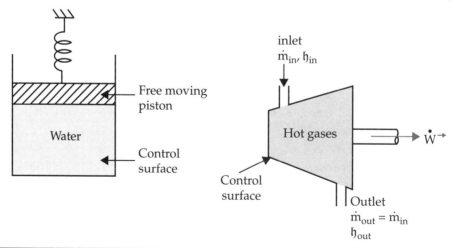

FIGURE 1.1 In the setup on the left, the piston-cylinder arrangement is perfectly sealed so that there is no mass exchange with the surrounding. The water entrapped inside is a control mass. In the setup on the right, the fixed control surface encases the turbine and cuts through the inlet, outlet, and shaft. This is an example of a control volume that exchanges mass with the environment through an inlet and an outlet.

is a control volume encasing a turbine with bulk flow crossing the boundary in and out of the control volume through the inlet and outlet of the turbine, as depicted in Figure 1.1 (right). The mass flowing in and out of the control volume carries with it momentum and energy. The boundary of the control mass or control volume is often referred as the control surface. Expressions for the conservation of mass, momentum, and energy for a control volume must capture what occurs over the control surface by bulk flow. In addition, bulk flow transfers work energy as it enters a control volume and alternatively, the control volume transfers work to the bulk flow that leaves the control volume. These bulk flow work interactions must also be included in the energy balance of a control volume.

Applying the conservation laws on a finite control mass or control volume yields the integral form of these laws. The integral form of a conservation law of a physical property relates the rate of change of the total amount of this physical property contained in the domain of interest to transfers of the property, by various physical interactions, across the boundary. Integral forms are useful when one is not after the knowledge of the detailed distributions of the thermophysical properties in space and time, but only interested in integral quantities. These may be volume quantities or surface quantities. Examples of volume quantities include the total mass contained in a control volume, total momentum of a control mass, or total internal energy in a control volume. Examples of surface quantities include the mass flow rate across the inlet to a control volume, the total shear forces acting on the surface of a control mass, or the rate of heat transfer by conduction across the control surface. When applied to an infinitesimal control mass or control volume, expressions of the conservation laws are in the form of partial differential equations (PDEs) whose solution yields the fields of the physical properties in space and time, i.e., $\phi(x, y, z, t) \in \{\rho, p, T, \vec{u}\}$, where ρ, p, T, and $\vec{u}$ are density, pressure, temperature, and velocity vector, respectively. These partial differential equations are generally nonlinear and coupled, and their solution requires knowledge of the initial and boundary conditions at all times. Naturally, solving this system of coupled nonlinear PDEs is more challenging than finding integral quantities using the integral forms, but it yields a detailed and complete knowledge of properties in space and time. Computational fluid dynamics is about predicting the evolution of the fields of properties $\phi(x, y, z, t) \in \{\rho, p, T, \vec{u}\}$ by solving the system of coupled PDEs that govern the conservation of mass, momentum, and energy numerically on a discrete representation of the domain called the grid or mesh.

We next present expressions for the conservation of mass, momentum, and energy for a control mass and a control volume, both in the integral as well as the differential form defined in Section 1.2.

1.1   Conservation Laws in Integral Form

1.1.1   Conservation of Mass

Control Mass

Since a control mass is made up of the same molecules at all times, its mass, m_{CM}, does not change. The conservation of mass of a control mass is then expressed as

$$\frac{\partial m_{CM}}{\partial t} = \frac{\partial}{\partial t} \int_{V_{CM}} \rho \, dV = 0 \tag{1.1}$$

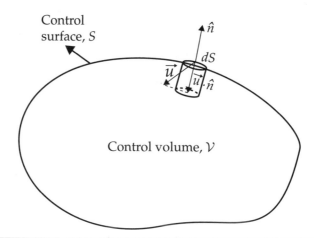

FIGURE 1.2 Control volume for deriving the equations in integral form.

where the mass of the control mass is the volume integral of the density, ρ, over the control mass volume, $\mathcal{V}_{CM}$. Note that in this case, we can replace the partial derivative $\partial/\partial t$ with the total derivative d/dt since the mass does not change in space.[1]

Control Volume

Consider the control volume, $\mathcal{V}$, shown in Figure 1.2. The conservation of mass for control volume states that the rate of increase of the mass of the control volume is balanced by the net mass flow rate into the control volume across its control surface, S, that is,

$$\frac{\partial m_{CV}}{\partial t} = \dot{m}_{in} - \dot{m}_{out} \tag{1.2}$$

In this book, we will use the notation $()_{in}$ and $()^{\leftarrow}$ interchangeably to refer to quantities flowing into a control mass or a control volume across the control surface. Similarly, we will use the notation $()_{out}$ and $()^{\rightarrow}$ interchangeably for out-flowing quantities.

The mass of fluid in the control volume, m_{CV}, is the integral of the density over the control volume. Denoting $\hat{n}$ as the unit vector normal to the control surface at dS and pointing away from the control volume, the mass leaving the control surface through dS per unit time is the mass of the volume swept per unit time by the velocity projected along $\hat{n}$, that is, $\rho\left((\vec{u}_r \cdot \hat{n})\,dS\right)$, where $\vec{u}_r$ is the flow velocity relative to the control surface, at the center of dS, that is, $\vec{u}_r = \vec{u} - \vec{u}_S$, where $\vec{u}$ and $\vec{u}_S$ are respectively the velocity of the fluid and the velocity of the control surface at the center of dS, measured in the same frame of reference. Dividing by dS yields the net mass flow rate into the control volume across its boundary to be equal to the negative of the integral over the control surface of the mass flux, $\rho\vec{u}_r \cdot \hat{n}$. Then, the conservation of mass for a control volume can be expressed as

$$\frac{\partial}{\partial t}\int_{\mathcal{V}} \rho\,d\mathcal{V} + \int_{S} \rho\vec{u}_r \cdot \hat{n}\,dS = 0 \tag{1.3}$$

---

[1]The partial derivative of a multivariable function is a derivative with respect to one variable with the other variables held constant. The total derivative is the derivative of a multivariable function where all the variables are allowed to vary. If the function depends on only one variable, the partial and total derivatives are equivalent.

1.1.2 Conservation of Momentum

Control Mass

The conservation of linear momentum states that the sum of forces, $\sum \vec{F}$, acting on the control mass is balanced by the rate of change of its momentum, $\vec{\mathcal{M}}_{CM}$:

$$\sum \vec{F} = \frac{d\vec{\mathcal{M}}_{CM}}{dt} = \frac{d}{dt} \int_{\mathcal{V}_{CM}} \rho \vec{u} \, d\mathcal{V} \tag{1.4}$$

The total momentum, $\vec{\mathcal{M}}_{CM}$, of the control mass is the sum of the momenta of very small masses that make up the control mass. Each small mass, δm_i, is chosen to be sufficiently small such that the density and velocity may be assumed uniform over its volume, $\delta \mathcal{V}_i$, in which case $\delta m_i = \rho_i \delta \mathcal{V}_i$. Then, $\vec{\mathcal{M}}_{CM} = \sum_i \delta \vec{\mathcal{M}}_{CM,i} = \sum_i \vec{u}_i \delta m_i = \sum_i \rho_i \vec{u}_i \delta \mathcal{V}_i$. In the limit $\delta m_i \to dm$ and $\delta \mathcal{V}_i \to d\mathcal{V}$, then $\vec{\mathcal{M}}_{CM} = \int_{\mathcal{V}_{CM}} \rho \vec{u} \, d\mathcal{V}$.

In fluids, the forces acting on the control mass (or control volume) may be categorized as body forces, surface forces, and line forces. A body force is a force that is proportional to the volume of the control mass (volume), i.e., it scales as L^3, where L is a characteristic length scale. The body force can be expressed as

$$\vec{F}_b = \int_{\mathcal{V}_{CM}} \rho \vec{f}_b \, d\mathcal{V} \tag{1.5}$$

where $\vec{f}_b$ is the body force per unit mass. The most common body force is the one due to gravity, i.e., the weight is $\vec{F}_g = \int_{\mathcal{V}_{CM}} \rho \vec{g} \, d\mathcal{V}$, for constant gravity $\vec{g}$, $\vec{F}_g = m_{CM} \vec{g}$.

Surface forces, which act on the control surface, arise from stresses. The surface force due to stress $\vec{\sigma}$ acting on $d\mathcal{S} \in \mathcal{S}_{CM}$ is $d\vec{F}_\sigma = \vec{\sigma} \, d\mathcal{S}$. The total stress force is then

$$\vec{F}_\sigma = \int_{\mathcal{S}_{CM}} \vec{\sigma} \, d\mathcal{S} \tag{1.6}$$

Unlike body forces, which scale as L^3, surface forces scale as L^2. For flow applications with very small characteristic length scale, as is the case in microflows, surface forces dominate body forces as the ratio $|\vec{F}_\sigma|/|\vec{F}_b| \sim L^{-1}$. For flows with large characteristic length scale, e.g., atmospheric and oceanic flows, body forces dominate.

As depicted in Figure 1.3, the stress acting on $d\mathcal{S}$ can be decomposed into a normal component, $\vec{\sigma}_n$, and a tangential component, $\vec{\tau}$, that is, $\vec{\sigma} = \vec{\sigma}_n + \vec{\tau}$ and upon further decomposition of the normal component into an isotropic (pressure) term,[2] $-p\hat{n}$, and a viscous term, $\sigma_{n,v}$, that is, $\vec{\sigma}_n = (-p + \sigma_{n,v})\hat{n}$, we can express

$$\vec{\sigma} = -p\hat{n} + \sigma_{n,v}\hat{n} + \vec{\tau} \tag{1.7}$$

so that

$$\vec{F}_\sigma = \int_{\mathcal{S}_{CM}} (-p\hat{n} + \sigma_{n,v}\hat{n}) \, d\mathcal{S} + \int_{\mathcal{S}_{CM}} \vec{\tau} \, d\mathcal{S} \tag{1.8}$$

where we used the conventional symbol for shear stress, $\vec{\tau}$, to denote the tangentia stress, $\vec{\sigma}_t$.

---

[2]Note that the pressure is a compressive force per unit area while, by convention, the normal stress points away from the control volume, which explains the minus sign.

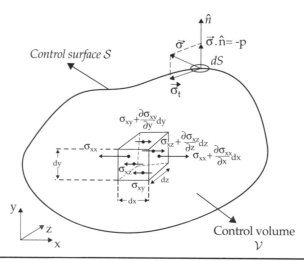

FIGURE 1.3 Decomposition of the stress force acting on a surface infinitesimal element into a normal and a tangential component. Also shown are the x–components of the stress force acting on the infinitesimal volume $(x, y, z) - (x + dx, y + dy, z + dz)$.

Line forces arise due to surface tension at the interface between two fluids or two phases of the same fluid. If the control mass (or control volume) contains a homogeneous single phase fluid, surface tension forces will be absent, and the conservation of momentum of a control mass, for incompressible flows, may be expressed as

$$\frac{d}{dt} \int_{\mathcal{V}_{CM}} \rho \vec{u} \, d\mathcal{V} = \int_{\mathcal{V}_{CM}} \rho \vec{g} \, d\mathcal{V} + \int_{\mathcal{S}_{CM}} (-p + \sigma_{n,v}) \hat{n} \, d\mathcal{S} + \int_{\mathcal{S}_{CM}} \vec{\tau} \, d\mathcal{S} + \vec{F}^* \qquad (1.9)$$

where $\vec{F}^*$ includes all forces other than gravity and stresses. For incompressible flow of a Newtonian fluid, we show in Section 1.2.2 that the only contribution to the normal stress is due to pressure, i.e., $\sigma_{n,v} = 0$ and $\vec{\sigma}_n = -p\hat{n}$.

Control Volume

The conservation of linear momentum for a control volume has to include the rate of momentum transferred by the bulk flow in and out of the control volume across the control surface, denoted as $\dot{\mathcal{M}}_{net}^{\leftarrow}$. The conservation of momentum, applied for a finite control volume, states that the rate of increase of the momentum in the control volume is balanced by the net rate of momentum transported by mass flow into the control volume across the control surface, in addition to all the external forces acting on the control volume:

$$\frac{\partial \vec{\mathcal{M}}_{CV}}{\partial t} = \dot{\mathcal{M}}_{net}^{\leftarrow} + \sum \vec{F} \qquad (1.10)$$

where the total momentum in the control volume is the sum of the momenta of very small masses constituting the control volume, which in the limit of differential elements becomes the integral $\vec{\mathcal{M}}_{CV} = \int_{\mathcal{V}} \rho \vec{u} \, d\mathcal{V}$. If the flow velocity at the center of $d\mathcal{S}$ relative to the control surface is $\vec{u}_r$, then the rate of momentum transport per unit area out of the control surface across $d\mathcal{S}$ will be the product of the momentum per unit mass (i.e., the velocity) and the mass flux at the center of $d\mathcal{S}$, i.e., $\vec{u} \, (\rho \vec{u}_r \cdot \hat{n})$. The net rate of momentum transfer by bulk flow across the control surface into the control volume is then the negative of the integral of the momentum flux over the control surface, i.e., $\dot{\mathcal{M}}_{net}^{\leftarrow} = -\int_S \rho\vec{u}\,\vec{u}_r \cdot \hat{n}\,dS$. This rate of momentum transfer by bulk flow into the control volume, in addition to the sum of forces acting on the control volume, balances the rate of increase of the total momentum of the control volume:

$$\frac{\partial}{\partial t}\int_V \rho\vec{u}\,dV = -\int_S \rho\vec{u}\,\vec{u}_r \cdot \hat{n}\,dS + \sum\vec{F} \tag{1.11}$$

The forces acting on the control volume typically include the gravity body force and the normal and tangential stress surface forces, $\vec{\sigma} = \vec{\sigma}_n + \vec{\sigma}_t$, in addition to other forces, whose sum is denoted as $\vec{F}^*$, so that

$$\frac{\partial}{\partial t}\int_V \rho\vec{u}\,dV + \int_S \rho\vec{u}\,\vec{u}_r \cdot \hat{n}\,dS = \int_V \rho\vec{g}\,dV + \int_S \vec{\sigma}\,dS + \vec{F}^* \tag{1.12}$$

where the stress may be expressed as the sum presented in Eq. (1.7).

1.1.3 Conservation of Energy

Control Mass

The conservation of energy for a control mass is an expression of the first law of thermodynamics, which states that the rate of change in total energy of the control mass is balanced by the rate of heat and work transfer into the control mass across the control surface

$$\frac{dE_{CM}}{dt} = \dot{Q}^{\leftarrow} + \dot{W}^{\leftarrow} \tag{1.13}$$

The total energy E_{CM} is the sum of internal energy U, kinetic energy KE, and potential energy PE, i.e., $E_{CM} = U+KE+PE$. If the total energy per unit mass (specific total energy) is e, then E_{CM} is the sum of the total energy of very small masses, each of mass δm_i and specific total energy e_i, constituting the control mass, i.e., $E_{CM} = \sum_i \delta E_i = \sum_i e_i\delta m_i = \sum \rho_i e_i\delta V_i$, where ρ_i and δV_i are respectively the density and volume of small mass δm_i. In the limit, $\delta m_i \rightarrow dm$ and $\delta V_i \rightarrow dV$ so that $E_{CM} = \int_{V_{CM}} e\rho\,dV$. The total specific energy may be expressed as the sum of the specific internal energy u, specific kinetic energy ke, and the specific potential energy pe, i.e., $e = u + ke + pe$, so that the first law of thermodynamics for a control mass may be expressed as

$$\frac{d}{dt}\int_{V_{CM}} (u + ke + pe)\,\rho\,dV = \dot{Q}^{\leftarrow} + \dot{W}^{\leftarrow} \tag{1.14}$$

The specific internal energy (internal energy per unit mass), u, is determined by the thermodynamic state and is a function of two independent intensive properties[3] for a simple compressible fluid. The specific kinetic and potential energy are respectively $ke = \frac{1}{2}|\vec{u}|^2$ and $pe = gz$, where z is the vertical z-axis coordinate. The rate of heat transfer into the control mass across the boundary can be expressed as the integral, over the control surface, of the component of the heat flux $\vec{q}$ at the boundary, along $-\hat{n}$:

$$\dot{Q}^{\leftarrow} = \int_{S_{CM}} \vec{q}\cdot(-\hat{n})\,dS \tag{1.15}$$

---

[3]Unlike extensive properties, which are proportional to the mass, intensive properties are independent of the mass. Example of extensive properties are V, E, U, KE, and PE. Examples of intensive properties are T, p, e, u, ke, and pe.

Since heat transfer across the boundary can generally take place by conduction and/or by radiation, we express $\vec{q} = \vec{q}_c + \vec{q}_r$. The conduction heat flux can be related to the temperature gradient according to Fourier's law:

$$\vec{q}_c = -\lambda \nabla T \tag{1.16}$$

where λ is the thermal conductivity. We may then write:

$$\dot{Q}^{\leftarrow} = \int_{\mathcal{S}_{CM}} \lambda \nabla T \cdot \hat{n}\, d\mathcal{S} - \int_{\mathcal{S}_{CM}} \vec{q}_r \cdot \hat{n}\, d\mathcal{S}$$

where $\nabla T \cdot \hat{n} = \partial T / \partial n$. In this book, we will not discuss radiation heat transfer further, and the interested reader is referred to references. [1]

Work exchanges across the control surface can be determined by identifying all the possible ways by which we can change the total energy of the control mass, other than by heat transfer. In the absence of surface tension, forms of work done on a fluid control mass can be explored by revisiting the forces acting on the control volume. Noting that the work done by gravity is already included in the potential energy term, the rate of work done by other body forces that may be present, such as electrostatic force in a charged fluid subject to an electric field, can be expressed as

$$\dot{W}_b^{\leftarrow} = \int_{\mathcal{V}_{CM}} \rho \vec{u} \cdot \vec{f}_b\, d\mathcal{V} \tag{1.17}$$

where $\vec{f}_b$ is the body force per unit mass. Note that the work done by conservative body forces is often accounted for in terms of a corresponding change in potential energy.[4]

The force acting on an infinitesimal control mass (of mass dm) is $\vec{f}_b\, dm$, where $dm = \rho\, d\mathcal{V}$, see Figure 1.4 (left). The work done by the body force over an infinitesimal thermodynamic process from time t to time $t + dt$ is $(\vec{f}_b\, dm) \cdot d\vec{x}_p$, where $d\vec{x}_p$ is the displacement vector of dm over the infinitesimal time interval dt. Dividing by dt yields the rate of work done on the infinitesimal control mass, $\frac{(\vec{f}_b\, dm) \cdot d\vec{x}_p}{dt}$. Noting that $\frac{d\vec{x}_p}{dt}$ is the velocity $\vec{u}_p$ at $(\vec{x}_p, t)$. Choosing $\vec{x} = \vec{x}_p$ (and $\vec{u} = \vec{u}_p$), the expression in Eq. (1.17) follows after integrating $(\vec{f}_b\, dm) \cdot \vec{u}$ over the volume of the control mass.

One can infer that work exchange by surface forces is caused by the stresses acting on the control surface. The stress force acting on an infinitesimal area $d\mathcal{S} \in \mathcal{S}_{CM}$ is $\vec{\sigma}\, d\mathcal{S}$, see Figure 1.4 (right). The work done onto the system as $d\mathcal{S}$ is deformed by $d\vec{x}_p$ over the time interval $t \to t + dt$ is then $(\vec{\sigma}\, d\mathcal{S}) \cdot d\vec{x}_p$, where $d\vec{x}_p$ is the displacement of the position vector of the center of $d\mathcal{S}$. Dividing by dt yields the rate of work done on $d\mathcal{S}$ as $\frac{(\vec{\sigma}\, d\mathcal{S}) \cdot d\vec{x}_p}{dt}$. Choosing $\vec{x} = \vec{x}_p$, it follows after integrating $\frac{(\vec{\sigma}\, d\mathcal{S}) \cdot d\vec{x}_p}{dt}$ over the area of the control surface that the rate of work done onto the control mass by the stress is

$$\dot{W}_{\sigma}^{\leftarrow} = \int_{\mathcal{S}_{CM}} \vec{\sigma} \cdot \vec{u}\, d\mathcal{S} \tag{1.18}$$

We express the stress as the sum of an isotropic normal pressure component, a viscous normal component, and a viscous tangential component, as presented in Eq. (1.7):

$$\dot{W}_{\sigma}^{\leftarrow} = \int_{\mathcal{S}_{CM}} -p\vec{u} \cdot \hat{n}\, d\mathcal{S} + \int_{\mathcal{S}_{CM}} \sigma_{n,v} \vec{u} \cdot \hat{n}\, d\mathcal{S} + \int_{\mathcal{S}_{CM}} \vec{\tau} \cdot \vec{u}\, d\mathcal{S} \tag{1.19}$$

---

[4]A conservative force is a force that can be expressed as the negative of the gradient of a potential, i.e., $\vec{f}_{cons} = -\nabla \Phi$, so that the work done by this force over a finite process $1 \to 2$ is $\int_1^2 -\nabla \Phi \cdot d\vec{x} = \Phi_1 - \Phi_2$.

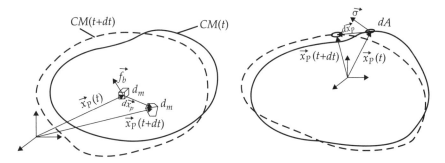

FIGURE 1.4 Left: work done by body force acting on an infinitesimal control mass (particle) over the time interval $t \rightarrow t + dt$. Right: work done by stress acting on an infinitesimal area of the control surface over the time interval $t \rightarrow t + dt$.

We note that the velocity $\vec{u}$ at the center of $dS \in S_{CM}$ is the same as the velocity of S_{CM}. This is because for a control mass, mass cannot be leaving or entering. It follows that for a control mass, the rate of work done by stresses is zero when the control surface is fixed. Expansion or compression due to deformation of the control surface at dS can only be caused by dS pulled/pushed along its normal direction. The term $\vec{u} \cdot \hat{n}$ can then be thought of as the outward distance, per unit time, by which dS is pulled or pushed. The energy equation for a control mass can then be expressed as

$$\frac{d}{dt} \int_{\mathcal{V}_{CM}} \left(u + \frac{1}{2}|\vec{u}|^2 + gz \right) \rho d\mathcal{V}$$

$$= \int_{S_{CM}} \lambda \frac{\partial T}{\partial n} dS + \int_{S_{CM}} (-p + \sigma_{n,v}) \vec{u} \cdot \hat{n} \, dS + \int_{S_{CM}} \vec{\tau} \cdot \vec{u} \, dS + \int_{\mathcal{V}_{CM}} \vec{f}^* \cdot \vec{u} \, d\mathcal{V} - \int_{S_{CM}} \vec{q}_r \cdot \hat{n} \, dS \tag{1.20}$$

where $\dot{W}_{\vec{F}*}^{\leftarrow} = \int_{\mathcal{V}_{CM}} \vec{f}^* \cdot \vec{u} \, d\mathcal{V}$ is the rate of work done by forces other than gravity and stresses and $\vec{f}^*$ is the force per unit volume.

Control Volume
The conservation of energy for a control volume follows from that for the control mass by including the total energy carried by bulk flow crossing the control surface. Using the Reynolds Transport theorem,[5] and starting with Eq. (1.20) for a control mass,

$$\frac{\partial}{\partial t} \int_{\mathcal{V}} \rho e \, d\mathcal{V} + \int_{S} \rho e \vec{u}_r \cdot \hat{n} \, dS$$

$$= \int_{S} \lambda \frac{\partial T}{\partial n} dS + \int_{S} (-p + \sigma_{n,v}) \vec{u} \cdot \hat{n} \, dS + \int_{S} \vec{\tau} \cdot \vec{u} \, dS + \int_{\mathcal{V}} \vec{f}^* \cdot \vec{u} \, d\mathcal{V} - \int_{S} \vec{q}_r \cdot \hat{n} \, dS \tag{1.21}$$

---

[5]Reynolds transport theorem states that the rate of change of a quantity f is equal to the sum of the rate of change of f within the control volume and the net rate of flux of f through the control surface. It is expressed mathematically as $\frac{d}{dt} \int_{\mathcal{V}} f d\mathcal{V} = \int_{\mathcal{V}} \frac{\partial f}{\partial t} d\mathcal{V} + \int_{S} f(\vec{u} \cdot \hat{n}) \, dS.$

where, as presented before, $\vec{u}_r = \vec{u} - \vec{u}_S$ is the flow velocity relative to the control surface, at the center of dS. Equation (1.21) may be alternatively expressed as

$$
\frac{\partial}{\partial t} \int_{\mathcal{V}} \rho e \, d\mathcal{V} + \int_{\mathcal{S}} \rho (e + p\upsilon - \sigma_{n,\upsilon}\upsilon) \vec{u}_r \cdot \hat{n} \, d\mathcal{S}
$$
$$
= \int_{\mathcal{S}} \lambda \frac{\partial T}{\partial n} \, d\mathcal{S} + \int_{\mathcal{S}} (-p + \sigma_{n,\upsilon}) \, \vec{u}_S \cdot \hat{n} \, d\mathcal{S} + \int_{\mathcal{S}} \vec{\tau} \cdot \vec{u} \, d\mathcal{S}
$$
$$
+ \int_{\mathcal{V}} \vec{f}^{*} \cdot \vec{u} \, d\mathcal{V} - \int_{\mathcal{S}} \vec{q}_r \cdot \hat{n} \, d\mathcal{S} \tag{1.22}
$$

The specific volume is denoted by υ. Noting that the specific enthalpy is $\mathfrak{h} = \mathfrak{u} + p\upsilon$, Eq. (1.22) is commonly expressed as

$$
\frac{\partial}{\partial t} \int_{\mathcal{V}} \rho e \, d\mathcal{V} + \int_{\mathcal{S}} \left(\mathfrak{h} + \frac{1}{2} |\vec{u}|^2 + gz \right) \rho \vec{u}_r \cdot \hat{n} \, d\mathcal{S} + \int_{\mathcal{S}} \rho(-\sigma_{n,\upsilon}\upsilon) \vec{u}_r \cdot \hat{n} \, d\mathcal{S}
$$
$$
= \int_{\mathcal{S}} \lambda \frac{\partial T}{\partial n} \, d\mathcal{S} + \int_{\mathcal{S}} (-p + \sigma_{n,\upsilon}) \, \vec{u}_S \cdot \hat{n} \, d\mathcal{S} + \int_{\mathcal{S}} \vec{\tau} \cdot \vec{u} \, d\mathcal{S}
$$
$$
+ \int_{\mathcal{V}} \vec{f}^{*} \cdot \vec{u} \, d\mathcal{V} - \int_{\mathcal{S}} \vec{q}_r \cdot \hat{n} \, d\mathcal{S} \tag{1.23}
$$

For incompressible flow of a Newtonian fluid, the normal viscous stress is zero, $\sigma_{n,\upsilon} = 0$. In the absence of radiative heat transfer, $\vec{q}_r = \vec{0}$. In the absence of forces other than stresses and gravity, $\vec{f}^{*} = \vec{0}$. Under these conditions

$$
\frac{\partial}{\partial t} \int_{\mathcal{V}} \rho e \, d\mathcal{V} + \int_{\mathcal{S}} \left(\mathfrak{h} + \frac{1}{2} |\vec{u}|^2 + gz \right) \rho \vec{u}_r \cdot \hat{n} \, d\mathcal{S}
$$
$$
= \int_{\mathcal{S}} \lambda \frac{\partial T}{\partial n} \, d\mathcal{S} + \int_{\mathcal{S}} (-p) \, \vec{u}_S \cdot \hat{n} \, d\mathcal{S} + \int_{\mathcal{S}} \vec{\tau} \cdot \vec{u} \, d\mathcal{S} \tag{1.24}
$$

Note that the second term on the right-hand side of Eq. (1.24) is the expansion or compression (or $pd\mathcal{V}$) work associated with motion of the control surface. This term is zero if the control surface is fixed, and Eq. (1.24) simplifies to

$$
\frac{\partial}{\partial t} \int_{\mathcal{V}} \rho e \, d\mathcal{V} + \int_{\mathcal{S}} \left(\mathfrak{h} + \frac{1}{2} |\vec{u}|^2 + gz \right) \rho \vec{u} \cdot \hat{n} \, d\mathcal{S} = \int_{\mathcal{S}} \lambda \frac{\partial T}{\partial n} \, d\mathcal{S} + \int_{\mathcal{S}} \vec{\tau} \cdot \vec{u} \, d\mathcal{S} \tag{1.25}
$$

The second term on the right-hand side of Eq. (1.25) is the work done by shear forces acting on the control surface. For those parts of the control surface where the fluid velocity is zero, as is the case with the no-slip boundary condition at a solid impermeable wall, the contribution from this term is zero. For other parts of the control surface not limited by a no-slip wall, e.g., an inlet or outlet, this term captures any work done by shear forces.

We finally note that the equations presented for conservation of energy applied for a control mass and a control volume do not include a heat source term, which may arise from internal heat sources such as electro-magnetic heating.

1.2 Conservation Laws in Differential Form

There is more than one way to arrive at the conservation laws in differential from. One way is to start with a control volume that consists of a fixed infinitesimal cube with faces orthogonal to the coordinates axes and extending from x to $x + dx$, y to $y + dy$ and from z

to $z+dz$. Alternatively, one can start with the integral forms in Eqs. (1.3), (1.12), and (1.21) for a fixed control volume. The surface integrals can be converted to volume integrals using Gauss' divergence theorem which, for vector $\vec{B}$, is expressed as

$$\int_S \hat{n} \cdot \vec{B} \, dS = \int_V \nabla \cdot \vec{B} \, dV \tag{1.26}$$

Note that $\hat{n} \cdot \vec{B} = \vec{B} \cdot \hat{n}$. For first (or second) order tensor $\mathbf{A}$, the divergence theorem is expressed as

$$\int_S \hat{n} \cdot \mathbf{A} \, dS = \int_V \nabla \cdot \mathbf{A} \, dV \tag{1.27}$$

It is important to respect the order in which $\hat{n}$ and $\mathbf{A}$ appear in the dot product. This is because $\hat{n} \cdot \mathbf{A}$ is not equal to $\mathbf{A} \cdot \hat{n}$. They are actually related according to $\hat{n} \cdot \mathbf{A} = \mathbf{A}^\mathsf{T} \cdot \hat{n}$, where the superscript T is the transpose.

1.2.1   The Continuity Equation

Noting that by Gauss' divergence theorem $\int_S \rho\vec{u} \cdot \hat{n} \, dS = \int_V \nabla \cdot (\rho\vec{u}) \, dV$ and that for a fixed control volume $\vec{u}_S = 0$ and $\vec{u}_r = \vec{u}$, the conservation of mass for a fixed finite control volume, given by Eq. (1.3), can be expressed as

$$\int_V \left(\frac{\partial \rho}{\partial t} + \nabla \cdot (\rho\vec{u}) \right) dV = 0 \tag{1.28}$$

Since the volume integral has to be equal to zero for any choice of control volume, then the integrand must necessarily be equal to zero, which yields the continuity equation in conservative form

$$\frac{\partial \rho}{\partial t} + \nabla \cdot (\rho\vec{u}) = 0 \tag{1.29}$$

which can also be expressed using the Lagrangian derivative

$$\frac{d\rho}{dt} + \rho\nabla \cdot \vec{u} = 0 \tag{1.30}$$

where the Lagrangian derivative is defined as

$$\frac{d}{dt} \equiv \frac{\partial}{\partial t} + \vec{u} \cdot \nabla \tag{1.31}$$

An important relation that allows us to switch between conservative forms of the continuity, momentum, and energy PDEs and the corresponding forms involving the Lagrangian derivative is

$$\rho\frac{d\phi}{dt} = \frac{\partial(\rho\phi)}{\partial t} + \nabla \cdot (\rho\vec{u}\phi) \tag{1.32}$$

The proof of Eq. (1.32) is as follows:

$$\begin{aligned} \rho\frac{d\phi}{dt} &= \rho \left(\frac{\partial \phi}{\partial t} + \vec{u} \cdot \nabla\phi \right) \\ &= \frac{\partial(\rho\phi)}{\partial t} - \phi\frac{\partial \rho}{\partial t} + \nabla \cdot (\rho\vec{u}\phi) - \phi\nabla \cdot (\rho\vec{u}) \\ &= \frac{\partial(\rho\phi)}{\partial t} + \nabla \cdot (\rho\vec{u}\phi) - \phi\left(\frac{\partial \rho}{\partial t} + \nabla \cdot (\rho\vec{u}) \right) \end{aligned}$$

By the continuity, expressed in Eq. (1.29), the term in parentheses on the right-hand side vanishes, and Eq. (1.32) follows.

1.2.2   Differential Form of the Conservation of Momentum

Noting that by Gauss' divergence theorem $\int_S b\,(\rho\vec{u}\cdot\hat{n})\,dS = \int_V \nabla\cdot(\rho b\vec{u})\,dV$, where b can be u, v, or w, which are the x, y, and z components of the velocity vector $\vec{u}$, then $\int_S \vec{u}\,(\rho\vec{u}\cdot\hat{n})\,dS = \int_V \nabla\cdot(\rho\vec{u}\vec{u})\,dV$. The vector product in $\rho\vec{u}\vec{u}$ is a dyadic product which results in a tensor. The conservation of momentum for a fixed finite control volume, given by Eq. (1.12), can be expressed as

$$\int_V \left[\frac{\partial(\rho\vec{u})}{\partial t} + \nabla\cdot(\rho\vec{u}\vec{u})\right] dV = \int_V \rho\vec{g}\,dV + \int_S \vec{\sigma}\,dS + \int_V \vec{f}^*\,dV \tag{1.33}$$

where $\vec{f}^*$ is the sum of all forces per unit volume, other than stresses and gravity, acting on the infinitesimal volume $dV \in V$.

In order to express the surface integral of the stress forces acting on the control surface as a volume integral, we first explore the nature of stress in a deforming fluid. The stress force acting on an infinitesimal area of the control surface, $dS \in S$, is simply the product of the stress vector and the area, i.e., $\vec{\sigma}\,dS$. The force due to stresses acting on the infinitesimal fixed control volume shown in Figure 1.3 arises from the combined effect of the stress forces acting on each of the six faces of the infinitesimal element. If σ_{xx} is the normal stress acting on face 1 in the $y-z$ plane at x, then, from infinitesimal calculus the normal stress acting on face 2 in the $y-z$ plane at $x+dx$ is $\sigma_{xx}+\frac{\partial\sigma_{xx}}{\partial x}dx$. Additional stress contributions to the $x-$component of the force arise from the $x-$components of the tangential stresses in the faces in the $x-z$ planes at y and $y+dy$ and in the faces in the $x-y$ planes at z and $z+dz$. Then, x force is the sum of these six components acting on the six faces,

$$F_{\sigma_x} = -\sigma_{xx}dydz + \left(\sigma_{xx} + \frac{\partial\sigma_{xx}}{\partial x}dx\right)dy\,dz - \sigma_{xy}dxdz + \left(\sigma_{xy} + \frac{\partial\sigma_{xy}}{\partial y}dy\right)dx\,dz$$

$$- \sigma_{xz}dxdy + \left(\sigma_{xz} + \frac{\partial\sigma_{xz}}{\partial z}dz\right)dx\,dy$$

$$= \left(\frac{\partial\sigma_{xx}}{\partial x} + \frac{\partial\sigma_{xy}}{\partial y} + \frac{\partial\sigma_{xz}}{\partial z}\right)dx\,dy\,dz \tag{1.34}$$

Similar expressions for the y and z components of the stress force acting on the infinitesimal element can be obtained:

$$F_{\sigma_y} = \left(\frac{\partial\sigma_{yx}}{\partial x} + \frac{\partial\sigma_{yy}}{\partial y} + \frac{\partial\sigma_{yz}}{\partial z}\right)dx\,dy\,dz$$

$$F_{\sigma_z} = \left(\frac{\partial\sigma_{zx}}{\partial x} + \frac{\partial\sigma_{zy}}{\partial y} + \frac{\partial\sigma_{zz}}{\partial z}\right)dx\,dy\,dz$$

so that the stress force per unit volume is $\vec{f}_\sigma = \frac{F_{\sigma_x}\hat{x}+F_{\sigma_y}\hat{y}+F_{\sigma_z}\hat{z}}{dV}$, $dV = dx\,dy\,dz$, or

$$\vec{f}_\sigma = \left(\frac{\partial\sigma_{xx}}{\partial x} + \frac{\partial\sigma_{xy}}{\partial y} + \frac{\partial\sigma_{xz}}{\partial z}\right)\hat{x} + \left(\frac{\partial\sigma_{yx}}{\partial x} + \frac{\partial\sigma_{yy}}{\partial y} + \frac{\partial\sigma_{yz}}{\partial z}\right)\hat{y}$$

$$+ \left(\frac{\partial\sigma_{zx}}{\partial x} + \frac{\partial\sigma_{zy}}{\partial y} + \frac{\partial\sigma_{zz}}{\partial z}\right)\hat{z}$$

which may be expressed as the divergence of the stress tensor.[6]

$$\vec{f}_\sigma = \nabla \cdot \boldsymbol{\sigma} \tag{1.35}$$

where

$$\boldsymbol{\sigma} = \begin{bmatrix} \sigma_{xx} & \sigma_{yx} & \sigma_{zx} \\ \sigma_{xy} & \sigma_{yy} & \sigma_{zy} \\ \sigma_{xz} & \sigma_{yz} & \sigma_{zz} \end{bmatrix} \tag{1.36}$$

Key characteristics of the stress tensor, $\boldsymbol{\sigma}$, which determines the state of stress at a given location and time in a deforming fluid, are

1. It is a 3×3 matrix, reflecting the fact that it is describing the components of a force (which is a vector of three components) acting on a surface of orientation defined by a unit normal vector (which is also made up of three components).

2. The indices are such that the first index denotes the direction of the stress component, and the second index denotes direction of the unit vector normal to the plane of action. For example, σ_{xz} denotes the x-component of the stress acting on the face in the $x - y$ plane normal to the z−coordinate.

3. The diagonal components are the normal stresses which, by convention, point away from the surface.

4. The off diagonal terms are the tangential stresses.

5. The stress tensor is symmetric, i.e., $\sigma_{xy} = \sigma_{yx}$, $\sigma_{xz} = \sigma_{zx}$ and $\sigma_{yz} = \sigma_{zy}$. This is a consequence of the local equilibrium condition on a small element. In the limit of $\delta x, \delta y, \delta z \to 0$, the force balance has to be established by stress forces only, because body force and inertia scale as $(\delta x)(\delta y)(\delta z)$.

6. The sum of the diagonal terms, called the trace of the tensor, is invariant under rotation of the axes. This is also a consequence of the local equilibrium condition on a small element in the limit of $\delta x, \delta y, \delta z \to 0$.

Using Eq. (1.35) for the stress force per unit volume, $\int_S \vec{\sigma} \, d\mathcal{S} = \int_V \vec{f}_\sigma \, d\mathcal{V}$, the momentum equation [Eq. (1.33)] is expressed in terms of volume integrals:

$$\int_V \left[\frac{\partial(\rho \vec{u})}{\partial t} + \nabla \cdot (\rho \vec{u} \vec{u}) \right] d\mathcal{V} = \int_V \left(\nabla \cdot \boldsymbol{\sigma} + \rho \vec{g} + \vec{f}^* \right) d\mathcal{V} \tag{1.37}$$

Noting that the equality must hold for all choices of control volume, then the integrands on either side of the equality must be equal:

$$\frac{\partial(\rho \vec{u})}{\partial t} + \nabla \cdot (\rho \vec{u} \vec{u}) = \nabla \cdot \boldsymbol{\sigma} + \rho \vec{g} + \vec{f}^* \tag{1.38}$$

---

[6]Tensors and matrices will appear in bold.

Stresses in a Deforming Fluid

The stresses acting at a point[7] in a deforming fluid may be decomposed into normal stresses and tangential stresses as

$$
\sigma =
\begin{bmatrix}
\sigma_{xx} & \sigma_{xy} & \sigma_{xz} \\
\sigma_{xy} & \sigma_{yy} & \sigma_{yz} \\
\sigma_{xz} & \sigma_{yz} & \sigma_{zz}
\end{bmatrix}
=
\begin{bmatrix}
\sigma_{xx} & 0 & 0 \\
0 & \sigma_{yy} & 0 \\
0 & 0 & \sigma_{zz}
\end{bmatrix}
+
\begin{bmatrix}
0 & \sigma_{xy} & \sigma_{xz} \\
\sigma_{xy} & 0 & \sigma_{yz} \\
\sigma_{xz} & \sigma_{yz} & 0
\end{bmatrix}
\tag{1.39}
$$

Note that in a static fluid, the normal stresses are equal and invariant under rotation of the axes, $\sigma_{xx} = \sigma_{yy} = \sigma_{zz} = -p$, which reflects the nature of static pressure in a static fluid. In the absence of relative motion in a static (or rigidly moving) fluid body, the tangential stresses do not exist, i.e., $\sigma_{xy} = \sigma_{xz} = \sigma_{yz} = 0$.

In a deforming fluid, the normal stresses change when the coordinate system is rotated from (x, y, z) to (x', y', z'), that is, $\sigma_{xx} \neq \sigma_{x'x'}$, $\sigma_{yy} \neq \sigma_{y'y'}$, and $\sigma_{zz} \neq \sigma_{z'z'}$. Their sum, however, is invariant under rotation of the coordinates, that is, $\sigma_{xx} + \sigma_{yy} + \sigma_{zz} = \sigma_{x'x'} + \sigma_{y'y'} + \sigma_{z'z'}$. This enables us to define an equivalent of the static pressure as the negative arithmetic mean of the sum, as $p_m = -\frac{\sigma_{xx} + \sigma_{yy} + \sigma_{zz}}{3}$. Note that the negative sign is because the pressure points into the surface whereas the normal stresses, by convention, point out of the surface, as seen in Figure 1.3. We introduce the mechanical pressure p_m as the equivalent of the thermodynamic pressure, where p_m acts isotropically normal to the surface of interest. This enables us to express the normal stress as the sum of an isotropic[8] component and a nonisotropic component as follows:

$$
\begin{bmatrix}
\sigma_{xx} & 0 & 0 \\
0 & \sigma_{yy} & 0 \\
0 & 0 & \sigma_{zz}
\end{bmatrix}
= -
\begin{bmatrix}
p_m & 0 & 0 \\
0 & p_m & 0 \\
0 & 0 & p_m
\end{bmatrix}
$$
$$
+
\begin{bmatrix}
\frac{2\sigma_{xx}-\sigma_{yy}-\sigma_{zz}}{3} & 0 & 0 \\
0 & \frac{2\sigma_{yy}-\sigma_{xx}-\sigma_{zz}}{3} & 0 \\
0 & 0 & \frac{2\sigma_{zz}-\sigma_{xx}-\sigma_{yy}}{3}
\end{bmatrix}
\tag{1.40}
$$

Substituting into Eq. (1.39) yields a decomposition of the stress as the sum of an isotropic component and a nonisotropic component:

$$
\sigma = -p_m \, I + \sigma_d
$$
$$
= -
\begin{bmatrix}
p_m & 0 & 0 \\
0 & p_m & 0 \\
0 & 0 & p_m
\end{bmatrix}
+
\begin{bmatrix}
\frac{2\sigma_{xx}-\sigma_{yy}-\sigma_{zz}}{3} & \sigma_{xy} & \sigma_{xz} \\
\sigma_{xy} & \frac{2\sigma_{yy}-\sigma_{xx}-\sigma_{zz}}{3} & \sigma_{yz} \\
\sigma_{xz} & \sigma_{yz} & \frac{2\sigma_{zz}-\sigma_{xx}-\sigma_{yy}}{3}
\end{bmatrix}
\tag{1.41}
$$

where I is the identity matrix and the nonisotropic component, expressed by the second tensor on the right-hand side of Eq. (1.41), is referred to as the deviatoric stress tensor, σ_d. Substituting Eq. (1.41) in Eq. (1.38) and using the fact that $\nabla \cdot (p_m I) = \nabla p_m$

$$
\frac{\partial(\rho \vec{u})}{\partial t} + \nabla \cdot (\rho \vec{u} \vec{u}) = -\nabla p_m + \nabla \cdot \sigma_d + \rho \vec{g} + \vec{f^*}
\tag{1.42}
$$

---

[7]Or on a vanishingly small cubic control volume aligned with the coordinates axes, $\delta x \ll dx$, $\delta y \ll dy$, $\delta z \ll dz$.

[8]Invariant under rotation of the coordinates axes.

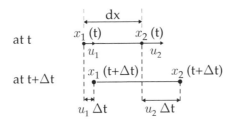

FIGURE 1.5 Deformation of an element in a one-dimensional flow.

The dyadic product $\rho \vec{u}\vec{u}$ results in a tensor, and its divergence, $\nabla \cdot (\rho \vec{u}\vec{u})$, is a vector. Equation (1.42) is a nonlinear PDE expressing the balance per unit volume of the flow acceleration (left-hand side) with the various body and surface forces (right-hand side), applicable at any time and at any location in the flow.

Local Deformation in Deforming Fluid

As opposed to solids, whose mechanical behavior is characterized by relating the stress tensor to the strain tensor, the mechanical behavior of fluids is characterized by relating the stress to the rate of deformation. This arises from the defining aspect of a fluid, in that it continuously deforms under the action of a tangential force. The rate of deformation tensor can be described by inspecting the relative deformation near a point. Consider the fluid element shown in Figure 1.5 in a one-dimensional unsteady flow. At time t, the fluid element extends from $x_1(t)$ to $x_2(t)$. At time $t + \Delta t$, it extends from $x_1(t + \Delta t)$ to $x_2(t + \Delta t)$. The change in length of the fluid element from t to $t + \Delta t$ is then $(x_2(t + \Delta t) - x_1(t + \Delta t)) - (x_2(t) - x_1(t))$ and the strain is the ratio of the change in length to the length at t,

$$\epsilon = \frac{(x_2(t + \Delta t) - x_1(t + \Delta t)) - (x_2(t) - x_1(t))}{x_2(t) - x_1(t)} \tag{1.43}$$

The rate of strain is then the limit $\lim_{\Delta t \to 0} \frac{\epsilon}{\Delta t}$, which may be expressed as

$$\lim_{\Delta t \to 0} \frac{\frac{x_2(t + \Delta t) - x_2(t)}{\Delta t} - \frac{x_1(t + \Delta t) - x_1(t)}{\Delta t}}{x_2(t) - x_1(t)} = \frac{u(x_2, t) - u(x_1, t)}{x_2(t) - x_1(t)} \tag{1.44}$$

where u is the velocity component in the x−direction. If the element length at t is chosen to be such that $x_2(t) = x_1(t) + dx$, then $u(x_2) = u(x_1 + dx) = u(x_1) + \frac{\partial u}{\partial x}\big|_{x_1} dx$ so that the rate of strain simplifies to

$$\frac{\frac{\partial u}{\partial x}\big|_{x_1} dx}{dx} = \frac{\partial u}{\partial x}\bigg|_{x_1} \tag{1.45}$$

which is the x-component of the gradient of u. In general three-dimensional flows, the rate of deformation in the neighborhood of a point represents the $x, y,$ and z components of the gradient of each of the velocity components $u, v,$ and w:

$$\epsilon = \nabla \vec{u} = \begin{bmatrix} \frac{\partial u}{\partial x} & \frac{\partial v}{\partial x} & \frac{\partial w}{\partial x} \\ \frac{\partial u}{\partial y} & \frac{\partial v}{\partial y} & \frac{\partial w}{\partial y} \\ \frac{\partial u}{\partial z} & \frac{\partial v}{\partial z} & \frac{\partial w}{\partial z} \end{bmatrix} \tag{1.46}$$

This arises from the second term of the left-hand side of Eq. (1.42) where the divergence operator acts on a dyadic product:

$$\nabla \vec{u} = \left(\frac{\partial}{\partial x} \hat{x} + \frac{\partial}{\partial y} \hat{y} + \frac{\partial}{\partial z} \hat{z} \right) (u\hat{x} + v\hat{y} + w\hat{z})$$

$$= \frac{\partial u}{\partial x} \hat{x}\hat{x} + \frac{\partial v}{\partial x} \hat{x}\hat{y} + \frac{\partial w}{\partial x} \hat{x}\hat{z} + \frac{\partial u}{\partial y} \hat{y}\hat{x} + \frac{\partial v}{\partial y} \hat{y}\hat{y} + \frac{\partial w}{\partial y} \hat{y}\hat{z} + \frac{\partial u}{\partial z} \hat{z}\hat{x}$$

$$+ \frac{\partial v}{\partial z} \hat{z}\hat{y} + \frac{\partial w}{\partial z} \hat{z}\hat{z}$$

$$= \begin{bmatrix} \frac{\partial u}{\partial x} & \frac{\partial v}{\partial x} & \frac{\partial w}{\partial x} \\ \frac{\partial u}{\partial y} & \frac{\partial v}{\partial y} & \frac{\partial w}{\partial y} \\ \frac{\partial u}{\partial z} & \frac{\partial v}{\partial z} & \frac{\partial w}{\partial z} \end{bmatrix}$$

The rate of deformation tensor may be expressed as the sum of a symmetric tensor, e, and an antisymmetric tensor, ξ,

$$\epsilon = e + \xi \tag{1.47}$$

$$= \frac{1}{2} \left(\nabla \vec{u} + (\nabla \vec{u})^{\mathsf{T}} \right) + \frac{1}{2} \left(\nabla \vec{u} - (\nabla \vec{u})^{\mathsf{T}} \right) \tag{1.48}$$

$$= \frac{1}{2} \begin{pmatrix} 2\frac{\partial u}{\partial x} & \frac{\partial v}{\partial x} + \frac{\partial u}{\partial y} & \frac{\partial w}{\partial x} + \frac{\partial u}{\partial z} \\ \frac{\partial u}{\partial y} + \frac{\partial v}{\partial x} & 2\frac{\partial v}{\partial y} & \frac{\partial w}{\partial y} + \frac{\partial v}{\partial z} \\ \frac{\partial u}{\partial z} + \frac{\partial w}{\partial x} & \frac{\partial v}{\partial z} + \frac{\partial w}{\partial y} & 2\frac{\partial w}{\partial z} \end{pmatrix} + \frac{1}{2} \begin{pmatrix} 0 & \frac{\partial v}{\partial x} - \frac{\partial u}{\partial y} & \frac{\partial w}{\partial x} - \frac{\partial u}{\partial z} \\ \frac{\partial u}{\partial y} - \frac{\partial v}{\partial x} & 0 & \frac{\partial w}{\partial y} - \frac{\partial v}{\partial z} \\ \frac{\partial u}{\partial z} - \frac{\partial w}{\partial x} & \frac{\partial v}{\partial z} - \frac{\partial w}{\partial y} & 0 \end{pmatrix} \tag{1.49}$$

where the superscript T is the transpose and e is the rate of strain tensor while the antisymmetric tensor, ξ, is the rate of rotation tensor. It can be shown that a fluid element, shown as a rectangle in Figure 1.6, undergoes, in addition to translation, deformation that can be decomposed into pure rotation, pure shear, and pure compression/expansion. The rate of strain tensor, e, captures the pure shear and pure compression/expansion deformations while ξ captures the pure rotation.

Relation between Stress and Rate of Deformation

Newtonian fluids are a special class of fluids whose mechanical behavior can be described by a linear relation between the deviatoric stress tensor, σ_d, and the rate of strain tensor, e,

$$\sigma_d = 2\mu \left(e - \frac{1}{3} (\nabla \cdot \vec{u}) I \right) \tag{1.50}$$

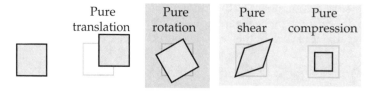

Pure translation   Pure rotation   Pure shear   Pure compression

FIGURE 1.6 Deformation of the motion of a fluid element.

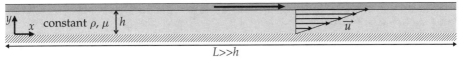

FIGURE 1.7 Couette flow is a shear-driven two-dimensional steady flow established by imparting shear onto a thin fluid film by dragging a plate in the x-direction with a constant speed V_p.

where μ is the coefficient of dynamic viscosity. Note that we have neglected the effect of the bulk viscosity,[9] which is a valid assumption for most flows (except in extreme cases of compressible flows). [2] For the Couette flow, depicted in Figure 1.7, μ is the ratio $\frac{F_p/A_p}{V_p/h}$. For a given plate velocity V_p, the plate must be dragged with a force, F_p, that is proportional to the viscosity of the fluid.

The Navier-Stokes Equations
Substituting the linear relation between rate of strain tensor and the stress tensor characterizing Newtonian fluids, expressed in Eq. (1.50), in Eq. (1.42) we get

$$\frac{\partial(\rho\vec{u})}{\partial t} + \nabla \cdot (\rho\vec{u}\vec{u}) = -\nabla p_m + \nabla \cdot \left(2\mu\left(e - \frac{1}{3}\nabla \cdot \vec{u}\,I\right)\right) + \rho\vec{g} + \vec{f}^* \qquad (1.51)$$

Substituting in the expression for e in the left term of Eq. (1.49), and noting that $\nabla \cdot \left(\frac{2\mu}{3}(\nabla \cdot \vec{u})\,I\right) = \nabla\left(\frac{2\mu}{3}(\nabla \cdot \vec{u})\right)$, we arrive at the Navier-Stokes equations expressed in conservative form as

$$\frac{\partial(\rho\vec{u})}{\partial t} + \nabla \cdot (\rho\vec{u}\vec{u}) = -\nabla p + \nabla \cdot \left(\mu\left(\nabla\vec{u} + (\nabla\vec{u})^{\mathsf{T}}\right)\right) - \frac{2}{3}\nabla(\mu\nabla \cdot \vec{u}) + \rho\vec{g} + \vec{f}^* \quad (1.52)$$

which for constant viscosity, μ, simplifies to

$$\frac{\partial(\rho\vec{u})}{\partial t} + \nabla \cdot (\rho\vec{u}\vec{u}) = -\nabla p + \mu\nabla^2\vec{u} + \frac{1}{3}\mu\nabla(\nabla \cdot \vec{u}) + \rho\vec{g} + \vec{f}^* \qquad (1.53)$$

where we used the identity $\nabla \cdot \left((\nabla\vec{u})^{\mathsf{T}}\right) - \frac{2}{3}\nabla(\nabla \cdot \vec{u}) = \frac{1}{3}\nabla(\nabla \cdot \vec{u})$. Note that in Eq. (1.53), the mechanical pressure p_m has been replaced by the thermodynamic pressure p, which is true if the condition of local thermodynamic equilibrium[10] holds, as is the case in most flows encountered in engineering applications and in natural phenomena.

Using Eq. (1.32), the inertia term on the left-hand side of the Navier-Stokes equation [Eq. (1.53)] may be replaced by a Lagrangian derivative to yield the following alternative form of the momentum equation:

$$\rho\frac{d\vec{u}}{dt} = -\nabla p + \mu\nabla^2\vec{u} + \frac{1}{3}\mu\nabla(\nabla \cdot \vec{u}) + \rho\vec{g} + \vec{f}^* \qquad (1.54)$$

---

[9]To account for bulk viscosity, Eq. (1.50) is replaced with $\sigma_d = 2\mu e + \left(\mu_B - \frac{2}{3}\mu\right)\nabla \cdot \vec{u}\,I$.

[10]Local thermodynamic equilibrium refers to local conditions where the thermodynamic processes proceed at time scales that are much larger than the time needed to reach equilibrium under hypothetical isolation. In thermodynamics, we refer to these processes as quasi-static processes.

1.2.3 The Energy Equation

Starting with the integral form of the energy equation for a fixed control volume, and using Gauss' divergence theorem to express $\int_S \lambda \nabla T \cdot \hat{n}\, dS = \int_V \nabla \cdot (\lambda \nabla T)\, dV$ and $\int_S e\,(\rho \vec{u} \cdot \hat{n})\, dS = \int_V \nabla \cdot (\rho \vec{u} e)\, dV$, Eq. (1.21) can be expressed as

$$\int_V \left[\frac{\partial(\rho e)}{\partial t} + \nabla \cdot (\rho \vec{u} e) \right] dV$$

$$= \int_V \nabla \cdot (\lambda \nabla T)\, dV + \int_S \vec{\sigma} \cdot \vec{u}\, dS + \int_V \vec{f}^* \cdot \vec{u}\, dV - \int_S \vec{q}_r \cdot \hat{n}\, dS \qquad (1.55)$$

where $\vec{\sigma} = (-p + \sigma_{n,v})\hat{n} + \vec{\tau}$, and for a stationary control volume, $\vec{u}_r = \vec{u}$. We now express the rate of work done by the stresses onto the control volume as a volume integral, $\int_S \vec{\sigma} \cdot \vec{u}\, dS = \int_V \dot{w}_\sigma\, dV$. We evaluate this rate of work for the face, in the $y - z$ plane at x, of the infinitesimal cubic element, where the stress vector is $-(\sigma_{xx}\hat{x} + \sigma_{yx}\hat{y} + \sigma_{zx}\hat{z})$. The term $\vec{\sigma} \cdot \vec{u}\, dS$ for the face is $-(\sigma_{xx}u + \sigma_{yx}v + \sigma_{zx}w)dydz$. The corresponding term for the face in the $y - z$ plane at $x + dx$ is

$$\left(\sigma_{xx}u + \frac{\partial(\sigma_{xx}u)}{\partial x}\,dx + \sigma_{yx}v + \frac{\partial(\sigma_{yx}v)}{\partial x}\,dx + \sigma_{zx}w + \frac{\partial(\sigma_{zx}w)}{\partial x}\,dx \right) dydz \qquad (1.56)$$

Summing these two terms yields $\frac{\partial}{\partial x}\left(\sigma_{xx}u + \sigma_{yx}v + \sigma_{zx}w\right)dV$, where $dV = dx\,dy\,dz$. Similarly, the rate of work done by stresses acting on the $x - z$ faces at y and $y + dy$ is $\frac{\partial}{\partial y}\left(\sigma_{xy}u + \sigma_{yy}v + \sigma_{zy}w\right)dV$, and the rate of work done by stresses acting on the $x - y$ faces at z and $z + dz$ is $\frac{\partial}{\partial z}\left(\sigma_{xz}u + \sigma_{yz}v + \sigma_{zz}w\right)dV$. Recalling that $\sigma_{ij} = \sigma_{ji}$ where $i, j, \in \{x, y, z\}$, the rate of work done per unit volume by all the stresses acting on all the faces is

$$\dot{w}_\sigma = \frac{\partial}{\partial x}\left(\sigma_{xx}u + \sigma_{xy}v + \sigma_{xz}w\right) + \frac{\partial}{\partial y}\left(\sigma_{xy}u + \sigma_{yy}v + \sigma_{yz}w\right)$$

$$+ \frac{\partial}{\partial z}\left(\sigma_{xz}u + \sigma_{yz}v + \sigma_{zz}w\right) \qquad (1.57)$$

which can be expressed as

$$\dot{w}_\sigma = u\left(\frac{\partial\sigma_{xx}}{\partial x} + \frac{\partial\sigma_{xy}}{\partial y} + \frac{\partial\sigma_{xz}}{\partial z} \right) + \left(\sigma_{xx}\frac{\partial u}{\partial x} + \sigma_{xy}\frac{\partial u}{\partial y} + \sigma_{xz}\frac{\partial u}{\partial z} \right)$$

$$+ v\left(\frac{\partial\sigma_{xy}}{\partial x} + \frac{\partial\sigma_{yy}}{\partial y} + \frac{\partial\sigma_{yz}}{\partial z} \right) + \left(\sigma_{xy}\frac{\partial v}{\partial x} + \sigma_{yy}\frac{\partial v}{\partial y} + \sigma_{yz}\frac{\partial v}{\partial z} \right)$$

$$+ w\left(\frac{\partial\sigma_{xz}}{\partial x} + \frac{\partial\sigma_{yz}}{\partial y} + \frac{\partial\sigma_{zz}}{\partial z} \right) + \left(\sigma_{xz}\frac{\partial w}{\partial x} + \sigma_{yz}\frac{\partial w}{\partial y} + \sigma_{zz}\frac{\partial w}{\partial z} \right)$$

or

$$\dot{w}_\sigma = \vec{u} \cdot (\nabla \cdot \boldsymbol{\sigma}) + \boldsymbol{\sigma} : \nabla \vec{u} \qquad (1.58)$$

Using the momentum equation [Eq. (1.38)], and noting that $\vec{u} \cdot \frac{\partial(\rho\vec{u})}{\partial t} = \frac{\partial\left(\rho\frac{|\vec{u}|^2}{2}\right)}{\partial t}$ and $\vec{u} \cdot \nabla \cdot (\rho\vec{u}\vec{u}) = \nabla \cdot \left(\rho\vec{u}\frac{|\vec{u}|^2}{2}\right)$,

$$\vec{u} \cdot \nabla \cdot \boldsymbol{\sigma} = \vec{u} \cdot \left(\frac{\partial(\rho\vec{u})}{\partial t} + \nabla \cdot (\rho\vec{u}\vec{u}) \right) - \vec{u} \cdot (\rho\vec{g}) - \vec{u} \cdot \vec{f}^*$$

$$= \left(\frac{\partial(\rho\frac{|\vec{u}|^2}{2})}{\partial t} + \nabla \cdot (\rho\vec{u}\frac{|\vec{u}|^2}{2}) \right) - \vec{u} \cdot (\rho\vec{g}) - \vec{u} \cdot \vec{f}^* \qquad (1.59)$$

The work done by stresses per unit volume and expressed in Eq. (1.58) becomes

$$\dot{w}_\sigma = \left(\frac{\partial(\rho \frac{|\vec{u}|^2}{2})}{\partial t} + \nabla \cdot (\rho \vec{u} \frac{|\vec{u}|^2}{2}) \right) - \vec{u} \cdot (\rho \vec{g}) - \vec{u} \cdot \vec{f}^* + \boldsymbol{\sigma} : \nabla \vec{u} \qquad (1.60)$$

Substituting in Eq. (1.55), noting that the integrands on either side of the equality must be equal, and neglecting the radiation term

$$\frac{\partial(\rho e)}{\partial t} + \nabla \cdot (\rho \vec{u} e) = \nabla \cdot (\lambda \nabla T) + \left(\frac{\partial(\rho \frac{|\vec{u}|^2}{2})}{\partial t} + \nabla \cdot (\rho \vec{u} \frac{|\vec{u}|^2}{2}) \right) - \vec{u} \cdot (\rho \vec{g}) + \boldsymbol{\sigma} : \nabla \vec{u} \quad (1.61)$$

Noting that $e = u + \frac{1}{2}|\vec{u}|^2 + gz$, then

$$\frac{\partial}{\partial t}\left(\rho(u + \frac{1}{2}|\vec{u}|^2 + gz) \right) + \nabla \cdot \left(\rho \vec{u}(u + \frac{1}{2}|\vec{u}|^2 + gz) \right)$$

$$= \nabla \cdot (\lambda \nabla T) + \left(\frac{\partial(\rho \frac{|\vec{u}|^2}{2})}{\partial t} + \nabla \cdot (\rho \vec{u} \frac{|\vec{u}|^2}{2}) \right) - \vec{u} \cdot (\rho \vec{g}) + \boldsymbol{\sigma} : \nabla \vec{u} \qquad (1.62)$$

The terms involving $|\vec{u}|^2$ cancel from both sides of the equation yielding:

$$\frac{\partial}{\partial t}\left(\rho(u + gz) \right) + \nabla \cdot (\rho \vec{u}(u + gz)) = \nabla \cdot (\lambda \nabla T) - \vec{u} \cdot (\rho \vec{g}) + \boldsymbol{\sigma} : \nabla \vec{u} \qquad (1.63)$$

Further simplification can be realized by noting that

$$\frac{\partial(\rho gz)}{\partial t} + \nabla \cdot (\rho \vec{u} gz) = \rho \frac{\partial(gz)}{\partial t} + \rho \vec{u} \cdot \nabla(gz) + gz \left[\frac{\partial \rho}{\partial t} + \nabla \cdot (\rho \vec{u}) \right] \qquad (1.64)$$

The term in square brackets on the right-hand side cancels by the continuity equation, and noting that gz is time independent, and that $\nabla(gz) = -\vec{g}$, then

$$\frac{\partial(\rho gz)}{\partial t} + \nabla \cdot (\rho \vec{u} gz) = -\rho \vec{u} \cdot \vec{g} \qquad (1.65)$$

Consequently, the potential energy terms also cancel in Eq. (1.63) so that

$$\frac{\partial(\rho u)}{\partial t} + \nabla \cdot (\rho \vec{u} u) = \nabla \cdot (\lambda \nabla T) + \boldsymbol{\sigma} : \nabla \vec{u} \qquad (1.66)$$

Substituting Eq. (1.41) into Eq. (1.66),

$$\frac{\partial(\rho u)}{\partial t} + \nabla \cdot (\rho \vec{u} u) = \nabla \cdot (\lambda \nabla T) - p\boldsymbol{I} : \nabla \vec{u} + \boldsymbol{\sigma}_d : \nabla \vec{u} \qquad (1.67)$$

Noting that $\boldsymbol{I} : \nabla \vec{u} = \nabla \cdot \vec{u}$, the differential form of the energy equation, expressed in terms of the specific internal energy is

$$\frac{\partial(\rho u)}{\partial t} + \nabla \cdot (\rho \vec{u} u) = \nabla \cdot (\lambda \nabla T) - p\nabla \cdot \vec{u} + \boldsymbol{\sigma}_d : \nabla \vec{u} \qquad (1.68)$$

For a Newtonian fluid,

$$\frac{\partial(\rho u)}{\partial t} + \nabla \cdot (\rho \vec{u} u) = \nabla \cdot (\lambda \nabla T) - p \nabla \cdot \vec{u} + \mu \Phi \tag{1.69}$$

where the viscous dissipation term, Φ, is given by

$$\Phi = 2 \left(e - \frac{1}{3} \nabla \cdot \vec{u} I \right) : \nabla \vec{u} \tag{1.70}$$

$$= 2e : \nabla \vec{u} - \frac{2}{3} (\nabla \cdot \vec{u})^2 \tag{1.71}$$

where $I : \nabla \vec{u} = \nabla \cdot \vec{u}$ and

$$2e : \nabla \vec{u} = 2 \left(\left(\frac{\partial u}{\partial x} \right)^2 + \left(\frac{\partial v}{\partial y} \right)^2 + \left(\frac{\partial w}{\partial z} \right)^2 \right)$$
$$+ \left(\frac{\partial u}{\partial y} + \frac{\partial v}{\partial x} \right)^2 + \left(\frac{\partial u}{\partial z} + \frac{\partial w}{\partial x} \right)^2 + \left(\frac{\partial v}{\partial z} + \frac{\partial w}{\partial y} \right)^2 \tag{1.72}$$

Then,

$$\Phi = \frac{2}{3} \left(\left(\frac{\partial u}{\partial x} - \frac{\partial v}{\partial y} \right)^2 + \left(\frac{\partial v}{\partial y} - \frac{\partial w}{\partial z} \right)^2 + \left(\frac{\partial u}{\partial x} - \frac{\partial w}{\partial z} \right)^2 \right)$$
$$+ \left(\frac{\partial u}{\partial y} + \frac{\partial v}{\partial x} \right)^2 + \left(\frac{\partial u}{\partial z} + \frac{\partial w}{\partial x} \right)^2 + \left(\frac{\partial v}{\partial z} + \frac{\partial w}{\partial y} \right)^2 \tag{1.73}$$

Using Eq. (1.32), the inertia term on the left-hand side of the energy equation [Eq. (1.69)] may be replaced by a Lagrangian derivative to yield the following alternative form of the energy equation:

$$\rho \frac{du}{dt} = \nabla \cdot (\lambda \nabla T) - p \nabla \cdot \vec{u} + \mu \Phi \tag{1.74}$$

Equation (1.74) states that the rate of change of internal energy of a small differential volume of fluid is balanced by heat transfer across its boundary, work done by/onto the particle as it expands/contracts, and a non-negative contribution from viscous effects ($\Phi \geq 0$). The viscous dissipation term accounts for the irreversible conversion into local heating of the work done by the fluid to overcome friction.

Expressing the energy equation in terms of temperature requires knowledge of the equation of state for the fluid over the operating pressure and temperature conditions. For a simple compressible material, one may express the specific internal energy as a function of two independent intensive properties. Choosing these two properties to be the temperature and the density, then $u = u(T, \rho)$ so that

$$du = \frac{\partial u}{\partial T} \bigg|_{\rho} dT + \frac{\partial u}{\partial \rho} \bigg|_{T} d\rho = c_v dT + \frac{\partial u}{\partial v} \bigg|_{T} dv = c_v dT + \left(-p + T \frac{\alpha}{\beta} \right) dv \tag{1.75}$$

The constant volume specific heat, $c_v = \frac{\partial u}{\partial T} \big|_{v}$, measures the increase in specific internal energy with increasing temperature in an isochoric process (constant volume). The coefficient of volume thermal expansion, $\alpha = \frac{1}{v} \frac{\partial v}{\partial T}\big|_p$, measures the relative increase of volume with increasing temperature in an isobaric process (constant pressure). The coefficient of isothermal compressibility, $\beta = -\frac{1}{v} \frac{\partial v}{\partial p}\big|_T$, measures the relative decrease of volume with increasing pressure in an isothermal process. For an ideal gas where $pv = RT$, $\alpha = 1/T$, and $\beta = 1/p$, the specific internal energy is a function of temperature only and $du = c_v dT$. For liquids, the specific internal energy can be approximated by the internal energy of the saturated liquid at the same temperature, i.e., $u = u_f(T)$. One may also use $du \simeq c_v dT$ for liquids since dv is typically very small due to the high incompressibility.

To conclude this section, we point out that one may alternatively express the energy equation in terms of the specific enthalpy, $\mathfrak{h}$, by using $u = \mathfrak{h} - \frac{p}{\rho}$. In this context, it is worth noting that for a simple compressible material, $\mathfrak{h}$ is a function of two independent intensive properties. Choosing temperature and pressure as the two properties, then $\mathfrak{h} = \mathfrak{h}(T, p)$

$$
d\mathfrak{h} = \frac{\partial \mathfrak{h}}{\partial T}\bigg|_p dT + \frac{\partial \mathfrak{h}}{\partial p}\bigg|_T dp
$$
$$
= c_p dT + (1 - \alpha T)v dp \tag{1.76}
$$

Note that for an ideal gas, $\alpha = 1/T$ and $d\mathfrak{h} = c_p dT$. For liquids, α is very small and $d\mathfrak{h} \simeq c_p dT + v dp$. Note for both liquids and ideal gases, the specific heat c_v is in general temperature dependent.

1.3 Special Cases

We note here special flow conditions which lead to simplification of the governing equations.

1.3.1 Steady Flow

A flow is *steady* when the macroscopic properties do not change in time at any location in the flow field, i.e., $\frac{\partial \phi(\vec{x},t)}{\partial t} = 0$, $\forall \vec{x} \in \mathcal{V}$, $\phi \in \{\rho, p, \vec{u}, T\}$ which implies that the properties are a function of space only, i.e., $\phi(\vec{x})$. For the flow to be steady, the boundary conditions must be time independent. In addition, source terms, such as production/consumption of species and heat due to chemical reactions must also be time independent. Note that these are necessary but not sufficient conditions for the flow to be steady after enough time has passed for the effect of the initial conditions to die out.

1.3.2 Two-Dimensional Flow

A flow can be modeled as *two-dimensional*, e.g., in the $x - y$ plane, if the flow variables do not vary in the third dimension, e.g., $\frac{\partial \phi}{\partial z} = 0$, $\phi \in \{\rho, p, \vec{u}, T\}$. An example is the flow in a rectangular channel depicted in Figure 1.8 where the flow conditions at the inlet of the channel do not depend on z. If the channel width, W, is much larger than the channel height H, that is, $W >> H$, then the flowfield in plane $x_1 - y_1$ is similar to that in plane $x_2 - y_2$ as long as the two planes are sufficiently far from the side walls, $z = \pm\frac{W}{2}$.

1.3.3 Inviscid Flow

An inviscid or frictionless flow is a flow in which the viscous effects are not present. For a Newtonian flow, this implies that the viscous tangential ($\nabla \cdot (\mu \nabla \vec{u}) = 0$) and normal stresses ($\frac{1}{3}\nabla (\mu \nabla \cdot \vec{u}) = 0$) are zero. The inviscid flow approximation should be

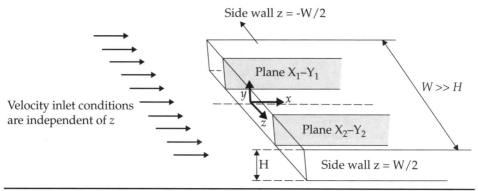

FIGURE 1.8 Flow in a wide rectangular channel, where the inlet conditions do not depend on z, can be modeled as two-dimensional in the $x - y$ plane.

taken with care as it may lead to nonphysical results. A condition for the viscous effect to be neglected with respect to inertia is that the Reynolds number must be very large, $\text{Re} = \frac{\rho_\infty V_\infty L}{\mu_\infty} \gg 1$. Whether a large Reynolds number flow can be modeled as inviscid depends on additional considerations. If the flow is turbulent due to growth of instabilities from boundaries or within the domain, then one cannot make the inviscid flow approximation. This is because turbulence is characterized by viscous effects that grow as large eddies break down into smaller eddies in a cascading fashion. Therefore, for the inviscid flow approximation to hold, it must apply to zones where the flow is streamlined with little turbulence, if any. In addition, viscous effects are always present near solid boundaries where the no-slip conditions hold. As such, the inviscid flow approximation should be made for flow regions outside the boundary layer (defined in Chapter 3). Another consideration is that the inviscid flow approximation should not be used to infer flow parameters that are strongly affected by the viscous effects that are present in isolated zones, as in the thin boundary layer of a streamlined flow over an airfoil, as depicted in the Figure 1.9. In the cases of streamlined flow over an airfoil, one can model the flow outside the boundary layer as an inviscid flow, but predicting the drag force would not be possible without accounting for the viscous effects inside the boundary layer.

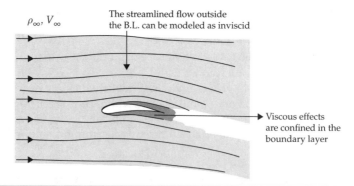

FIGURE 1.9 Two-dimensional large Reynolds number steady flow over an airfoil. If the flow is streamlined, one can model the flow region outside the boundary layer as inviscid.

1.3.4 Inertia-free (Stokes) Flow

Flows characterized by a very small Reynolds number, $\text{Re} = \frac{\rho VL}{\mu} \ll 1$, can be modeled as inertia-free flows. The assumption of inertia-free flows implies that one can neglect the inertia terms of the momentum equation, i.e., $\nabla \cdot (\rho \vec{u}\vec{u}) \simeq \vec{0}$ and depending on the problem, $\frac{\partial(\rho \vec{u})}{\partial t} \simeq \vec{0}$. Inertia-free flows arise in many microfluidics applications, where the length scale is of the order of 1 micrometer, so that over the ranges of velocities encountered in most of these applications, the condition $\text{Re} = \frac{\rho VL}{\mu} \ll 1$ holds. Note that in many micro and nano flows, especially in gas flows, the continuum hypothesis breaks for $\text{Kn} > 0.01$, where the Knudsen number $\text{Kn} = \frac{l_m}{L}$ and l_m is a microscopic length scale. For gases, l_m is the mean free path. For flow conditions where the continuum hypothesis condition does not hold, one cannot use the governing equations in the form presented in this chapter to simulate the flow. Instead, alternative approaches are adopted such as Molecular Dynamics (MD) and Direct Simulation Monte Carlo (DSMC).

1.3.5 Constant Density Flow

If the flow conditions are such that the density remains constant, then the flow is a constant density flow $\rho(\vec{x}, t) = \rho_0$. An example is water flow under thermodynamic conditions that are in the compressed liquid region but not far from a reference state, at which the density is ρ_0. It is important to note the difference between a constant density flow and an incompressible flow, which is described next.

1.3.6 Incompressible Flow

Incompressible flow is a flow that satisfies the condition $|\nabla \cdot \vec{u}| \ll |\frac{\partial u}{\partial x}| + |\frac{\partial v}{\partial y}| + |\frac{\partial w}{\partial z}|$, which upon employing the conservation of mass, implies that the relative rate of change of the density of a particle is very small, i.e., $\left|\frac{1}{\rho}\frac{d\rho}{dt}\right| = |\nabla \cdot \vec{u}| \ll \frac{U}{L}$, where the velocity and length scales are chosen to capture the order of magnitude of $|\frac{\partial u}{\partial x}| + |\frac{\partial v}{\partial y}| + |\frac{\partial w}{\partial z}|$. Further development of the incompressible flow condition $\left|\frac{1}{\rho}\frac{d\rho}{dt}\right| \ll \frac{U}{L}$ depends on whether the flow is a large inertia flow or a low inertia flow. For large inertia flows, one may invoke the inviscid steady flow limit in the momentum equation, which in the absence of a body force, simplifies to $\rho V \frac{dV}{ds} = -\frac{\partial p}{\partial s}$, where s is the coordinate along a streamline. If we assume that the flow is isentropic, which is in accordance with the inviscid (frictionless) flow limiting case of large inertia flows, then the change of pressure with density while holding the entropy constant is related to the speed of sound in the medium according to $dp = a_s^2 \, d\rho$. Then, $\frac{d\rho}{\rho}$ scales as $\frac{VdV}{a_s^2}$ and $\left|\frac{1}{\rho}\frac{d\rho}{dt}\right|$ scales as $\frac{V}{a_s^2}\frac{dV}{dt}$. Scaling V with U and t with $\frac{L}{U}$, $\left|\frac{1}{\rho}\frac{d\rho}{dt}\right|$ scales as $\frac{U^2}{a_s^2}\frac{U}{L}$. The incompressible flow condition for large inertia flows is then $\text{Ma}^2 \ll 1$, where Mach number, Ma, is the ratio of the characteristic flow speed, U, to the speed of sound, a_s in the medium, i.e., $\text{Ma} = \frac{U}{a_s}$. In physical terms, if the flow conditions are such that changes in the flow speed yield changes in the pressure ($dp \sim \rho VdV$) that are sufficiently large to inflict considerable changes in the density ($d\rho \sim \frac{dp}{a_s^2}$), then the flow is a compressible flow, otherwise it is an incompressible flow. Note that density can still vary in an incompressible flow, but its variation does not originate from larger pressure changes over small distances. Examples of incompressible flows with variable density are large scale oceanic and atmospheric flows. Buoyancy driven flows are other examples of incompressible flows with variable density.

1.4   General Form of the Conservation Laws

The equations governing the conservation of mass, momentum, and energy may be expressed in the following form:

$$\frac{\partial(\rho\Lambda\phi)}{\partial t} + \nabla \cdot (\rho\Lambda\phi\vec{u}) = \nabla \cdot (\Gamma\nabla\phi) + \dot{S} \tag{1.77}$$

where t is time (sec), ρ is the density (kg/m^3), $\vec{u}$ is the velocity vector, ϕ is the intensive macroscopic property carried by mass. The diffusion coefficient, Γ, the source term, $\dot{S}$, the coefficient, Λ, and ϕ are listed in Table 1.1 for the different conservation laws, where the momentum equation is expressed in terms of its components along Cartesian coordinates. In Table 1.1, μ is the dynamic viscosity (kg/(m.s)), p is the absolute thermodynamic pressure (Pa), $\vec{g}$ is the gravitational acceleration (m/s^2), c_v is the constant volume specific heat (J/kg.K), c_p is the constant pressure specific heat (J/kg.K), and λ is thermal conductivity (W/(m.K)). Note that we also added $\dot{q}_g$ to the energy equation, which is the rate of heat production per unit volume (J/(s.m^3)) due to sources inside the domain, such as electro-magnetic heating. Note that expressions for u and ħ in terms of temperature, specific volume, and pressure for a simple compressible fluid are presented in Eqs. (1.75) and (1.76), respectively.

The finite volume method, which is the subject of Chapter 2, is based on numerical representation of the integral form of the conservation laws over a control volume, $\mathcal{V}$, taken to be a computational cell (or grid cell). As done before, we can arrive at the integral by integrating Eq. (1.77) over $\mathcal{V}$, bounded by surface $\mathcal{S}$:

$$\int_{\mathcal{V}} \left(\frac{\partial(\rho\Lambda\phi)}{\partial t} + \nabla \cdot (\Lambda\phi\,\rho\vec{u}) = \nabla \cdot (\Gamma\nabla\phi) + \dot{S} \right) d\mathcal{V} \tag{1.78}$$

We then employ Gauss' divergence theorem which enables representing the volume integral of the divergence of a vector quantity as a surface integral of the flux across the bounding surface,

$$\int_{\mathcal{V}} \frac{\partial(\rho\Lambda\phi)}{\partial t} d\mathcal{V} + \int_{\mathcal{S}} \Lambda\phi\,\rho\vec{u} \cdot \hat{n}\, d\mathcal{S} = \int_{\mathcal{S}} \Gamma\nabla\phi \cdot \hat{n}\, d\mathcal{S} + \int_{\mathcal{V}} \dot{S}\, d\mathcal{V} \tag{1.79}$$

TABLE 1.1 Expressions for ϕ, Λ, Γ, and $\dot{S}$ for the various conservation laws. In the energy equation, it is assumed that the specific internal energy and enthalpy depend only on the temperature[11], and the contribution of the bulk viscosity is ignored.

Conservation law	ϕ	Λ	Γ	$\dot{S}$
Mass	1	1	0	0
Linear momentum x	u	1	μ	$-\frac{\partial p}{\partial x} + \nabla \cdot \left(\mu(\nabla u)^{\mathsf{T}}\right) - \frac{2}{3}\frac{\partial p}{\partial x}(\mu\nabla \cdot \vec{u}) + \rho\vec{g} \cdot \hat{x} + \vec{f}* \cdot \hat{x}$
Linear momentum y	v	1	μ	$-\frac{\partial p}{\partial y} + \nabla \cdot \left(\mu(\nabla v)^{\mathsf{T}}\right) - \frac{2}{3}\frac{\partial p}{\partial y}(\mu\nabla \cdot \vec{u}) + \rho\vec{g} \cdot \hat{y} + \vec{f}* \cdot \hat{y}$
Linear momentum z	w	1	μ	$-\frac{\partial p}{\partial z} + \nabla \cdot \left(\mu(\nabla w)^{\mathsf{T}}\right) - \frac{2}{3}\frac{\partial p}{\partial z}(\mu\nabla \cdot \vec{u}) + \rho\vec{g} \cdot \hat{z} + \vec{f}* \cdot \hat{z}$
Energy	T	c_v	λ	$-p\nabla \cdot \vec{u} + \mu\Phi + \dot{q}_g$
Energy	T	c_p	λ	$\rho\frac{d}{dt}\left(\frac{p}{\rho}\right) - (1-\alpha T)\frac{dp}{dt} - p\nabla \cdot \vec{u} + \mu\Phi + \dot{q}_g$

---

[11]Which is generally a good approximation for gases and liquids.

If the control volume $\mathcal{V}$ is fixed, $\int_{\mathcal{V}} \frac{\partial(\rho\Lambda\phi)}{\partial t} d\mathcal{V} = \frac{\partial}{\partial t} \int_{\mathcal{V}} \rho\Lambda\phi \, d\mathcal{V}$, so that

$$\frac{\partial}{\partial t} \int_{\mathcal{V}} \rho\Lambda\phi \, d\mathcal{V} + \int_{\mathcal{S}} \vec{J}_{\phi} \cdot \hat{n} \, d\mathcal{S} = \int_{\mathcal{V}} \dot{S} \, d\mathcal{V} \tag{1.80}$$

where the total flux, $\vec{J}_{\phi}$, of the physical quantity ϕ across the control surface is the sum of the advection flux, $\vec{J}_{adv} = \Lambda\phi \, \rho\vec{u}$, and the diffusion flux, $\vec{J}_{dif} = -\Gamma\nabla\phi$

$$\vec{J}_{\phi} = \Lambda\phi \, \rho\vec{u} - \Gamma\nabla\phi \tag{1.81}$$

Equation (1.80) is the starting point for numerical representation of the conservation laws on a grid representing the spatial domain for the various applications presented in Chapters 3 to 7.

1.5 Cartesian, Cylindrical, and Spherical Coordinates

Choice of the coordinate system is often dictated by the problem being studied. For example, the cylindrical coordinate system is an appropriate choice for studying the pipe flow discussed in Chapters 3 to 6. In what follows, we present the representation of vectors and vector operators in Cartesian coordinates, Cylindrical coordinates, and Spherical coordinates.

1.5.1 Cartesian Coordinates

In Cartesian coordinates, the unit vectors, $\hat{x}, \hat{y}$, and $\hat{z}$ along the x, y, and z coordinates, illustrated in Figure 1.10, are orthogonal to each other. This makes Cartesian coordinates the easiest to deal with. Considering the vectors $\vec{u} = u_x\hat{x} + u_y\hat{y} + u_z\hat{z}$ and $\vec{v} = v_x\hat{x} + v_y\hat{y} + v_z\hat{z}$ and the scalar T, then

$$\vec{u} \cdot \vec{v} = u_x v_x + u_y v_y + u_z v_z \tag{1.82}$$

$$\vec{u} \times \vec{v} = (u_y v_z - u_z v_y)\hat{x} + (u_z v_x - u_x v_z)\hat{y} + (u_x v_y - u_y v_x)\hat{z} \tag{1.83}$$

$$\nabla = \frac{\partial}{\partial x}\hat{x} + \frac{\partial}{\partial y}\hat{y} + \frac{\partial}{\partial z}\hat{z} \tag{1.84}$$

$$\nabla \cdot \vec{u} = \frac{\partial u_x}{\partial x} + \frac{\partial u_y}{\partial y} + \frac{\partial u_z}{\partial z} \tag{1.85}$$

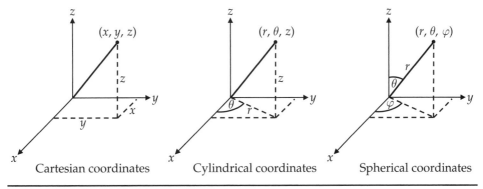

FIGURE 1.10 Cartesian, cylindrical and spherical coordinates.

$$\nabla \times \vec{u} = \left(\frac{\partial u_z}{\partial y} - \frac{\partial u_y}{\partial z} \right) \hat{x} + \left(\frac{\partial u_x}{\partial z} - \frac{\partial u_z}{\partial x} \right) \hat{y} + \left(\frac{\partial u_y}{\partial x} - \frac{\partial u_x}{\partial y} \right) \hat{z} \tag{1.86}$$

$$\vec{u} \cdot \nabla = u_x \frac{\partial}{\partial x} + u_y \frac{\partial}{\partial y} + u_z \frac{\partial}{\partial z} \tag{1.87}$$

$$\nabla^2 T = \nabla \cdot (\nabla T) = \frac{\partial^2 T}{\partial x^2} + \frac{\partial^2 T}{\partial y^2} + \frac{\partial^2 T}{\partial z^2} \tag{1.88}$$

$$\nabla^2 \vec{u} = \nabla \cdot (\nabla \vec{u}) = \frac{\partial^2 \vec{u}}{\partial x^2} + \frac{\partial^2 \vec{u}}{\partial y^2} + \frac{\partial^2 \vec{u}}{\partial z^2} = (\nabla^2 u_x) \hat{x} + (\nabla^2 u_y) \hat{y} + (\nabla^2 u_z) \hat{z} \tag{1.89}$$

1.5.2 Cylindrical Coordinates

In Cylindrical coordinates, r, θ, and z, illustrated in Figure 1.10, the unit vectors, $\hat{r}$ and $\hat{\theta}$ are dependent on θ according to $\frac{\partial \hat{r}}{\partial \theta} = \hat{\theta}$ and $\frac{\partial \hat{\theta}}{\partial \theta} = -\hat{r}$. This demands care when dealing with these coordinates, especially when taking the spatial derivatives of a vector. Considering the vectors $\vec{u} = u_r \hat{r} + u_\theta \hat{\theta} + u_z \hat{z}$ and $\vec{v} = v_r \hat{r} + v_\theta \hat{\theta} + v_z \hat{z}$ and the scalar T, then

$$\vec{u} \cdot \vec{v} = u_r v_r + u_\theta v_\theta + u_z v_z \tag{1.90}$$

$$\vec{u} \times \vec{v} = (u_\theta v_z - u_z v_\theta) \hat{r} + (u_z v_r - u_r v_z) \hat{\theta} + (u_r v_\theta - u_\theta v_r) \hat{z} \tag{1.91}$$

$$\nabla = \frac{\partial}{\partial r} \hat{r} + \frac{1}{r} \frac{\partial}{\partial \theta} \hat{\theta} + \frac{\partial}{\partial z} \hat{z} \tag{1.92}$$

$$\nabla \cdot \vec{u} = \frac{1}{r} \frac{\partial (r u_r)}{\partial r} + \frac{1}{r} \frac{\partial u_\theta}{\partial \theta} + \frac{\partial u_z}{\partial z} \tag{1.93}$$

$$\nabla \times \vec{u} = \left(\frac{1}{r} \frac{\partial u_z}{\partial \theta} - \frac{\partial u_\theta}{\partial z} \right) \hat{r} + \left(\frac{\partial u_r}{\partial z} - \frac{\partial u_z}{\partial r} \right) \hat{\theta} + \frac{1}{r} \left(\frac{\partial (r u_\theta)}{\partial r} - \frac{\partial u_r}{\partial \theta} \right) \hat{z} \tag{1.94}$$

$$\vec{u} \cdot \nabla = u_r \frac{\partial}{\partial r} + \frac{u_\theta}{r} \frac{\partial}{\partial \theta} + u_z \frac{\partial}{\partial z} \tag{1.95}$$

$$\nabla^2 T = \frac{1}{r} \frac{\partial}{\partial r} \left(r \frac{\partial T}{\partial r} \right) + \frac{1}{r^2} \frac{\partial^2 T}{\partial \theta^2} + \frac{\partial^2 T}{\partial z^2} \tag{1.96}$$

$$\nabla^2 \vec{u} = \left[\nabla^2 u_r - \frac{u_r}{r^2} - \frac{2}{r^2} \frac{\partial u_\theta}{\partial \theta} \right] \hat{r} + \left[\nabla^2 u_\theta - \frac{u_\theta}{r^2} + \frac{2}{r^2} \frac{\partial u_r}{\partial \theta} \right] \hat{\theta} + (\nabla^2 u_z) \hat{z} \tag{1.97}$$

In axisymmetric cylindrical coordinates, the properties are independent of θ, so that $\frac{\partial}{\partial \theta}(T, u_r, u_x, u_\theta) = 0$. Additionally, in the absence of a swirl velocity component, $u_\theta = 0$.

1.5.3 Spherical Coordinates

In spherical coordinates, r, θ, φ, illustrated in Figure 1.10, the unit vectors are dependent on θ and φ according to $\frac{\partial \hat{r}}{\partial \theta} = \hat{\theta}$, $\frac{\partial \hat{\theta}}{\partial \theta} = -\hat{r}$, $\frac{\partial \hat{\varphi}}{\partial \theta} = 0$, $\frac{\partial \hat{r}}{\partial \varphi} = \sin \theta \, \hat{\varphi}$, $\frac{\partial \hat{\theta}}{\partial \varphi} = \cos \theta \, \hat{\varphi}$ and $\frac{\partial \hat{\varphi}}{\partial \varphi} = -\sin \theta \, \hat{r} - \cos \theta \, \hat{\theta}$. This demands care when dealing with these coordinates, especially when taking the spatial derivatives of a vector. Considering the vectors $\vec{u} = u_r \hat{r} + u_\theta \hat{\theta} + u_\varphi \hat{\varphi}$ and $\vec{v} = v_r \hat{r} + v_\theta \hat{\theta} + v_\varphi \hat{\varphi}$ and the scalar T, then

$$\vec{u} \cdot \vec{v} = u_r v_r + u_\theta v_\theta + u_\varphi v_\varphi \tag{1.98}$$

$$\vec{u} \times \vec{v} = (u_\theta v_\varphi - u_\varphi v_\theta) \hat{r} + (u_\varphi v_r - u_r v_\varphi) \hat{\theta} + (u_r v_\theta - u_\theta v_r) \hat{\varphi} \tag{1.99}$$

$$\nabla = \frac{\partial}{\partial r}\hat{r} + \frac{1}{r}\frac{\partial}{\partial \theta}\hat{\theta} + \frac{1}{r\sin\theta}\frac{\partial}{\partial \varphi}\hat{\varphi} \tag{1.100}$$

$$\nabla \cdot \vec{u} = \frac{1}{r^2}\frac{\partial(r^2 u_r)}{\partial r} + \frac{1}{r\sin\theta}\frac{\partial(u_\theta \sin\theta)}{\partial \theta} + \frac{1}{r\sin\theta}\frac{\partial u_\varphi}{\partial \varphi} \tag{1.101}$$

$$\nabla \times \vec{u} = \frac{1}{r\sin\theta}\left(\frac{\partial(u_\varphi \sin\theta)}{\partial \theta} - \frac{\partial u_\theta}{\partial \varphi}\right)\hat{r} + \frac{1}{r}\left(\frac{1}{\sin\theta}\frac{\partial u_r}{\partial \varphi} - \frac{\partial(r u_\varphi)}{\partial r}\right)\hat{\theta}$$
$$+ \frac{1}{r}\left(\frac{\partial(r u_\theta)}{\partial r} - \frac{\partial u_r}{\partial \theta}\right)\hat{z} \tag{1.102}$$

$$\vec{u} \cdot \nabla = u_r\frac{\partial}{\partial r} + \frac{u_\theta}{r}\frac{\partial}{\partial \theta} + \frac{u_\varphi}{r\sin\theta}\frac{\partial}{\partial \varphi} \tag{1.103}$$

$$\nabla^2 T = \frac{1}{r^2}\frac{\partial}{\partial r}\left(r^2\frac{\partial T}{\partial r}\right) + \frac{1}{r^2\sin\theta}\frac{\partial}{\partial \theta}\left(\sin\theta\frac{\partial T}{\partial \theta}\right) + \frac{1}{r^2\sin^2\theta}\frac{\partial^2 T}{\partial \varphi^2} \tag{1.104}$$

$$\nabla^2 \vec{u} = \left[\nabla^2 u_r - \frac{2 u_r}{r^2} - \frac{2}{r^2\sin\theta}\frac{\partial(\sin\theta\, u_\theta)}{\partial \theta} - \frac{2}{r^2\sin\theta}\frac{\partial u_\varphi}{\partial \varphi}\right]\hat{r}$$
$$+ \left[\nabla^2 u_\theta - \frac{u_\theta}{r^2\sin^2\theta} - \frac{2\cos\theta}{r^2\sin^2\theta}\frac{\partial u_\varphi}{\partial \varphi} + \frac{2}{r^2}\frac{\partial u_r}{\partial \theta}\right]\hat{\theta}$$
$$+ \left[\nabla^2 u_\varphi - \frac{u_\varphi}{r^2\sin^2\theta} + \frac{2\cos\theta}{r^2\sin^2\theta}\frac{\partial u_\theta}{\partial \varphi} + \frac{2}{r^2\sin\theta}\frac{\partial u_r}{\partial \varphi}\right]\hat{\varphi}. \tag{1.105}$$

CHAPTER 2

Introduction to Computational Fluid Dynamics Using the Finite Volume Method

List of Symbols	
Reynolds number	Re
Density	ρ
Error	e
Grid size	Δx
Time step	Δt
Volume	$\mathcal{V}$
Centroid of the cell under consideration	$\mathbb{O}_i$
Set of neighbors sharing a face with cell i	$\mathcal{N}_i$
Face of a cell i	$\Bbbk$
List of faces	$\mathcal{F}$
Area of face $\Bbbk$	$\mathcal{A}_{\Bbbk}$
Centroid of face $\Bbbk$	$O_{\Bbbk}$
Time	t
Area	$\mathcal{S}$
Diffusion coefficient	Γ
Normal vector	$\hat{n}$
Source term per unit volume	$\dot{S}$
Advection flux	$\vec{J}_{adv}$
Diffusion flux	$\vec{J}_{dif}$
Velocity vector	$\vec{u}$
Mass flow rate leaving across face $\Bbbk$	$\dot{m}_{\Bbbk}^{\rightarrow}$
Pressure	p

Continued

List of Symbols	
Dynamic viscosity	μ
Gravitational acceleration	$\vec{g}$
Discretization error	ϵ_h
Global cell size	h
Grid convergence index	GCI
Global order of accuracy	$\not{p}_{av}$

The theoretical solutions to fluid dynamics problems are available to specific problems, such as the Poiseuille and Couette flows. For other engineering applications, the experimental approach is used to acquire measurements of flow properties. In many cases, it is difficult and/or expensive to replicate the actual problem in the laboratory. With the emergence of computers and their rapid advancements enabled continuous improvement in processing power and storage capacity, numerical methods for solving a discretized version of the Navier-Stokes equations to yield a detailed description of the flow field offered a powerful venue that not only complemented the experimental approach, but also enabled the investigation of new problems. Moreover, computational fluid dynamics (CFD) offered new insights and deeper and more comprehensive understanding of many fluid phenomena arising in nature, such as understanding of the flight dynamics of insects and birds, fire dynamics, and atmospheric and ocean dynamics. Investigation, using computational methods, of new problems or old challenging problems enabled advancement of many technological fields.

2.1   What Is Computational Fluid Dynamics?

CFD numerically solves a discretized form of the mathematical model representing the physical laws governing fluid flows in a given domain divided into cells, subject to prescribed boundary conditions. The discretized system is the outcome of employing truncated Taylor series expansion to approximate the governing equations at each cell of the grid, which is a discrete spatial representation of the domain of interest. This approximation yields a system of algebraic equations governing the unknown quantities at discrete locations, often selected to be the centroids of the grid cells. The unknown quantities are then determined by solving the system of algebraic equations on a computer. The building blocks that constitute the CFD solution method are presented in the following section, along with the criteria that govern its viability and accuracy. The applications of CFD are many and span many engineering and other fields like the design of microfluidic devices, lab-on-chip, and micro-total analysis systems.

CFD not only complements the experimental approach but also enables exploration of problems that cannot be studied or are too expensive to study in a laboratory. CFD also, in principle, allows simulating a flow in the actual prototype as opposed to a scaled down or scaled up model. The computational laboratory enabled by CFD does not put a constraint on how small or large the study domain is or how short or long the observation time is. Indeed, CFD has been used to study microflows over short observation time as well as in climate investigation of planetary scale atmospheric and ocean dynamics over a very long observation time (order of 100 years). In CFD, one can also implement boundary conditions that are, in many cases, closer to those in the actual flow than the boundary conditions that can be implemented in the laboratory. In addition, one can easily change the working fluid, the operating conditions, and the dimensions by simply changing the corresponding values in the model. Modern CFD tools also allow effortless sweeping of design variables and even embedding a model within optimization tools that yield improved design.

One, however, should be careful when using CFD to study complex flows with multi-physics. For these problems, there are always approximations involved for the purpose of making the computational cost affordable without jeopardizing the accuracy of the predicted quantities of interest. In addition to the errors incurred in the numerical solution of the mathematical system, which are discussed in Section 2.2, there are errors associated with the mathematical system being not an accurate representation of the actual flow. This is either because an accurate representation requires an unfeasible numerical solution due to the prohibitively expensive associated computational cost (e.g., turbulence) or because such an accurate representation is not possible (e.g., combustion).

A major challenge in CFD is simulating turbulent flows to an acceptable degree of accuracy. This is because resolving the length and time scales of the small eddies in turbulent flows is impossible in most engineering problems. Resolving these scales is possible using direct numerical simulation, but this is viable only for very small problems due to the associated cost. To render the computational cost of simulating three-dimensional turbulent flows affordable, the grid size is chosen to be significantly larger than the smallest eddy scales in the flow. In order to resolve the sub-grid scale, turbulence modeling is used. There is no single turbulence model that is suitable for all applications, and each turbulence model is characterized by parameters that should be tuned or determined empirically. Complex physics, such as combustion, also add to the challenges in obtaining accurate CFD solutions. This is because of the difficulty in accurately representing the associated chemical reactions. Even when an adequate representation of the chemical reactions is attained, the cost of handling all the species involved along with accurately modeling all the reactions, which take place at different rates, is too large. To make CFD affordable in this case, a subset of the reactions is used where the reactions that proceed at a very large rate may be approximated as infinitely fast. Another challenge is properly and accurately describing the boundary conditions on the surface of the grid that cuts through the flow in unbounded or semi-bounded domains. Furthermore, expressing the conservation laws using differential calculus is only valid for flows in the continuum regime.[1] As such, computational solution methodologies that are based on truncated Taylor series approximations of derivatives, such as the finite volume method (FVM), are not suitable for non-continuum flows.

2.2   The Building Blocks of a CFD Solution Method

The building blocks of CFD solution methods are depicted in the diagram of Figure 2.1. The first step is carefully identifying the most suitable mathematical model of the actual flow (or prototype) to be studied in an appropriate coordinate system. In fluid flow

---

[1]The continuum hypothesis holds if the flow time and length scales are much larger than the time and length scales characterizing molecular interactions, which gives rise to the Knudsen number condition $Kn \ll 1$ introduced in Section 1.3.4. In gas flow in a pipe, $Kn = \frac{\lambda}{D}$, where λ is the mean free path and D is the pipe diameter.

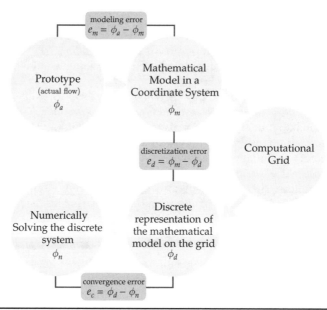

FIGURE 2.1 Schematic of the building Blocks of a CFD solution and their associated errors.

problems, the Navier-Stokes equations (or some simplification thereof) are generally used because they contain key features of real-world fluids. For example, if we want to numerically study the steady flow of water in a cylindrical pipe of circular cross-section under conditions where $\mathtt{Re} = 500$, then the most suitable mathematical model is the Navier-Stokes equations simplified for steady ($\frac{\partial}{\partial t} = 0$), constant-density ($\rho$), laminar pipe flow using the axisymmetric cylindrical coordinates, (r, θ, z), described in Section 1.5.2. The second step is to assign a computational grid to the real-world geometry of our problem. The choice of the grid takes into account the geometry of the domain and the level of detail needed in different parts of the domain. Different grid types are discussed in Section 2.3. The choice of grid type and cell-size is also influenced by physical effects we wish to capture in the simulation. For example, grid refinement near solid walls allows the method to more accurately resolve the large variations in the velocity field within the boundary layer, especially for turbulent flow where we need to resolve the viscous sublayer. Another example of flow physics influencing the grid choice appears in unbounded or semi-bounded flows. In these cases, an "artificial" boundary is introduced to truncate the problem to a reasonable global size. This boundary and its boundary conditions must be selected such that its presence does not compromise the model. The answer is a trade off between the cost arising from choosing too large of a computational domain and the loss in accuracy associated with choosing too small of a computational domain.In the third step, the discretization method is chosen. Although in this book we only cover the FVM, commonly used methods are the finite difference method and the finite element method [3]. We will demonstrate in Sections 2.4.1 to 2.4.3 that the FVM requires numerical representation of the volume integrals and surface integrals of diffusion and advection fluxes over a grid cell, taken as a control volume. Numerical approximations of these integrals involve computing approximations of the physical quantities (see Section 2.4.5) and their gradients (see Section 2.4.4) at the boundary (i.e., the faces of the grid cell) in terms of the grid points (often selected as the centroids of the grid cells). For unsteady problems, integrating in time may be accomplished in different ways, resulting in explicit, semi-implicit, and fully implicit time integration schemes, each with its convergence conditions that put an upper bound on the time step in relation to the grid size. Time integration schemes are discussed in Section 2.6. In the absence of an explicit equation that governs the pressure, solving for the pressure field is an additional challenge in CFD which requires a special treatment as discussed in Section 2.7. Discretizing the governing equations onto the computational grid results in a system of algebraic equations which must then be solved numerically. For most problems, the system is solved using iterative methods, as opposed to direct methods such as Gaussian elimination due to the large computational cost associated with the latter method. Iterative methods stop when a pre-specified convergence criterion is met. Numerical methods for solving the discretized equations governing unsteady flows commonly employ at least two nested iteration loops, with the outer loop for marching in time and the inner loop for reaching a solution that satisfies the algebraic equations at a given time.

2.2.1   Sources of Numerical Error

The diagram of Figure 2.1 helps in identifying various sources of errors between the actual flow solution, represented by ϕ_a, and the numerical solution obtained at the end of the iterations, denoted by ϕ_n. Since the mathematical model is not necessarily an exact representation of the actual flow, the difference between the two is captured by the modeling error, $e_m = \phi_a - \phi_m$, where ϕ_m is the exact solution of the mathematical model. Inaccurate representation of boundary conditions contributes to modeling errors. In some numerical methods, the discrete representation of the mathematical model is obtained by expressing each term in the mathematical model as a truncated Taylor series expansion up to a prespecified order. The aggregation of the truncation errors of the different terms gives rise to the discretization error, $e_d = \phi_m - \phi_d$, where ϕ_d is the exact numerical solution of the discrete representation. Since the discrete system is solved using iterative methods, the numerical solution obtained at the end of the iterations, ϕ_n, differs from ϕ_d by the convergence error $e_c = \phi_d - \phi_n$. The number of iterations is decided by the convergence criteria set by the user. In many cases, the number of iterations is additionally constrained by an upper bound to limit the computational cost.

2.2.2   Assessment of a CFD Solution Method

Assessing a CFD solution to a problem requires us to check different criteria including consistency, stability, convergence, and conservation.

Consistency A key property of a numerical method is that the discretized representation converges to the exact solution of the mathematical model as the grid size and time step tend to zero, that is, $\phi_d \to \phi_m$ as $\Delta x \to 0, \Delta t \to 0$. Since the discretized representation is obtained by representing the different terms in the mathematical model as truncated Taylor series, the discretization error may be expressed in terms of powers of Δx and Δt, that is, $e_d \sim \Delta x^n \Delta t^m$. For consistency the orders of the discretization in time and space must be positive, that is, $m, n > 0$.

Stability is another essential requirement for the viability of a numerical method. A numerical method is stable if it does not amplify the numerical errors that arise in the course of the numerical solution of the discrete system. Stability of the method is a necessary condition for the iterations to converge and for the boundedness of the solution. Time integration schemes, discussed in Section 2.6, may be conditionally or unconditionally stable. A conditionally stable scheme constraints the time step to be less than a certain value; otherwise the scheme will be unstable. One of the most widely used methods to study the stability of numerical schemes is the Von Neumann method [4].

Convergence of a numerical scheme may be assessed by inspecting the behavior of the discretization and convergence errors $|e_d| + |e_c|$ as $\Delta x, \Delta t \to 0$. A necessary condition for convergence is consistency of the numerical method, which requires $e_d \to 0$ as $\Delta x, \Delta t \to 0$. In addition, convergence of the numerical method requires the convergence error to tend to zero as $\Delta x, \Delta t \to 0$. This requires the method to be stable. The relation between consistency, stability, and convergence is captured by the *Lax equivalence theorem*, [5] which states that "given a properly posed linear initial value problem and a finite difference approximation to it that satisfies the consistency condition, stability is the necessary and sufficient condition for convergence." For more complex problems, convergence is assessed by inspecting $|e_c|$ as the grid is successively refined. If the solutions for grids $1, 2, 3, \dots$ characterized by $\Delta x_1 > \Delta x_2 > \Delta x_3 > \dots$ are $\phi_1, \phi_2, \phi_3, \dots$, then the convergence criterion is met if a grid-independent solution is obtained, that is, $\frac{|\phi(\Delta x_{i+1}) - \phi(\Delta x_i)|}{|\phi(\Delta x_{i+1})|} < \epsilon$, where the value of ϵ ($0 < \epsilon \ll 1$) is chosen by the user as an acceptable relative difference between solutions with differing grids. One way to assess the convergence using three grids of different sizes and to arrive at an improved solution is the Richardson extrapolation method, discussed in Section 2.9.

Conservation A numerical scheme designed to numerically solve a conservation law governing a conserved quantity Π is conservative if it conserves the quantity Π for each grid cell as well as for the entire domain. A key advantage of the FVM is that it is conservative by construction, which implies that the rate of change of Π in a grid cell is balanced by the advection and diffusion fluxes of Π across its boundary, in addition to contribution for physical sources within the cell, if any. As such, no artificial or nonphysical sources or sinks can manifest in a conservative scheme. This property, which sets the FVM apart from other methods such as the finite element and finite difference methods, made it the method of choice for most engineers.

2.3   Numerical Representation of the Domain

As depicted in the diagram of Figure 2.1, once a mathematical model in a properly selected coordinate system is set up, the next step in grid-based numerical methods, including the FVM, is to divide the domain into a finite number of control volumes (grid cells). The number of grid cells representing the domain is denoted by N. To simplify things, we use a single index notation. The subscript i refers to a single grid cell, where $i \in [1, N]$. The global index list of the grid cells is then $1, 2, \dots, N$. The volume of cell i is denoted by $\mathcal{V}_i$. Each grid cell is represented by a finite number of grid points, which can be thought of as points used for a finite integral. In the simplest representation, the grid cell is represented by a single grid or quadrature point, located at the centroid of the cell, denoted by $\mathbb{O}_i$, as depicted in Figure 2.2. This figure is an example for a regular two-dimensional rectilinear grid. The set of neighboring cells sharing a face with cell i is denoted $\mathcal{N}(i)$. The bounding surface of cell i is made up of the faces it shares with its neighbors. The global index of a face of cell i, denoted by $\mathbb{k}$, is a member of the list of faces $\mathcal{F}(i)$, that is, $\mathbb{k} \in \mathcal{F}(i)$. The area of face $\mathbb{k}$ is denoted $\mathcal{A}_{\mathbb{k}}$ and its centroid by $\mathbb{O}_{\mathbb{k}}$.

The choice of the grid is critical to the computational cost, convergence and accuracy of the solution. The factors that should be taken into account when choosing grid type

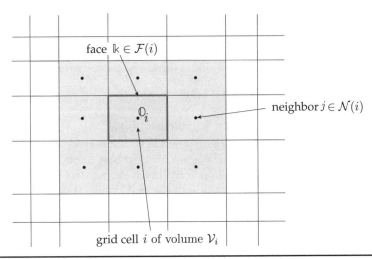

FIGURE 2.2 Notation used for indexing of grid cells, neighbors, and faces.

and spatial resolution are the geometry of the problem and the complexity of the flow and physics in time and space.

The discretized form of the governing equation is obtained by integrating the conservative form of the relevant conservation law over each grid cell, taken as a control volume. In carrying out the volume integrals, a profile of the dependent variables is assumed, which leads to a set of algebraic equations governing the values of the dependent variables at the grid points. The grid that spatially represents the computational domain of study can be structured, unstructured, or hybrid, as depicted in Figure 2.3.

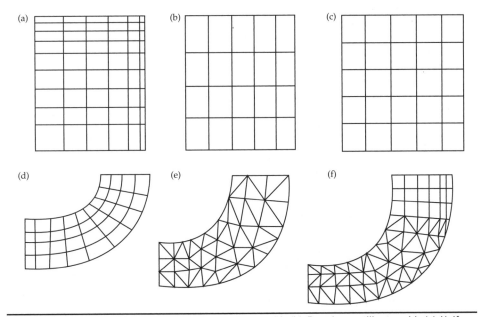

FIGURE 2.3 Structured grids: (a) General rectilinear grid. (b) Regular rectilinear grid. (c) Uniform cartesian rectilinear grid. (d) Curvilinear grid. (e) Unstructured tetrahedral grid. (f) Hybrid grid.

A structured grid (Figure 2.3[a–d]) is characterized by regular or simple connectivity which can be implicitly represented in the ordered spatial indexing of the arrays of the dependent variables. For example, a structured rectilinear grid (Figure 2.3[a–c]) representing a two-dimensional domain may employ two indices, $1 \leq i \leq N_x$ and $1 \leq j \leq N_y$, representing the indices of the grid cells ordered by position along each of the coordinates. This allows for representation of the dependent variables using two-dimensional arrays. In this representation, $\phi(i_x, i_y)$ is the value of property ϕ at the centroid of grid cell (i_x, i_y), which is located at column i_x and row i_y. Conventionally, the rows are indexed from bottom to top and the columns from left to right, in accordance with the coordinates directions. Because the grid is structured, the west, south, north, and east neighbors of grid cell (i_x, i_y) are respectively cells (i_x-1, i_y), (i_x, i_y-1), (i_x, i_y+1), and $(i_x + 1, i_y)$ and the value of ϕ at the centroids of these neighboring cells are respectively $\phi(i_x - 1, i_y)$, $\phi(i_x, i_y - 1)$, $\phi(i_x, i_y + 1)$, and $\phi(i_x + 1, i_y)$. Such simple regular connectivity saves on storage requirements because neighbors of a given cell do not need to be stored separately. For a structured grid, conversion from double indexing, (i_x, i_y), of grid cells representing a two-dimensional domain to single indexing, i, is accomplished by numbering the cells, starting from the bottom left cell, in the order they are traversed, one column at a time, as shown in Figure 2.6. In this case, $i = (i_x - 1) N_y + i_y$. Similar conversion from triple indexing of grid cells in three-dimensional domains to a single index can be done. A special case of the structured rectilinear grid is when $\Delta x, \Delta y$, and Δz are the same for all grid cells, in which case the grid is called a *regular grid*, as depicted in Figure 2.3(b). Because a rectilinear structured grid is made up of rectangles (2D) or rectangular cuboids (3D), its applicability is limited to simple domain geometries with boundary segments aligned with one of the coordinate axes. For example, a rectilinear structured grid cannot be used to represent a curved domain, such as a bend. In this case, a curvilinear grid must be used, as depicted in Figure 2.3(d). A curvilinear grid is a structured grid with the same combinatorial structure as a regular grid, in which the cells are quadrilaterals (2D) or general cuboids or hexagonal (3D).

For complex domain geometries, a more appropriate representation is accomplished using an unstructured grid. Unlike structured grids, unstructured grids, such as the one shown in Figure 2.3(e), require extra storage for the connectivity of cells. The connectivity of each grid cell consists of an ordered list of the indices of its nodes. The nodes' coordinates are stored in a global list (or array). For unstructured grids, commonly used cells are the triangular elements for two-dimensional domains and the tetrahedral elements for three-dimensional domains.

A hybrid grid is composed of structured and unstructured blocks of elements. An example is shown in Figure 2.3(f), where a structured grid is used for the rectangular part of the domain and an unstructured grid for the curved part.

2.4   The Finite Volume Method

The FVM numerically approximates the conservation law by applying the conservation form of the governing equation to a control volume, taken to be the grid cell. Each grid cell is represented by a finite number of grid points, referred to as computational nodes, at which the numerical solution is to be determined. In the simplest representation, the grid cell is represented by a single node, located at the centroid of the cell, as discussed in Section 2.3. Numerical representation of the general conservation law, given by Eq. (1.79), for each grid cell yields an equation governing the unknown quantities at the computational nodes (commonly the centroids) in that cell and its neighbors. Upon combining the linear equations for all the grid cells and incorporating the boundary conditions, the mathematical model is numerically represented by a linear system governing the unknown quantities at all the computational nodes. In addition to its key advantage of being a conservative method, the FVM can be used with any grid type, and is as such suitable for complex geometries.

For the grid cell i shown in Figure 2.2, we present a numerical approximation of each of the terms of the integral form of the conservation law, given by Eq. (1.80). Selecting grid cell i to be the control volume and recalling that the diffusion coefficient Γ is constant, Eq. (1.80) can be written as

$$\frac{\partial}{\partial t} \int_{\mathcal{V}_i} \rho \Lambda \phi \, d\mathcal{V} + \int_{\mathcal{S}_i} (\rho \Lambda \phi \vec{u} - \Gamma \nabla \phi) \cdot \hat{n} \, d\mathcal{S} = \int_{\mathcal{V}_i} \dot{S} \, d\mathcal{V} \tag{2.1}$$

Next, we present the simplest numerical approximation of the volume and surface integrals over grid cell i as quadrature formulae in terms of nodal values. Numerical approximation of the surface integrals of the advection and diffusion fluxes, $\vec{J}_{adv}$ and $\vec{J}_{dif}$, require numerical integration of quantities on the faces of the grid cell, which are in turn interpolated from nodal values. This double approximation (integration and interpolation) makes it difficult to develop a FVM of order higher than two for three-dimensional flows.

2.4.1 Evaluation of the Volume Integral

Numerical representation (computation) of volume integrals is generally accomplished as a sum of weighted evaluations of the integrand at quadrature points [6].

$$\int_{\mathcal{V}} f \, d\mathcal{V} \simeq \sum_{q=1}^{N_q} w_q f_q \tag{2.2}$$

where N_q is the number of quadrature points, f_q is the value of the function f at quadrature point $\vec{x}_q$, and w_q is the associated weight. The simplest representation uses one quadrature point at the centroid of the volume and therefore the weight w_q is simply 1. For grid cell i, this representation for the two volume integrals of Eq. (2.1) is

$$\int_{\mathcal{V}_i} \rho \Lambda \phi \, d\mathcal{V} \simeq \rho_i \Lambda_i \phi_i \mathcal{V}_i \tag{2.3}$$

and

$$\int_{\mathcal{V}_i} \dot{S} \, d\mathcal{V} \simeq \dot{S}_i \mathcal{V}_i \tag{2.4}$$

where the subscript i refers to evaluation of the relevant property at the centroid, $\mathbb{O}_i$, of cell i. Approximation of the volume integrals that uses one quadrature point at the centroid is a second-order approximation. We present a proof for the one-dimensional grid-cell, with centroid $\mathbb{O}_i$. Expressing $f(x)$ as a Taylor series expansion about the centroid, $f(x) = f_i + \frac{\partial f}{\partial x}\big|_i (x - x_i) + \frac{\partial^2 f}{\partial x^2}\big|_i \frac{(x-x_i)^2}{2} + h.o.t.$, then

$$\int_{\mathcal{V}_i} f \, d\mathcal{V} = \int_{x_i - \Delta x_i/2}^{x_i + \Delta x_i/2} \left(f_i + \frac{\partial f}{\partial x}\bigg|_i (x - x_i) + \frac{\partial^2 f}{\partial x^2}\bigg|_i \frac{(x - x_i)^2}{2} + h.o.t. \right) dx \tag{2.5}$$

where $h.o.t.$ refers to the higher order terms of the expansion. Since $\mathbb{O}_i$ is the centroid of the grid cell, it follows that $\int_{x_i - \Delta x_i/2}^{x_i + \Delta x_i/2} (x - x_i) \, dx = 0$, yielding a second-order accurate approximation for the average value of f over the cell volume

$$\frac{1}{\mathcal{V}_i} \int_{\mathcal{V}_i} f \, d\mathcal{V} = f_i + O(\Delta x_i^2) \tag{2.6}$$

2.4.2   Evaluation of the Surface Integral of the Diffusion Flux

The surface integral of the diffusion flux, $\vec{J}_{dif}$, normal to the control surface, $\mathcal{S}_i$, is expressed as the sum over all the cell faces that make up the control surface

$$\int_{\mathcal{S}_i} \vec{J}_{dif} \cdot \hat{n} \, d\mathcal{S} = \sum_{\Bbbk \in \mathcal{F}_i} \int_{\mathcal{A}_{\Bbbk}} \vec{J}_{dif} \cdot \hat{n} \, d\mathcal{S} \tag{2.7}$$

where $\mathcal{A}_{\Bbbk}$ is the area of face $\Bbbk$. Noting that $\nabla \phi \cdot \hat{n} = \frac{\partial \phi}{\partial n}$, then $\vec{J}_{dif} \cdot \hat{n} = -\Gamma \frac{\partial \phi}{\partial n}$ so that

$$\int_{\mathcal{S}_i} \vec{J}_{dif} \cdot \hat{n} \, d\mathcal{S} = \sum_{\Bbbk \in \mathcal{F}_i} \int_{\mathcal{A}_{\Bbbk}} \Gamma \left(-\frac{\partial \phi}{\partial n} \right) d\mathcal{S} \tag{2.8}$$

Discrete representation of the surface integral of the diffusion flux commonly involves two approximations. The first approximation expresses the surface integral using nodal quantities, and in its simplest form employs the centroid of face $\Bbbk$ as the only node so that

$$\int_{\mathcal{S}_i} \vec{J}_{dif} \cdot \hat{n} \, d\mathcal{S} \simeq \sum_{\Bbbk \in \mathcal{F}_i} \Gamma_{\Bbbk} \left(-\frac{\partial \phi}{\partial n} \right)_{\Bbbk} \mathcal{A}_{\Bbbk} \tag{2.9}$$

In a *collocated grid*, all the dependent variables, denoted by ϕ, are numerically represented as discrete values at the centroids of the grid cells, which means that the values of ϕ and $\frac{\partial \phi}{\partial n}$ at the faces' centroids need to be expressed in terms of values of ϕ at the centers of the grid cells. The second approximation expresses $\frac{\partial \phi}{\partial n}$ at the centroid of face $\Bbbk$ in terms of these discrete values at the cells centroids. This approximation is presented in Section 2.4.4 for a two-dimensional rectilinear grid.

2.4.3   Evaluation of the Surface Integral of the Advection Flux

The surface integral of the advection flux of $\Lambda \phi$ transported by mass across the control surface bounding grid cell i is expressed as a sum over the faces of cell i as follows:

$$\int_{\mathcal{S}_i} \vec{J}_{adv} \cdot \hat{n} \, d\mathcal{S} = \sum_{\Bbbk \in \mathcal{F}_i} \int_{\mathcal{A}_{\Bbbk}} \vec{J}_{adv} \cdot \hat{n} \, d\mathcal{S} \tag{2.10}$$

where $\vec{J}_{adv} = \rho \Lambda \phi \vec{u}$, so that

$$\int_{\mathcal{S}_i} \vec{J}_{adv} \cdot \hat{n} \, d\mathcal{S} = \sum_{\Bbbk \in \mathcal{F}_i} \int_{\mathcal{A}_{\Bbbk}} \rho \Lambda \phi \vec{u} \cdot \hat{n} \, d\mathcal{S} \tag{2.11}$$

Discrete representation of the surface integral of the advection flux commonly involves two approximations. The first approximation expresses the surface integral using nodal quantities, and in its simplest form employs the centroid of face $\Bbbk$ as the only node so that

$$\int_{\mathcal{S}_i} \vec{J}_{adv} \cdot \hat{n} \, d\mathcal{S} \simeq \sum_{\Bbbk \in \mathcal{F}_i} \Lambda_{\Bbbk} \phi_{\Bbbk} \left(\rho_{\Bbbk} \vec{u}_{\Bbbk} \cdot \hat{n}_{\Bbbk} \mathcal{A}_{\Bbbk} \right) \tag{2.12}$$

Denoting $\rho_{\Bbbk}\vec{u}_{\Bbbk} \cdot \hat{n}_{\Bbbk}\mathcal{A}_{\Bbbk} = \dot{m}_{\Bbbk}^{\rightarrow}$ as an approximation of the mass flow rate leaving control volume i across face $\Bbbk$, then

$$\int_{\mathcal{S}_i} \vec{J}_{adv} \cdot \hat{n}\, d\mathcal{S} \simeq \sum_{\Bbbk \in \mathcal{F}_i} \dot{m}_{\Bbbk}^{\rightarrow} \Lambda_{\Bbbk}\phi_{\Bbbk} \tag{2.13}$$

The second approximation expresses $\Lambda_{\Bbbk}\phi_{\Bbbk}$ and $\dot{m}_{\Bbbk}^{\rightarrow}$ at the centroid of face $\Bbbk$ in terms of these discrete values at the cells centroids. A second-order approximation of the advection term, for a two-dimensional rectilinear grid, is presented in Section 2.4.5.

Using the discretizations of Eqs. (2.3), (2.4), (2.9), and (2.13), a discrete spatial representation of the general conservation law of Eq. (2.1) applied to cell i follows as:

$$\frac{\partial}{\partial t}\left(\rho_i \Lambda_i \phi_i \mathcal{V}_i\right) + \sum_{\Bbbk \in \mathcal{F}_i}\left(\dot{m}_{\Bbbk}^{\rightarrow} \Lambda_{\Bbbk}\phi_{\Bbbk} - \Gamma_{\Bbbk}\mathcal{A}_{\Bbbk}\left(\frac{\partial\phi}{\partial n}\right)_{\Bbbk}\right) = \dot{S}_i \mathcal{V}_i \tag{2.14}$$

As mentioned above, in a collocated grid, all the dependent variables are represented at the centroids of the grid cells, which means that the values of ϕ and $\frac{\partial\phi}{\partial n}$ at the faces' centroids need to be expressed in terms of values of ϕ at the centers of the grid cells. In what follows, we present approximations for a two-dimensional rectilinear grid.

Figure 2.4 shows a schematic of a grid cell and its neighbors. We refer to the centroids of the cell (i) and its neighbors $i + 1$, $i - 1$, $i + N_y$, and $i - N_y$, respectively as $\mathbb{O}$, and $\mathbb{N}$, $\mathbb{S}$, $\mathbb{E}$, and $\mathbb{W}$, denoting north, south, east, and west. The size of control volume i is $\Delta x_i \times \Delta y_i$ or $\Delta x \times \Delta y$ when we are referring to the control volume under consideration. The sizes of its east, west, north, and south neighbors are $\Delta x_{\mathbb{E}}\Delta y_{\mathbb{E}}$, $\Delta x_{\mathbb{W}}\Delta y_{\mathbb{W}}$, $\Delta x_{\mathbb{N}}\Delta y_{\mathbb{N}}$, and $\Delta x_{\mathbb{S}}\Delta y_{\mathbb{S}}$, respectively. The distances between the cell's centroid, $\mathbb{O}$, and the $\mathbb{E}$, $\mathbb{W}$, $\mathbb{N}$, and $\mathbb{S}$ neighbors' centroids are respectively $\delta x_{\mathbb{E}}$, $\delta x_{\mathbb{W}}$, $\delta y_{\mathbb{N}}$, and $\delta y_{\mathbb{S}}$. Note that $\delta x_{\mathbb{E}} = \frac{\Delta x + \Delta x_{\mathbb{E}}}{2}$, $\delta x_{\mathbb{W}} = \frac{\Delta x + \Delta x_{\mathbb{W}}}{2}$, $\delta y_{\mathbb{N}} = \frac{\Delta y + \Delta y_{\mathbb{N}}}{2}$, and $\delta y_{\mathbb{S}} = \frac{\Delta y + \Delta y_{\mathbb{S}}}{2}$. The unit outward normal vectors to faces e, w, n, and s are respectively $\hat{n}_{\mathrm{e}} = \hat{x}$, $\hat{n}_{\mathrm{w}} = -\hat{x}$, $\hat{n}_{\mathrm{n}} = \hat{y}$, and $\hat{n}_{\mathrm{s}} = -\hat{y}$ so that $\left.\frac{\partial\phi}{\partial n}\right|_{\mathrm{e}} = \left.\frac{\partial\phi}{\partial x}\right|_{\mathrm{e}}$, $\left.\frac{\partial\phi}{\partial n}\right|_{\mathrm{w}} = -\left.\frac{\partial\phi}{\partial x}\right|_{\mathrm{w}}$, $\left.\frac{\partial\phi}{\partial n}\right|_{\mathrm{n}} = \left.\frac{\partial\phi}{\partial y}\right|_{\mathrm{n}}$, and $\left.\frac{\partial\phi}{\partial n}\right|_{\mathrm{s}} = -\left.\frac{\partial\phi}{\partial y}\right|_{\mathrm{s}}$.

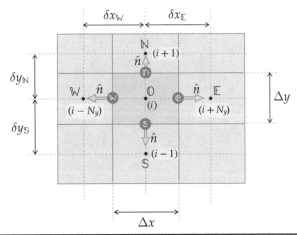

FIGURE 2.4 Schematic of a control volume and its neighbors in a two-dimensional rectilinear grid. The grid cells are ordered using a single global index that starts from the bottom right most cell and sweeps the grid to the right column by column.

2.4.4   Evaluation of $\partial\phi/\partial n$ at the Face

The spatial derivatives of ϕ at the different faces of a control volume of the two-dimensional rectilinear mesh of Figure 2.4 can be approximated using linear interpolation of ϕ between the centroid of the control volume and the centroid of the neighbor sharing the face so that

$$\left.\frac{\partial\phi}{\partial n}\right|_{e} = \left.\frac{\partial\phi}{\partial x}\right|_{e} = \frac{\phi_E - \phi_O}{\delta x_E} \tag{2.15}$$

$$\left.\frac{\partial\phi}{\partial n}\right|_{w} = -\left.\frac{\partial\phi}{\partial x}\right|_{w} = -\frac{\phi_O - \phi_W}{\delta x_W} \tag{2.16}$$

$$\left.\frac{\partial\phi}{\partial n}\right|_{n} = \left.\frac{\partial\phi}{\partial y}\right|_{n} = \frac{\phi_N - \phi_O}{\delta y_N} \tag{2.17}$$

$$\left.\frac{\partial\phi}{\partial n}\right|_{s} = -\left.\frac{\partial\phi}{\partial y}\right|_{s} = -\frac{\phi_O - \phi_S}{\delta y_S} \tag{2.18}$$

Approximations (Eqs. (2.15)–(2.18)) are accurate locally to second-order. Expressing ϕ_O and ϕ_E as Taylor series expansions about the centroid of face e,

$$\phi_O = \phi_e + \left.\frac{\partial\phi}{\partial n}\right|_{e}\left(-\frac{\Delta x}{2}\right) + \left.\frac{\partial^2\phi}{\partial n^2}\right|_{e}\frac{1}{2}\left(-\frac{\Delta x}{2}\right)^2 + h.o.t. \tag{2.19}$$

$$\phi_E = \phi_e + \left.\frac{\partial\phi}{\partial n}\right|_{e}\left(\frac{\Delta x_E}{2}\right) + \left.\frac{\partial^2\phi}{\partial n^2}\right|_{e}\frac{1}{2}\left(\frac{\Delta x_E}{2}\right)^2 + h.o.t. \tag{2.20}$$

Taking the difference between Eqs. (2.19) and (2.20) yields approximation (2.15) with a leading error term of the order $O(|\Delta x_E - \Delta x|)$. Note that the approximations become second-order accurate for a regular rectilinear grid, where $\Delta x = \Delta x_E = \Delta x_W$, and $\Delta y = \Delta y_N = \Delta y_S$.

Using approximations (2.15)–(2.18), the rate of transport of ϕ by diffusion is then approximated as

$$\sum_{k \in \mathcal{F}_i}\left(-\Gamma A\frac{\partial\phi}{\partial n}\right)_{k} \simeq$$
$$-\left(\Gamma_e A_e\frac{\phi_E - \phi_O}{\delta x_E} - \Gamma_w A_w\frac{\phi_O - \phi_W}{\delta x_W} + \Gamma_n A_n\frac{\phi_N - \phi_O}{\delta y_N} - \Gamma_s A_s\frac{\phi_O - \phi_S}{\delta y_S}\right) \tag{2.21}$$

For a rectilinear grid, $A_e = A_w = \Delta y$ and $A_s = A_n = \Delta x$, and assuming constant diffusion coefficient Γ, then the diffusion term may be expressed as

$$\sum_{k \in \mathcal{F}_i}\left(-\Gamma A\frac{\partial\phi}{\partial n}\right)_{k} \simeq c_{O,d}\phi_O + c_{E,d}\phi_E + c_{W,d}\phi_W + c_{N,d}\phi_N + c_{S,d}\phi_S \tag{2.22}$$

where

$$c_{0,d} = -(c_{E,d} + c_{W,d} + c_{N,d} + c_{S,d}) \tag{2.23}$$

$$c_{E,d} = -\Gamma \frac{\Delta y}{\delta x_E} \tag{2.24}$$

$$c_{W,d} = -\Gamma \frac{\Delta y}{\delta x_W} \tag{2.25}$$

$$c_{N,d} = -\Gamma \frac{\Delta x}{\delta y_N} \tag{2.26}$$

$$c_{S,d} = -\Gamma \frac{\Delta x}{\delta y_S} \tag{2.27}$$

and the subscript d refers to diffusion.

2.4.5 Evaluation of ϕ at the Face

For the two-dimensional rectilinear grid of Figure 2.4, the value of a dependent variable at the centroid of a face is commonly approximated using linear distance-weighted interpolation between the values at the centroids of the sharing cells,

$$\phi_e \simeq \alpha_E \phi_0 + (1 - \alpha_E)\phi_E, \quad \alpha_E = \frac{\Delta x_E/2}{\delta x_E} \tag{2.28}$$

$$\phi_w \simeq \alpha_W \phi_0 + (1 - \alpha_W)\phi_W, \quad \alpha_W = \frac{\Delta x_W/2}{\delta x_W} \tag{2.29}$$

$$\phi_n \simeq \alpha_N \phi_0 + (1 - \alpha_N)\phi_N, \quad \alpha_N = \frac{\Delta y_N/2}{\delta y_N} \tag{2.30}$$

$$\phi_s \simeq \alpha_S \phi_0 + (1 - \alpha_S)\phi_S, \quad \alpha_S = \frac{\Delta y_S/2}{\delta y_S} \tag{2.31}$$

By dividing Eq. (2.19) by Δx and Eq. (2.20) by Δx_E, and then taking the sum, approximation (Eq. (2.28)) can be shown to be second-order accurate, with leading error term $O(\Delta x_E \Delta x)$.

These approximations enable us to approximate the mass flow rates leaving the control volume across the four faces. For example, the mass flow rate leaving through face e is $\dot{m}_e^{\rightarrow} \simeq (\rho_e \vec{u}_e \cdot \hat{n}_e)\mathcal{A}_e$. Noting that $\hat{n}_e = \hat{x}$, then $\dot{m}_e^{\rightarrow} \simeq (\rho u)_e \mathcal{A}_e$, where $\mathcal{A}_e = \Delta y$ and $(\rho u)_e$ can be approximated using Eq. (2.28), so that

$$\dot{m}_e^{\rightarrow} \simeq [\alpha_E (\rho u)_0 + (1 - \alpha_E)(\rho u)_E] \Delta y \tag{2.32}$$

Similarly,

$$\dot{m}_w^{\rightarrow} \simeq -[\alpha_W (\rho u)_0 + (1 - \alpha_W)(\rho u)_W] \Delta y \tag{2.33}$$

$$\dot{m}_n^{\rightarrow} \simeq [\alpha_N (\rho v)_0 + (1 - \alpha_N)(\rho v)_N] \Delta x \tag{2.34}$$

$$\dot{m}_s^{\rightarrow} \simeq -[\alpha_S (\rho v)_0 + (1 - \alpha_S)(\rho v)_S] \Delta x \tag{2.35}$$

Note that a positive $\dot{m}_k^{\rightarrow}$ indicates flow leaving the control volume through face $k \in \{e, w, n, s\}$ and a negative value indicates an entering flow.

2.4.6 Evaluation of the Advection Term

According to Eq. (2.14), the advection term associated with conservation of property ϕ in the control volume consisting of grid cell i is $\sum_{k=1}^{\mathcal{F}_i} \dot{m}_k^{\rightarrow} \Lambda_k \phi_k$, where $\dot{m}_k^{\rightarrow}$, the mass flow rate leaving the control volume across face k, is approximated using Eqs. (2.32–2.35).

At first, one may think of using the central difference scheme (CDS), expressed by Eqs. (2.28–2.31) to approximate $\Lambda_{\Bbbk}\phi_{\Bbbk}$ at face $\Bbbk$ in terms of nodal values. It turns out that this approximation, although second-order accurate, can yield oscillatory solutions if the grid cell size is not sufficiently small [7]. In addition, Pantakar [7] showed that the size of the grid for large Reynolds number flows is computationally prohibitive, especially for two-dimensional problems and much more so for three-dimensional problems. To circumvent this limitation, the upwind scheme was proposed by Courant et al. [8, 9]. The upwind scheme is a physically motivated scheme that accounts for the fact that the upstream conditions affect the value of ϕ at the face more than the downstream conditions. The order of approximation of the convection term is decided by which is smaller, the order of the surface integral approximation or the order of approximating ϕ at the quadrature points. Using the face centroid as the only quadrature point, the surface integral approximation is second order. The order of approximating ϕ at the face centroid is decided by the upwind scheme used. The first-order upwind scheme (or upwind differencing scheme [UDS]) approximates $\phi_{\Bbbk}$ in the convection term $\dot{m}_{\Bbbk}^{\rightarrow}\phi_{\Bbbk}$ using the upstream value of ϕ as follows:

$$\phi_{\Bbbk} \simeq \begin{cases} \phi_{\mathbb{O}} & \text{for } \dot{m}_{\Bbbk}^{\rightarrow} > 0 \\ \phi_{\mathbb{K}} & \text{for } \dot{m}_{\Bbbk}^{\rightarrow} < 0 \end{cases} \qquad (2.36)$$

where $\Bbbk \in \{\mathbb{e}, \mathbb{w}, \mathbb{n}, \mathbb{s}\}$ and $\mathbb{K} \in \{\mathbb{E}, \mathbb{W}, \mathbb{N}, \mathbb{S}\}$. A schematic for approximating $\phi_{\mathbb{e}}$ is shown in Figure 2.5. Using the Heaviside function, $\mathcal{H}$,

$$\phi_{\Bbbk} \simeq \phi_{\mathbb{O}}\mathcal{H}(\dot{m}_{\Bbbk}^{\rightarrow}) + \phi_{\mathbb{K}}\mathcal{H}(-\dot{m}_{\Bbbk}^{\rightarrow}) \qquad (2.37)$$

where by definition $\mathcal{H}(x) = 1$ for $x > 0$ and 0 otherwise. This scheme is accurate to first order since the associated truncation error is decided by the leading term of the truncated Taylor series expansion, which is $O(\Delta x)$. The scheme is unconditionally stable, and as such puts no restriction on how large the grid size can be for a given size of the time step. On the downside, the first-order upwind scheme introduces numerical diffusion, which arises from the leading term of the truncation error. For the steady one-dimensional advection diffusion problem, the leading term in approximating $\dot{m}_{\Bbbk}^{\rightarrow}\phi_{\Bbbk}$ using (2.37) is

$$\dot{m}_{\Bbbk}^{\rightarrow}\left.\frac{\partial\phi}{\partial x}\right|_{\Bbbk}\left[-\frac{\Delta x}{2}\mathcal{H}(\dot{m}_{\Bbbk}^{\rightarrow}) + \frac{\Delta x_{\mathbb{K}}}{2}\mathcal{H}(-\dot{m}_{\Bbbk}^{\rightarrow})\right] \qquad (2.38)$$

which is effectively a diffusion flux, albeit an artificial one, with a diffusion coefficient $\frac{\dot{m}_{\Bbbk}^{\rightarrow}}{2}\left[\Delta x\,\mathcal{H}(\dot{m}_{\Bbbk}^{\rightarrow}) + \Delta x_{\mathbb{K}}\mathcal{H}(-\dot{m}_{\Bbbk}^{\rightarrow})\right]$. Upwind schemes of higher order have been proposed and some are presented in Appendix A.

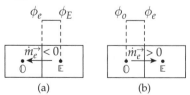

FIGURE 2.5 Schematic depicting approximation of ϕ at the center of face $\mathbb{e}$ using a first-order upwind scheme for (a) the mass flow rate across face $\mathbb{e}$ is out of cell $\mathbb{O}$ and (b) the mass flow rate across face $\mathbb{e}$ is into cell $\mathbb{O}$.

All these schemes for approximating ϕ at the face centroid in terms of value of ϕ at the centroids of the cells involved in the interpolation can be expressed as

$$\phi_\Bbbk \simeq c_{\mathbb{0},a}\phi_\mathbb{0} + c_{\mathbb{L},a}\phi_\mathbb{L} + c_{\mathbb{K},a}\phi_\mathbb{K} + c_{\mathbb{KK},a}\phi_{\mathbb{KK}} \tag{2.39}$$

Expressions for the coefficients $c_{\mathbb{0},a}$, $c_{\mathbb{L},a}$, $c_{\mathbb{K},a}$, and $c_{\mathbb{KK},a}$ are listed in Table A.5 for the UDS schemes and the higher order schemes presented Appendix A.

2.4.7 Numerically Solving the Steady General Transport Equation

At steady state, the term $\frac{\partial}{\partial t}(\rho_i \Lambda_i \phi_i \mathcal{V}_i)$ vanishes, and the discrete spatial representation of the general conservation law of Eq. (2.1) applied to cell i is

$$\sum_{\Bbbk \in \mathcal{F}_i} \left(\dot{m}_\Bbbk^{\rightarrow} \Lambda_\Bbbk \phi_\Bbbk - \Gamma_\Bbbk \mathcal{A}_\Bbbk \left(\frac{\partial \phi}{\partial n} \right)_\Bbbk \right) = \dot{S}_i \mathcal{V}_i \tag{2.40}$$

Expressing the diffusion and advection fluxes in terms of values at the cells' centroids for a rectilinear grid using Eqs. (2.22) and (2.39) respectively, Eq. (2.40) is written as

$$\sum_{\Bbbk \in \mathcal{F}_i} \dot{m}_\Bbbk^{\rightarrow} \Lambda_\Bbbk \left(c_{\mathbb{0},a}\phi_\mathbb{0} + c_{\mathbb{L},a}\phi_\mathbb{L} + c_{\mathbb{K},a}\phi_\mathbb{K} + c_{\mathbb{KK},a}\phi_{\mathbb{KK}} \right)$$
$$+ \left(c_{\mathbb{0},d}\phi_\mathbb{0} + c_{\mathbb{E},d}\phi_\mathbb{E} + c_{\mathbb{W},d}\phi_\mathbb{W} + c_{\mathbb{N},d}\phi_\mathbb{N} + c_{\mathbb{S},d}\phi_\mathbb{S} \right) = \dot{S}_i \mathcal{V}_i \tag{2.41}$$

In general, $\dot{m}_\Bbbk^{\rightarrow}$, $\Lambda_\Bbbk$, and $\Gamma_\Bbbk$, in addition to ϕ_i are not known at faces $\Bbbk$ and cell centroids i, which makes Eq. (2.41) nonlinear, and as such difficult to solve. To overcome this obstacle, Eq. (2.41) is numerically solved for ϕ_i at the centroids of the grid cells using an iterative scheme, where it is assumed at an iteration (k) that all quantities, except ϕ_i, are known and, if they depend on ϕ, assume their latest available values or their values from the previous iteration $(k-1)$. Thus, $\dot{m}_\Bbbk^{\rightarrow}$, $\Lambda_\Bbbk$, and $\Gamma_\Bbbk$ are computed using the expressions listed in Table 2.1.

In this case, Eq. (2.41) represents a linear system whose solution are the values of ϕ at all the cell centroids at the current iteration (k), that is, $\phi_i^{(k)}$, which can be cast as

$$\sum_{j \in \mathcal{I}_i} A_{ij}\phi_i^{(k)} = b_i, \quad i \ldots N \tag{2.42}$$

where the interaction list, $\mathcal{I}_i$ is the list of all cells involved in the discrete representation of the advection and diffusion fluxes across the faces of cell i. Using the first-order upwind scheme for the rectilinear grid shown in Figure 2.4, the nonzero coefficients of row i of

TABLE 2.1 Subscript 0 refers to variables that are known a priori and do not change their values in the iterations. The coefficients μ, c_v, and λ are assumed to be constant.

ϕ	Description	$\dot{m}_\Bbbk^{\rightarrow}$	$\Lambda_\Bbbk$	$\Gamma_\Bbbk$
$\vec{u}$	Velocity	$\left(\rho \vec{u}^{(k-1)} \cdot \hat{n}\mathcal{A} \right)_\Bbbk$	1	μ
T	Temperature	$\left(\dot{m}_\Bbbk^{\rightarrow} \right)_0$	c_v	λ

matrix $\mathbf{A}$ are

$$A_O = A_{i,i} = \sum_{\mathbb{k} \in \{e,w,n,s\}} \dot{\vec{m}}_{\mathbb{k}} \Lambda_{\mathbb{k}} \mathcal{A}_{\mathbb{k}} \mathcal{H}(\dot{\vec{m}}_{\mathbb{k}}) + c_{O,d} \tag{2.43}$$

$$A_{\mathbb{E}} = A_{i,i+N_y} = \dot{\vec{m}}_e \Lambda_e \mathcal{A}_e \mathcal{H}(-\dot{\vec{m}}_e) + c_{\mathbb{E},d} \tag{2.44}$$

$$A_{\mathbb{W}} = A_{i,i-N_y} = \dot{\vec{m}}_w \Lambda_w \mathcal{A}_w \mathcal{H}(-\dot{\vec{m}}_w) + c_{\mathbb{W},d} \tag{2.45}$$

$$A_{\mathbb{S}} = A_{i,i-1} = \dot{\vec{m}}_s \Lambda_s \mathcal{A}_s \mathcal{H}(-\dot{\vec{m}}_s) + c_{\mathbb{S},d} \tag{2.46}$$

$$A_{\mathbb{N}} = A_{i,i+1} = \dot{\vec{m}}_n \Lambda_n \mathcal{A}_n \mathcal{H}(-\dot{\vec{m}}_n) + c_{\mathbb{N},d} \tag{2.47}$$

and

$$b_i = \dot{S}_i \mathcal{V}_i \tag{2.48}$$

Note that conservation of mass requires $\sum_{\mathbb{k} \in \{e,w,n,s\}} \dot{\vec{m}}_{\mathbb{k}} = 0$. Since the number of cells is, in general, much smaller than N, it follows that matrix $\mathbf{A}$ is highly sparse, i.e., most of its coefficients are zero. So it only makes sense to employ solvers of the linear system, presented in Eq. (2.42), that take advantage of the sparsity of $\mathbf{A}$. Methods for solving linear systems, with focus on sparse matrices, are discussed next.

2.5   Solving of the Linear System of Equations

In this section, we discuss numerical methods for solving the linear system of equations, expressed in matrix form as

$$\mathbf{A}\vec{\phi} = \vec{b} \tag{2.49}$$

where $\mathbf{A}$ in an invertible $N \times N$ square matrix and $\vec{\phi}$ and $\vec{b}$ are column vectors of size N each. Methods for numerically solving the linear system for $\vec{\phi}$ can be direct or iterative. For large problems, where $\mathbf{A}$ is sparse, iterative methods are often used and direct methods such as the Gauss elimination and lower-upper (LU) decomposition are not used due to the associated prohibitive cost. As such, direct methods are not covered in this book, and the reader is referred to Ref [10].

Starting with the solution at iteration k, $\vec{\phi}^{(k)}$, an iterative method for solving the linear system is as follows:

$$\mathbf{A}\vec{\phi} = \vec{b} \tag{2.50}$$

splits the matrix $\mathbf{A}$ according to

$$\mathbf{A} = \mathbf{M} - \mathbf{N} \tag{2.51}$$

and arrives at an improved solution, $\vec{\phi}^{(k+1)}$, at iteration $k+1$ according to the iteration

$$\mathbf{M}\vec{\phi}^{(k+1)} = \mathbf{N}\vec{\phi}^{(k)} + \vec{b} \tag{2.52}$$

or

$$\vec{\phi}^{(k+1)} = \mathbf{M}^{-1}\mathbf{N}\vec{\phi}^{(k)} + \mathbf{M}^{-1}\vec{b} \tag{2.53}$$

where $\mathbf{M}^{-1}\mathbf{N}$ is the iteration matrix. The iterative method converges when $\vec{\phi}^{(k+1)} = \vec{\phi}^{(k)} = \vec{\phi}$, which implies that the Iteration (Eq. (2.52)) seeks to solve the original system, Eq. (2.50). Subtracting $\mathbf{M}\vec{\phi}^{(k)}$ from both sides of Eq. (2.52) yields

$$\mathbf{M}\vec{\varsigma}^{(k)} = \vec{\varepsilon}^{(k)} \tag{2.54}$$

where the correction update, $\vec{\varsigma}^{(k)}$, is

$$\vec{\varsigma}^{(k)} = \vec{\phi}^{(k+1)} - \vec{\phi}^{(k)} \tag{2.55}$$

and the corresponding residual is

$$\vec{\varepsilon}^{(k)} = \vec{b} - \mathbf{A}\vec{\phi}^{(k)} \tag{2.56}$$

Convergence depends on the spectral radius of the iteration matrix $\varrho(\mathbf{M}^{-1}\mathbf{N})$, which is the maximum absolute eigenvalue. Specifically, if $\mathbf{M}$ is invertible and $\varrho(\mathbf{M}^{-1}\mathbf{N}) < 1$, then $\vec{\phi}^{(k)}$ in Iteration (Eq. (2.52)) converges to the exact solution $\vec{\phi} = \mathbf{A}^{-1}\vec{b}$ for all choices of the initial guess $\vec{\phi}_0$. In numerical implementations of the algorithm, the iterations stop and convergence is reached if the norm of the residual is smaller than a preselected small positive number, that is, $\left\| \vec{\varepsilon}^{(k)} \right\| < \varsigma$. The number of iterations required to reach convergence depends on how close the spectral radius is to unity. The closer the spectral radius is to unity, the more iterations are needed.

There are two main classes of iterative methods. The first class is the stationary Jacobi-based iterative methods, presented below. Most iterative methods that have been developed over the years for solving linear systems with large sparse matrices belong to the second class, which covers the more general Krylov subspace projection-based methods, such as the bi-conjugate gradient method [11]. These methods are beyond the scope of this textbook, and, as such, will be not be covered here. Interested readers may refer to Refs. [12, 13].

2.5.1 Jacobi-Based Iterative Methods

Table 2.2 summarizes choices of $\mathbf{M}$ and $\mathbf{N}$ for different stationary iterative methods (SIM), where matrix $\mathbf{A}$ is expressed as the sum of a lower diagonal $\mathbf{L}$, diagonal $\mathbf{D}$, and upper diagonal $\mathbf{U}$ matrices, that is, $\mathbf{A} = \mathbf{L} + \mathbf{D} + \mathbf{U}$. If $\mathbf{A}$ is a symmetric positive definite matrix (SPD), the Gauss-Seidel method converges for all choices of $\vec{\phi}_0$. The method, however, becomes very slow when $\varrho(\mathbf{M}^{-1}\mathbf{N})$ is close to unity. To overcome this difficulty, the successive over-relaxation (SOR) method introduces the relaxation parameter ω, which is, ideally, chosen to minimize $\varrho(\mathbf{M}^{-1}\mathbf{N})$. In practice, it is hard to find the optimal value of ω and trial and error is often used. In what follows, we present a detailed description of the SOR method.

The Successive Over-Relaxation Method

One of the early widely used iterative methods for solving the linear system of discretized equations is the successive over-relaxation (SOR) method, which is based on the Gauss-Seidel iterative procedure. The linear equation arising from discretization of

TABLE 2.2 Stationary Iterative Methods—$\mathbf{A} = \mathbf{L} + \mathbf{D} + \mathbf{U}$. For all cases, $\mathbf{N} = \mathbf{A} - \mathbf{M}$

Method	M	ω
Jacobi	**D**	
Gauss-Seidel	**D** + **L**	
Successive Over-relaxation	$\omega^{-1}\mathbf{D} + \mathbf{L}$	$1 < \omega < 2$
Successive under-relaxation	$\omega^{-1}\mathbf{D} + \mathbf{L}$	$0 < \omega < 1$

the conservative form of the conservation law for a given grid cell, i, may be expressed as

$$A_{ii}\phi_i + \sum_{j\in\mathcal{I}_i, j\neq i} A_{ij}\phi_j = b_i \tag{2.57}$$

where $\mathcal{I}_i$ is the set of neighboring cells involved in the discretization. The Gauss-Seidel iterative procedure updates ϕ at the cells centroids sequentially according to the order in which the coefficients are stored in $\mathbf{A}$. So when it is the turn of cell i to update its solution from iteration k to iteration $k+1$, the solutions of cells $1, 2, \ldots, i-1$ would have already been updated. The SOR scheme takes advantage of this by updating ϕ_i according to

$$\phi_i^{(k+1)} = \omega A_{ii} \left(b_i - \sum_{j\in\mathcal{I}_i, j<i} A_{ij}\phi_j^{(k+1)} - \sum_{j\in\mathcal{I}_i, j>i} A_{ij}\phi_j^{(k)} \right) + (1-\omega)\phi_i^{(k)} \tag{2.58}$$

where ω, the over-relaxation factor, must be greater than one. For example, for a structured grid with second-order discretization schemes, $\mathcal{I}_i$ consists of $\mathbb{O}$, $\mathbb{E}$, $\mathbb{EE}$, $\mathbb{W}$, $\mathbb{WW}$, $\mathbb{N}$, $\mathbb{NN}$, $\mathbb{S}$, and $\mathbb{SS}$ cells relative to cell i. If, as depicted in Figure 2.6, $\mathbf{A}$ and $\vec{b}$ are filled by processing the grid cells column by column, starting from the eastern most column, and for each column northwards starting from the southern most cell, then the solutions at the centroids of cells $\mathbb{W}$, $\mathbb{WW}$, $\mathbb{S}$, and $\mathbb{SS}$ are updated before $\phi_{\mathbb{O}}$, and are used to update $\phi_{\mathbb{O}}$ according to

$$\phi_{\mathbb{O}}^{(k+1)} = \frac{\omega}{A_{\mathbb{O}}} \left(b_{\mathbb{O}} - (A_{\mathbb{W}}\phi_{\mathbb{W}}^{(k+1)} + A_{\mathbb{S}}\phi_{\mathbb{S}}^{(k+1)} + A_{\mathbb{N}}\phi_{\mathbb{N}}^{(k)} + A_{\mathbb{E}}\phi_{\mathbb{E}}^{(k)} \right.$$
$$\left. + A_{\mathbb{WW}}\phi_{\mathbb{WW}}^{(k+1)} + A_{\mathbb{SS}}\phi_{\mathbb{SS}}^{(k+1)} + A_{\mathbb{NN}}\phi_{\mathbb{NN}}^{(k)} + A_{\mathbb{EE}}\phi_{\mathbb{EE}}^{(k)}) \right) + (1-\omega)\phi_{\mathbb{O}}^{(k)} \tag{2.59}$$

Finding the optimal value of ω for maximum convergence speed is not straightforward. Too small of a value reduces the convergence speed while too large of a value causes

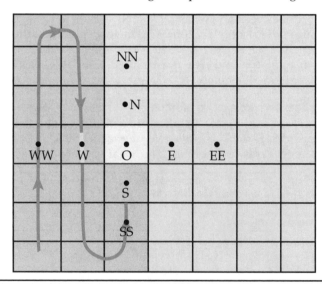

FIGURE 2.6 Schematic of the cells involved in the SOR method showing the direction in which the cells are updated.

oscillatory convergence behavior. For unsteady flows, this challenge can be overcome by searching for an updated optimal value of ω every few time steps, starting from the previous optimal value. In many cases, convergence requires under-relaxation ($\omega < 1$) to limit the changes in the variables in a time step or iteration. In summary, over-relaxation speeds up convergence as long as stability of the numerical system allows, and under-relaxation is used to stabilize the numerical scheme at the expense of the convergence speed.

2.6   Integration in Time for Unsteady Flow

In this section, we discuss the different methods for marching (or integrating) the solution of the spatially discretized Eq. (2.14) discretely in time from $t_1 = \ell\Delta t$ to $t_2 = (\ell + 1)\Delta t$ given past solutions $\phi(b\Delta t), b \le \ell$, where Δt is the time step and b is a positive integer. Equation (2.14) can be expressed as

$$\frac{\partial}{\partial t}\left(\rho_i \Lambda_i \phi_i \mathcal{V}_i\right) = g(t, \phi(t)), \tag{2.60}$$

where

$$g(t, \phi(t)) = -\sum_{k \in \mathcal{F}_i}\left(\dot{m}_k^{\rightarrow} \Lambda_k \phi_k - \Gamma_k \mathcal{A}_k \left(\frac{\partial \phi}{\partial n}\right)_k\right) + \dot{S}_i \mathcal{V}_i \tag{2.61}$$

In a collocated grid, the advection and diffusion flux terms are expressed in terms of values at the centroids of the cells. Before discussing the different methods for time-marching Eq. (2.60), we first present different methods for numerically solving the initial value problem:

$$\frac{d\phi}{dt} = f(t, \phi(t)) \tag{2.62}$$

subject to the initial condition $\phi(t = 0) = \phi_0$. The exact solution at $t_2 = (\ell + 1)\Delta t$ is obtained by integrating Eq. (2.62) from t_1 to t_2,

$$\phi^{(\ell+1)} = \phi^{(\ell)} + \int_{\ell\Delta t}^{(\ell+1)\Delta t} f(t', \phi(t'))dt' \tag{2.63}$$

where $\phi^{(\ell)}$ refers to the solution at $t_1 = \ell\Delta t$. To numerically approximate this solution, the integral at the right hand side of Eq. (2.63) can be expressed in terms of the values of ϕ at discrete times $b\Delta t$, where b is a positive integer. Noting that when marching from $t_1 = \ell\Delta t$ to $t_2 = (\ell + 1)\Delta t$, past solutions $\phi(b\Delta t), b \le \ell$ are available, and future solutions $\phi(b\Delta t), b > \ell$ are not available, the numerical methods that have been developed are classified into two categories: *explicit* and *implicit* schemes [11, 14]. Explicit schemes approximate the integral at the right hand side of Eq. (2.63) in terms of past discrete values $\phi(b\Delta t), b \le \ell$. Implicit schemes can be *fully implicit* or *semi-implicit*. While fully implicit schemes approximate the integral at the right hand side of Eq. (2.63) in terms of the future discrete values $\phi^{(\ell+1)}$, semi-implicit schemes employ both past and future values.

Numerical approximation using the discrete values can be generally expressed as

$$\phi^{(\ell+1)} = \phi^{(\ell)} + \sum_b a_b f(b\Delta t, \phi(b\Delta t))\Delta t \tag{2.64}$$

TABLE 2.3 Two-Level Methods

Method	Type	a_ℓ	$a_{\ell+1}$	Order	Stability
Forward Euler	explicit	1	0	1	$\Delta t < 2 \left(\dfrac{\partial f}{\partial \phi} \right)^{-1}$
Backward Euler	implicit	0	1	1	unconditional
Trapezoidal rule	semi-implicit	$\frac{1}{2}$	$\frac{1}{2}$	2	unconditional*

*May be unstable for nonlinear problems.

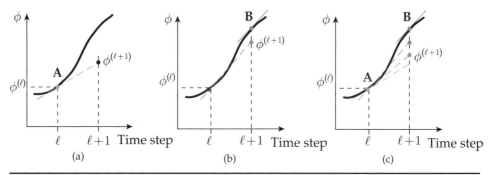

FIGURE 2.7 Plots of ϕ versus time depicting (a) the Forward Euler, (b) Backward Euler, and (c) Trapezoidal time integration schemes.

where the values of b used in the summation depend on the time integration scheme used. Two level schemes employ the past ($\ell\Delta t$) and future (($\ell+1)\Delta t$) time steps so that

$$\phi^{(\ell+1)} = \phi^{(\ell)} + a_\ell f(\ell\Delta t, \phi^{(\ell)})\Delta t + a_{\ell+1} f((\ell+1)\Delta t, \phi^{(\ell+1)})\Delta t \qquad (2.65)$$

The values of b and a_b for the different two-level schemes are listed in Table 2.3. Figure 2.7 shows how ϕ is updated from time step ℓ to time step $\ell + 1$ using the three schemes listed in the table. The Forward Euler explicit scheme integrates ϕ along a linear profile with slope at point A, as shown in Figure 2.7(a). The Backward Euler implicit scheme integrates ϕ along a linear profile with slope at point B, as shown in Figure 2.7(b). Figure 2.7(c) shows that the Trapezoidal scheme integrates ϕ along a linear profile with a slope equal to the arithmetic average of the slopes at A and B. Note that the explicit scheme is simple to implement as it readily provides the solution at ($\ell+1)\Delta t$. It is, however, conditionally stable, and as such restricts the size of the time step to the condition shown in the Table 2.3. In contrast, the implicit and semi-implicit schemes are more challenging. This is because Eq. (2.64) for control volume i involves future unknown values of ϕ at the centroids of neighbor cells belonging to the interaction list. Computing future values at all the cells centroids would then require solving a linear system of the form (2.50), which adds to the computational cost in terms of CPU time and memory storage, especially when f is a nonlinear function of ϕ, in which case a nonlinear equation has to be iteratively solved for $\phi^{(\ell+1)}$. The main advantage of the implicit schemes is that they are unconditionally stable and as such put no restriction on the choice of the size of the time step. Selected higher level schemes for solving the initial value problem are also discussed in Appendix B. A discussion on the stability conditions for various schemes is also presented in Appendix B.

2.7   The Navier-Stokes Equations

The Navier-Stokes equations refer to the differential form of the conservation of mass and momentum, expressed respectively in Eqs. (1.29) and (1.53) for a Newtonian fluid. Under conditions where the thermodynamic state equation may be expressed as the density being constant or depends uniquely on the pressure, the Navier-Stokes equations are a system of coupled nonlinear-partial differential equations, where the unknowns are the velocity vector field and the pressure (scalar) field. The Navier-Stokes equations are difficult to solve numerically for the following reasons:

- Since the equations are coupled, we need to solve for all unknowns at once.

- While the continuity and momentum equations govern the velocity vector field, there is no explicit equation that governs the pressure field. For compressible flows, the continuity equation serves as a dynamic equation for updating the density. The pressure field can then be obtained from the equation of state. In incompressible flows, the continuity equation serves as a kinematic constraint on the velocity field. In this case, the density is virtually constant and the equation of state cannot be used to determine the pressure. The divergence-free condition of the velocity is exploited to derive a Poisson equation governing the pressure field by taking the divergence of the momentum equation.

- The advection term $\nabla \cdot (\rho \vec{u}\vec{u})$ is nonlinear. Since most numerical schemes are based on discretizations that ultimately yield a linear system, discretization of this nonlinear term not only introduces a numerical error but also amplifies this error, depending on the discretization scheme, where initially small numerical errors grow uncontrollably resulting in an unstable numerical solution. As such, higher order discretizations are not always an improvement over lower order discretizations.

- A major challenge is that the spatial and temporal discretization schemes preserve the conservative nature of the original equations and do not introduce nonphysical effects.

In what follows, we present solution methods for a collocated grid, which became popular in the 1980s, after overcoming the difficulty in the pressure-velocity coupling and the resulting nonphysical oscillations in pressure. The advantages of using a collocated grid are (i) it is simple to implement, (ii) it requires less storage, (iii) it is appropriate for complex boundaries, and (iv) it is easier to deploy within a multi-grid scheme. To simplify the exposition, we consider flows where the fluid has a constant density and a constant viscosity.

In the absence of forces other than gravity and stress, the Navier-Stokes equations governing a constant density flow ($\rho = \rho_0$), Eqs. (1.29) and (1.53) reduce to

$$\nabla \cdot \vec{u} = 0 \tag{2.66}$$

$$\frac{\partial (\rho_0 \vec{u})}{\partial t} + \nabla \cdot (\rho_0 \vec{u}\vec{u}) = -\nabla p + \mu_0 \nabla^2 \vec{u} + \rho_0 \vec{g} \tag{2.67}$$

where the coefficient of dynamic viscosity is assumed to be constant, $\mu = \mu_0$. In order to derive the Poisson equation governing the pressure, we take the divergence of the momentum Eq. (2.67), yielding

$$\nabla \cdot \left[\frac{\partial(\rho_0 \vec{u})}{\partial t} + \nabla \cdot (\rho_0 \vec{u}\vec{u}) \right] = -\nabla \cdot \nabla p + \mu_0 \nabla \cdot (\nabla^2 \vec{u}) + \nabla \cdot (\rho_0 \vec{g}) \tag{2.68}$$

Equation (2.68) may be rearranged as

$$\frac{\partial}{\partial t} (\nabla \cdot (\rho_0 \vec{u})) = -\nabla^2 p + \nabla \cdot \left[-\nabla \cdot (\rho_0 \vec{u}\vec{u}) + \mu_0 \nabla^2 \vec{u} + \rho_0 \vec{g} \right] \tag{2.69}$$

Since the density is constant, the continuity equation simplifies to $\nabla \cdot \vec{u} = 0$. It then follows that the unsteady term, the viscous term and the gravity term in Eq. (2.69) vanish. This implies that to maintain the velocity field divergence-free as it evolves over time, the following condition must be satisfied

$$\nabla^2 p = -\nabla \cdot [\nabla \cdot (\rho_0 \vec{u}\vec{u})] \tag{2.70}$$

In the FVM, the discretized form of the Navier-Stokes equations is obtained by first expressing the Navier-Stokes Eqs. (2.66) and (2.67) in integral form applied to a control volume, intended to be a grid cell of volume $\mathcal{V}_i$ bounded by surface $\mathcal{S}_i$. Accordingly,

$$\int_{\mathcal{S}_i} \rho_0 \vec{u} \cdot \hat{n} \, d\mathcal{S} = 0 \tag{2.71}$$

$$\frac{\partial}{\partial t} \int_{\mathcal{V}_i} \rho_0 \vec{u} \, d\mathcal{V} + \int_{\mathcal{S}_i} \rho_0 \vec{u}\vec{u} \cdot \hat{n} \, d\mathcal{S} = -\int_{\mathcal{S}_i} p\hat{n} \, d\mathcal{S} + \mu_0 \int_{\mathcal{S}_i} \hat{n} \cdot \nabla \vec{u} \, d\mathcal{S} + \int_{\mathcal{V}_i} \rho_0 \vec{g} \, d\mathcal{V} \tag{2.72}$$

Integrating Eq. (2.70) over the volume of a grid cell of boundary $\mathcal{S}_i$, and using the divergence theorem, we obtain

$$\int_{\mathcal{S}_i} \hat{n} \cdot \nabla p \, d\mathcal{S} = -\int_{\mathcal{S}_i} \hat{n} \cdot [\nabla \cdot (\rho_0 \vec{u}\vec{u})] \, d\mathcal{S} \tag{2.73}$$

Using the FVM, the discretized form of Eq. (2.67) is given by Eq. (2.14), with $\phi = u, v, w$, $\Gamma_{\Bbbk} = \mu$, $\Lambda_{\Bbbk} = 1$, and $\dot{S} = -\nabla p + \rho \vec{g}$. In Cartesian coordinates, we obtain the following equations for $u, v,$ and w:

$$\frac{\partial}{\partial t} (\rho_0 u_i \mathcal{V}_i) + \sum_{\Bbbk \in \mathcal{F}_i} \left(\dot{m}_{\Bbbk}^{\rightarrow} u_{\Bbbk} - \mu_0 A_{\Bbbk} \left(\frac{\partial u}{\partial n} \right)_{\Bbbk} \right) = \sum_{\Bbbk \in \mathcal{F}_i} -p_{\Bbbk} \hat{n}_{\Bbbk} \cdot \hat{x} A_{\Bbbk} \tag{2.74}$$

$$\frac{\partial}{\partial t} (\rho_0 v_i \mathcal{V}_i) + \sum_{\Bbbk \in \mathcal{F}_i} \left(\dot{m}_{\Bbbk}^{\rightarrow} v_{\Bbbk} - \mu_0 A_{\Bbbk} \left(\frac{\partial v}{\partial n} \right)_{\Bbbk} \right) = \sum_{\Bbbk \in \mathcal{F}_i} -p_{\Bbbk} \hat{n}_{\Bbbk} \cdot \hat{y} A_{\Bbbk} \tag{2.75}$$

$$\frac{\partial}{\partial t} (\rho_0 w_i \mathcal{V}_i) + \sum_{\Bbbk \in \mathcal{F}_i} \left(\dot{m}_{\Bbbk}^{\rightarrow} w_{\Bbbk} - \mu_0 A_{\Bbbk} \left(\frac{\partial w}{\partial n} \right)_{\Bbbk} \right) = \sum_{\Bbbk \in \mathcal{F}_i} -p_{\Bbbk} \hat{n}_{\Bbbk} \cdot \hat{z} A_{\Bbbk} - \rho_0 g \mathcal{V}_i \tag{2.76}$$

where $\vec{g} = -g\hat{z}$. The values of the pressure, $p_{\Bbbk}$, the normal derivatives of the velocity components, $\left(\frac{\partial}{\partial n}(u, v, w) \right)_{\Bbbk}$, and the velocity, $(u, v, w)_{\Bbbk}$, at the centroids of the faces $\Bbbk$ are approximated in terms of values at the cells centroids according to the schemes presented in Section 2.4. In compact notation, Eqs. (2.74–2.76) above are written as

$$\frac{\partial}{\partial t} (\rho_0 \vec{u}_i \mathcal{V}_i) + \sum_{\Bbbk \in \mathcal{F}_i} \left(\dot{m}_{\Bbbk}^{\rightarrow} \vec{u}_{\Bbbk} - \mu_0 A_{\Bbbk} \left(\frac{\partial \vec{u}}{\partial n} \right)_{\Bbbk} \right) = \sum_{\Bbbk \in \mathcal{F}_i} -p_{\Bbbk} \hat{n}_{\Bbbk} A_{\Bbbk} + \rho_0 \vec{g} \mathcal{V}_i \tag{2.77}$$

Expressing the surface integral as the sum of integrals over the faces constituting S_i, and approximating each surface integral using the value at the face centroid, the discretized form of the Poisson equation governing the pressure, Eq. (2.73), is

$$\sum_{k \in \mathcal{F}_i} A_k \left(\frac{\partial p}{\partial n} \right)_k \simeq - \sum_{k \in \mathcal{F}_i} \hat{n}_k \cdot [\nabla \cdot (\rho_0 \vec{u}\vec{u})]_k \, A_k \qquad (2.78)$$

Equation (2.78) serves as a means to compute the pressure field from the velocity field while satisfying the Navier-Stokes equations for a constant density fluid.

In what follows, we present numerical methods for solving the Navier-Stokes equations for unsteady and steady flows using explicit and implicit time stepping schemes.

2.7.1   Unsteady Flows

Explicit Time Stepping Scheme

When using a first-order explicit time stepping scheme to arrive at the solution of Eq. (2.67) at time step $\ell + 1$ starting from the solution obtained previously at the time step ℓ, the velocities at the centroids of the computational cells at time step $\ell + 1$ are determined according to

$$\vec{u}^{(\ell+1)} = \vec{u}^{(\ell)} + \frac{\Delta t}{\rho_0} \left(-\nabla \cdot (\rho_0 \vec{u}\vec{u}) + \mu_0 \nabla^2 \vec{u} - \nabla p + \rho_0 \vec{g} \right)^{(\ell)} \qquad (2.79)$$

where to simplify the exposition, the density is taken to be constant, that is, $\rho = \rho_0$. Equation (2.79) is expressed in semi-discrete form, i.e., discrete in time and continuous in space. If the velocity field is divergence-free at time step ℓ, then it will be divergence-free at step $\ell + 1$ if the pressure, $p^{(\ell)}$, satisfies the Poisson equation (Eq. (2.70)).

$$\nabla^2 p^{(\ell)} = -\nabla \cdot [\nabla \cdot (\rho_0 \vec{u}\vec{u})]^{(\ell)} \qquad (2.80)$$

Upon integrating Eq. (2.79) over a control volume consisting of grid cell i, the discretized forms of the equation is

$$\vec{u}_i^{(\ell+1)} = \vec{u}_i^{(\ell)} - \frac{\Delta t}{\rho_0 \mathcal{V}_i} \left[\sum_{k \in \mathcal{F}_i} \left(\dot{m}_k \vec{u}_k - \mu_0 A_k \left(\frac{\partial \vec{u}}{\partial n} \right)_k \right) + \sum_{k \in \mathcal{F}_i} p_k \hat{n}_k A_k - \rho_0 \vec{g} \mathcal{V}_i \right]^{(\ell)} \qquad (2.81)$$

where the advection term is approximated using one of the discretization schemes presented in Section (2.4.6) and the diffusion term using one of the discretization schemes presented in Section (2.4.4). Recall that choice of these two discretization schemes decides not only on the spatial discretization accuracy but also on the size of the time step as constrained by the corresponding stability requirement. The pressure at the centroid of the faces, that appear in Eq. (2.81), is interpolated from the values at the cells centroids obtained by numerically solving Eq. (2.80), which, upon integration over control volume i, is expressed in the discrete form

$$\sum_{k \in \mathcal{F}_i} A_k \left(\frac{\partial p}{\partial n} \right)_k^{(\ell)} = - \sum_{k \in \mathcal{F}_i} \hat{n}_k \cdot [\nabla \cdot (\rho_0 \vec{u}\vec{u})]_k^{(\ell)} \, A_k \qquad (2.82)$$

Implicit Time Stepping Scheme

A first-order implicit time stepping scheme reaches the solution at time step $\ell + 1$ starting from the solution obtained previously at the time step ℓ by solving the following equation:

$$\vec{u}^{(\ell+1)} = \vec{u}^{(\ell)} + \frac{\Delta t}{\rho_0} \left(-\nabla \cdot (\rho_0 \vec{u}\vec{u}) + \mu_0 \nabla^2 \vec{u} - \nabla p + \rho_0 \vec{g} \right)^{(\ell+1)} \tag{2.83}$$

expressed in semi-discrete form. Unlike the explicit scheme, where the right hand side of Eq. (2.81) is readily available from the solution at the previous time step, the right side of Eq. (2.83) is a function of the velocity and pressure at the new time step, which are to be determined. As before, and in order to satisfy the continuity equation, the pressure satisfies the Poisson equation (Eq. (2.70)), which for constant density ($\rho = \rho_0$), is expressed as

$$\nabla^2 p^{(\ell+1)} = -\nabla \cdot [\nabla \cdot (\rho_0 \vec{u}\vec{u})]^{(\ell+1)} \tag{2.84}$$

Upon inspecting Eqs. (2.83) and (2.84), one can identify two key challenges that arise in the implicit scheme. The first challenge is that the pressure and velocity are coupled. In other words, Eqs. (2.83) and (2.84) have to be solved simultaneously for $\vec{u}^{(\ell+1)}$ and $p^{(\ell+1)}$. The second challenge is the nonlinearity arising from the term $\nabla \cdot (\vec{u}\vec{u})$, which does not allow for a system of linear equations governing the unknowns p and $\vec{u}$. These two challenges are largely overcome by linearizing the two equations by expressing:

$$p^{(\ell+1)} = p^* + p' \tag{2.85}$$

$$\vec{u}^{(\ell+1)} = \vec{u}^* + \vec{u}' \tag{2.86}$$

where p^* is chosen to be $p^{(\ell)}$, and $\vec{u}^*$ is the solution of the

$$\vec{u}^* = \vec{u}^{(\ell)} + \frac{\Delta t}{\rho_0} \left(-\nabla \cdot (\rho_0 \vec{u}^* \vec{u}^*) + \mu_0 \nabla^2 \vec{u}^* - \nabla p^{(\ell)} + \rho_0 \vec{g} \right) \tag{2.87}$$

Note that Eq. (2.87) differs from Eq. (2.83) by having p evaluated at time step ℓ instead of at time step $\ell + 1$, which decouples the pressure from the velocity. The solution $\vec{u}^*$, however, is not divergence-free. Taking the difference between Eqs. (2.83) and (2.87), we arrive at a relation between p' and $\vec{u}'$,

$$\vec{u}' = -\frac{\Delta t}{\rho_0} \left(\vec{B}' + \nabla p' \right) \tag{2.88}$$

where

$$\vec{B}' = \vec{B}^{(\ell+1)} - \vec{B}^*$$

and

$$\vec{B} = \nabla \cdot (\rho_0 \vec{u}\vec{u}) - \mu_0 \nabla^2 \vec{u} \tag{2.89}$$

The solution at the new time step is divergence-free if $\nabla \cdot \vec{u}^{(\ell+1)} = 0$, under which conditions Eqs. (2.86) and (2.88) lead to

$$\nabla \cdot \vec{u}^* = -\nabla \cdot \vec{u}' = \frac{\Delta t}{\rho_0} \left(\nabla \cdot \vec{B}' + \nabla \cdot \nabla p' \right) \tag{2.90}$$

or

$$\nabla \cdot \nabla p' = \frac{\rho_0}{\Delta t} \nabla \cdot \vec{u}^* - \nabla \cdot \vec{B}' \tag{2.91}$$

Different methods have been proposed for solving Eqs. (2.88) and (2.91) for p' and $\vec{u}'$. Since the two equations are coupled, simplification is often made by approximating the term $\nabla \cdot \vec{B}'$. For example, the semi-implicit method for pressure linked equations (SIMPLE) algorithm [15] assumes $\vec{B}' = \vec{0}$. In this case, the pressure correction can be readily obtained from $\vec{u}^*$ by solving the the Poisson equation:

$$\nabla^2 p' \simeq \frac{\rho_0}{\Delta t} \nabla \cdot \vec{u}^* \tag{2.92}$$

which follows from Eq. (2.91). An approximation for the the velocity correction follows from Eq. (2.88), which simplifies to

$$\vec{u}' \simeq -\frac{\Delta t}{\rho_0} \nabla p' \tag{2.93}$$

The discretized forms of Eqs. (2.87), (2.92), and (2.93), obtained by integrating over a control volume consisting of grid cell i, and carrying out the subsequent approximations for the surface and volume integrals, are

$$\vec{u}_i^* = \vec{u}_i^{(\ell)} - \frac{\Delta t}{\rho_0 \mathcal{V}_i} \left\{ \sum_{\Bbbk \in \mathcal{F}_i} \left(\dot{m}_{\Bbbk} \vec{} \vec{u}_{\Bbbk}^* - \mu_0 \mathcal{A}_{\Bbbk} \left(\frac{\partial \vec{u}}{\partial n} \right)_{\Bbbk}^* \right) + \sum_{\Bbbk \in \mathcal{F}_i} p_{\Bbbk}^{(\ell)} \hat{n}_{\Bbbk} \mathcal{A}_{\Bbbk} - \rho_0 \vec{g} \mathcal{V}_i \right\} \tag{2.94}$$

$$\sum_{\Bbbk \in \mathcal{F}_i} \left(\frac{\partial p'}{\partial n} \right)_{\Bbbk} \mathcal{A}_{\Bbbk} = \frac{\rho_0}{\Delta t} \sum_{\Bbbk \in \mathcal{F}_i} \left(\vec{u}^* \cdot \hat{n} \right)_{\Bbbk} \mathcal{A}_{\Bbbk} \tag{2.95}$$

$$\vec{u}_i^{(\ell+1)} = \vec{u}_i^* - \frac{\Delta t}{\rho_0 \mathcal{V}_i} \left(\sum_{\Bbbk \in \mathcal{F}_i} p'_{\Bbbk} \hat{n}_{\Bbbk} \mathcal{A}_{\Bbbk} \right) \tag{2.96}$$

Note that the errors that arise from the linearization are proportional to p' and $\vec{u}'$, and as such they are larger for larger values of the time step, Δt.

2.7.2 Steady Flow

For steady flows, the discretized momentum equation follows from Eq. (2.77) by setting the time derivative to zero, so that

$$\sum_{\Bbbk \in \mathcal{F}_i} \left(\dot{m}_{\Bbbk} \vec{} \vec{u}_{\Bbbk} - \mu_0 \mathcal{A}_{\Bbbk} \left(\frac{\partial \vec{u}}{\partial n} \right)_{\Bbbk} \right) + \sum_{\Bbbk \in \mathcal{F}_i} p_{\Bbbk} \hat{n}_{\Bbbk} \mathcal{A}_{\Bbbk} - \rho_0 \vec{g} \mathcal{V}_i = \vec{0} \tag{2.97}$$

Employing the interpolations in terms of cell centroid quantities in Eq. (2.22) and using Eq. (2.39)

$$\left(\sum_{\Bbbk \in \mathcal{F}_i} \dot{m}_{\Bbbk} \vec{} \left(c_{\mathbb{O},a} \phi_{\mathbb{O}} + c_{\mathbb{L},a} \phi_{\mathbb{L}} + c_{\mathbb{K},a} \phi_{\mathbb{K}} + c_{\mathbb{K}\mathbb{K},a} \phi_{\mathbb{K}\mathbb{K}} \right) \right)$$
$$+ \left(c_{\mathbb{O},d} \phi_{\mathbb{O}} + c_{\mathbb{E},d} \phi_{\mathbb{E}} + c_{\mathbb{W},d} \phi_{\mathbb{W}} + c_{\mathbb{N},d} \phi_{\mathbb{N}} + c_{\mathbb{S},d} \phi_{\mathbb{S}} \right) = - \sum_{\Bbbk \in \mathcal{F}_i} p_{\Bbbk} \hat{n}_{\Bbbk} \cdot \hat{i} \mathcal{A}_{\Bbbk} + \dot{S}_i^{(\phi)} \tag{2.98}$$

Algorithm 1 Using an outer and an inner iteration to solve the nonlinear equation (Eq. (2.100))

1: $o = 0$
2: $k = 0$
3: initialize $\vec{u}^{(k)}$
4: **repeat**
5:     $o \leftarrow o + 1$
6:     **for** $i = 1 \ldots N$ **do**
7:         Update $c_{ij}^{(o)}, j \in \mathcal{I}(i)$ from nodal velocities $\vec{u}_{i'}^{(k)}, i' = 1 \ldots N$
8:     **end for**
9:     $k = 0$
10:     **repeat**
11:         $k \leftarrow k + 1$
12:         Solve Eq. (2.100) for $\vec{u}_{i'}^{(k)}, i' = 1 \ldots N$
13:     **until** Convergence
14: **until** Convergence

---

where $\phi = \{u, v, w\}, \hat{i} = \{\hat{x}, \hat{y}, \hat{z}\}$, and the source term, here, does not include the pressure force, so that $\dot{S}_i^{(\phi)} = \{0, 0, -\rho_0 g \mathcal{V}_i\}$. Note that the coefficients c in Eq. (2.98) are different for different choices of ϕ. Equation (2.98) can be expressed in the following compact form:

$$c_{ii}^{(\phi)} \phi_i + \sum_{j \in \mathcal{I}(i), j \neq i} c_{ij}^{(\phi)} \phi_j = - \sum_{k \in \mathcal{F}_i} p_k \hat{n}_k \cdot \hat{i} \, \mathcal{A}_k + \dot{S}_i^{(\phi)} \qquad (2.99)$$

The nonlinear term is handled by linearization in the context of an iterative scheme which employs an outer iteration, referred to by the superscript o, and an inner iteration, referred to by the superscript k. The inner iteration updates ϕ from iteration $k - 1$ to iteration k while using the values for c_{ii} and c_{ij} from the outer iteration:

$$c_{ii}^{(o)} \phi_i^{(k)} + \sum_{j \in \mathcal{I}(i), j \neq i} c_{ij}^{(o)} \phi_j^{(k)} = - \sum_{k \in \mathcal{F}_i} p_k^{(k)} \hat{n}_k \cdot \hat{i} \, \mathcal{A}_k + \dot{S}_i^{(o)} \qquad (2.100)$$

where the superscript (ϕ) was dropped from the coefficients for simplicity of exposition. Note that the coefficients c_{ii} and c_{ij} contain the mass flow rates across the faces, which in turn requires knowledge of the velocities. To deal with the nonlinearity that arises, the velocities from the outer iteration,[2] o, are used to evaluate the coefficients in Eq. (2.100). A pseudocode of the inner and outer loops is presented in Algorithm 1.

Algorithm 1 does not show how the pressure at the cells faces is determined at iteration k. To this end, we present in what follows commonly used schemes for decoupling the pressure and velocity in Eq. (2.100).

Decoupling the pressure and velocity may be accomplished by using the previous iteration value for the pressure

$$c_{ii}^{(o)} \phi_i^* + \sum_{j \in \mathcal{I}(i), j \neq i} c_{ij}^{(o)} \phi_j^* = - \sum_{k \in \mathcal{F}_i} p_k^{(k-1)} \hat{n}_k \cdot \hat{i} \, \mathcal{A}_k + \dot{S}_i \qquad (2.101)$$

---

[2]Or previous time step when solving for temperature.

It is obvious that the solution, $\vec{u}^*$, to Eq. (2.101) does not satisfy the original Eq. (2.99) and is not divergence-free. Different methods have been devised to improve the solution. These are based on correcting the pressure and the velocity fields to make the velocity field divergence-free and, as such, a closer approximation of the solution of Eq. (2.99). The velocity (recall that $\phi \in \{u, v, w\}$) and pressure at the end of iteration k are then expressed as

$$\phi^{(k)} = \phi^* + \phi' \tag{2.102}$$

$$p^{(k)} = p^{(k-1)} + p' \tag{2.103}$$

where ϕ^* is the solution of Eq. (2.101). Since the solution at iteration k is very close to that at iteration $(k-1)$, it follows that the velocity and pressure corrections ϕ' and p' satisfy the conditions $|\phi'| << |\phi^{(k)}|$ and $|p'| << |p^{(k)}|$. Now we turn attention to deriving the equations governing ϕ' and p', followed by presenting the mainstream methods used to numerically solve these equations. The first equation may be obtained by taking the difference between Eqs. (2.100) and (2.101):

$$\phi'_i = -\frac{\sum_{j \in \mathcal{I}(i), j \neq i} c_{ij}^{(o)} \phi'_j}{c_{ii}^{(o)}} - \frac{1}{c_{ii}^{(o)}} \sum_{\Bbbk \in \mathcal{F}_i} p'_\Bbbk \hat{n}_\Bbbk \cdot \hat{i} \, \mathcal{A}_\Bbbk \tag{2.104}$$

Expressing in terms of the velocity vector, Eq. (2.104) may be expressed as

$$\vec{u}'_i = \vec{r}'_i - \sum_{\Bbbk \in \mathcal{F}_i} p'_\Bbbk \hat{n}_\Bbbk^{(s)} \, \mathcal{A}_\Bbbk \tag{2.105}$$

where $\vec{r}'_i$ is the first term on the right-hand side of Eq. (2.105), and the scaled unit vector is $\hat{n}_\Bbbk^{(s)} = \frac{\hat{n}_\Bbbk \cdot \hat{x}}{c_{ii,u}^{(o)}} \hat{x} + \frac{\hat{n}_\Bbbk \cdot \hat{y}}{c_{ii,v}^{(o)}} \hat{y} + \frac{\hat{n}_\Bbbk \cdot \hat{z}}{c_{ii,w}^{(o)}} \hat{z}$, where $c_{ii,u}$, $c_{ii,v}$, and $c_{ii,w}$ are the values of c_{ii} corresponding to $\phi = u, v$, and w, respectively. As mentioned before, the velocity is corrected so that it remains divergence-free, that is, $(\nabla \cdot \vec{u}^{(k)})_i = 0$. Using Eq. (2.102),

$$(\nabla \cdot \vec{u}^*)_i + (\nabla \cdot \vec{u}')_i = 0 \tag{2.106}$$

Taking the divergence of Eq. (2.105), and using Eq. (2.106), we arrive at the second equation that couples p' and ϕ'

$$(\nabla \cdot \vec{u}^*)_i + (\nabla \cdot \vec{r}')_i = \sum_{\Bbbk \in \mathcal{F}_i} \hat{n}_\Bbbk^{(s)} \cdot (\nabla p')_\Bbbk \, \mathcal{A}_\Bbbk \tag{2.107}$$

Different solution approaches are based on different approximations that will further simplify Eqs. (2.104) and (2.107) governing p' and $\vec{u}'$. The SIMPLE algorithm [15] solves Eq. (2.107) for the pressure correction, p', while neglecting the term $\vec{r}'$

$$\sum_{\Bbbk \in \mathcal{F}_i} \hat{n}_\Bbbk^{(s)} \cdot (\nabla p')_\Bbbk \, \mathcal{A}_\Bbbk \simeq (\nabla \cdot \vec{u}^*)_i \tag{2.108}$$

The term $\vec{r}'$ is, however, retained when solving Eq. (2.104) for $\vec{u}'$. A pseudocode for the SIMPLE algorithm is presented in Algorithm 2. Other algorithms for decoupling the pressure and velocity fields, based on the SIMPLE algorithm, include the SIMPLER algorithm [7, 16], SIMPLEC algorithm [17], and the PISO algorithm [18].

For compressible flows, density-based algorithms are preferred over pressure-based algorithms such as SIMPLE. In density-based algorithms, the density is computed from the continuity equation and the pressure is then computed from the state equation.

Algorithm 2 The SIMPLE Algorithm

1: Evaluate the coefficients $c_{ii}^{(o)}$, $c_{ij}^{(o)}$, $j \in \mathcal{I}(i)$, and $S_i^{(o)}$ using the solution at o.
2: $k = 1$
3: **repeat**
4:      $k \leftarrow k + 1$
5:      Iterate from $k - 1$ to k, given $\vec{u}_i^{(k-1)}$, $p_i^{(k-1)}$, $i \ldots N$
6:      Solve Eq. (2.101) for $\vec{u}^*$
7:      Solve Eq. (2.108) for p'
8:      Solve Eq. (2.104) for $\vec{u}'$
9:      $p^{(k)} = p^{(k-1)} + p'$
10:     $\vec{u}^{(k)} = \vec{u}^* + \vec{u}'$
11: **until** Convergence

2.8 Boundary Conditions

When representing the flow domain of interest using a collocated grid, the domain boundary is made up of the faces of the computational cells neighboring the boundary. Boundary conditions generally entail specification of the variable ϕ, or its flux, or a relation between the two at the boundary faces. In what follows, we present implementations in the FMV of common boundary conditions using the two-dimensional domain shown in Figure 2.8. The flow considered is a steady incompressible flow in a channel with sudden expansion. Since the flow is symmetric with respect to the symmetry plane shown in the figure, only half of the channel is taken to be the computational domain. The boundary of the computational domain is made up of (1) an inflow boundary,

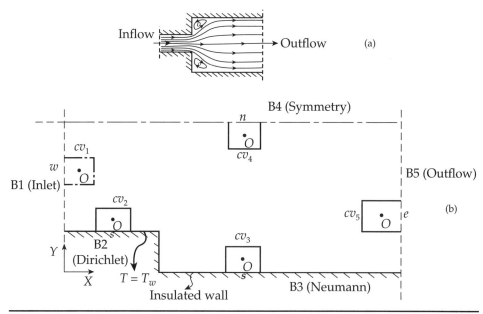

FIGURE 2.8 (a) Two-dimensional steady flow in a channel with sudden expansion. (b) Flow boundary conditions.

(2) a solid impermeable fixed wall with temperature held constant, (3) a solid impermeable thermally insulated fixed wall, (4) a plane of symmetry, and (5) an outflow boundary. These boundary parts are referred to in Figure 2.8(b) respectively as B1, B2, B3, B4, and B5. For a steady, two-dimensional, incompressible flow, the discretized momentum equation is given (Eq. (2.97)):

$$\sum_{\Bbbk \in \mathcal{F}_i} \left(\dot{m}_{\Bbbk} \vec{u}_{\Bbbk} - \mu \mathcal{A}_{\Bbbk} \left(\frac{\partial \vec{u}}{\partial n} \right)_{\Bbbk} \right) + \sum_{\Bbbk \in \mathcal{F}_i} p_{\Bbbk} \hat{n}_{\Bbbk} \mathcal{A}_{\Bbbk} - \rho \vec{g} \mathcal{V}_i = \vec{0} \tag{2.109}$$

Starting with Eq. (2.40), the discretized energy equation, in the absence of a heat source, is expressed as

$$\sum_{\Bbbk \in \mathcal{F}_i} \left(\dot{m}_{\Bbbk} c_v T_{\Bbbk} - \lambda \mathcal{A}_{\Bbbk} \left(\frac{\partial T}{\partial n} \right)_{\Bbbk} \right) = 0 \tag{2.110}$$

where viscous dissipation has been neglected. In Eqs. (2.109) and (2.110), the viscosity μ, density ρ, constant volume specific heat c_v, and thermal conductivity λ are assumed to be constants. In what follows, we discuss incorporation of the boundary conditions (1)–(5) in Eqs. (2.109) and (2.110).

2.8.1 Inflow Boundary Condition

The most common inflow boundary condition entails specification of the velocity vector and temperature at the boundary faces. For example, at face $\Bbbw$ of CV1, the mass flux $\dot{m}_{\Bbbw}$, momentum flux $\dot{m}_{\Bbbw} \vec{u}_{\Bbbw}$, and internal energy flux $\dot{m}_{\Bbbw} c_v T_{\Bbbw}$ are prescribed by the inflow boundary condition. To compute the diffusion flux at face $\Bbbw$ of CV1, $(\partial \phi / \partial n)_{\Bbbw}$ is approximated using one-sided interpolation schemes, the simplest of which is the first-order approximation

$$\left(\frac{\partial \phi}{\partial n} \right)_{\Bbbw} \simeq -\frac{\phi_{\mathbb{O}} - \phi_{\Bbbw}}{\frac{1}{2} \Delta x} \text{ for } \Bbbw \in \text{CV1} \tag{2.111}$$

Accordingly, Eqs. (2.109) and (2.110) may be expressed for CV1 as

$$\sum_{\Bbbk \in \{s,n,e\}} \left(\dot{m}_{\Bbbk} \vec{u}_{\Bbbk} - \mu \mathcal{A}_{\Bbbk} \left(\frac{\partial \vec{u}}{\partial n} \right)_{\Bbbk} \right) + \mu \mathcal{A}_{\Bbbw} \frac{\vec{u}_{\mathbb{O}}}{\frac{1}{2} \Delta x}$$

$$+ \sum_{\Bbbk \in \mathcal{F}_i} p_{\Bbbk} \hat{n}_{\Bbbk} \mathcal{A}_{\Bbbk} = \rho \vec{g} \mathcal{V}_i - \dot{m}_{\Bbbw} \vec{u}_{\Bbbw} + \mu \mathcal{A}_{\Bbbw} \frac{\vec{u}_{\Bbbw}}{\frac{1}{2} \Delta x} \tag{2.112}$$

$$\sum_{\Bbbk \in \{s,n,e\}} \left(\dot{m}_{\Bbbk} c_v T_{\Bbbk} - \lambda \mathcal{A}_{\Bbbk} \left(\frac{\partial T}{\partial n} \right)_{\Bbbk} \right) + \lambda \mathcal{A}_{\Bbbw} \frac{T_{\mathbb{O}}}{\frac{1}{2} \Delta x} = -\dot{m}_{\Bbbw} c_v T_{\Bbbw} + \lambda \mathcal{A}_{\Bbbw} \frac{T_{\Bbbw}}{\frac{1}{2} \Delta x} \tag{2.113}$$

where the right-hand sides of the two equations are known quantities, that are fed into the source term or the right hand side of the system of equations to be solved, expressed in Eq. (2.42). The pressure at the inflow boundary face $\Bbbw$ is not known, but can be related to interior nodes using extrapolation. A second-order extrapolation yields

$$p_{\Bbbw} \simeq p_{\mathbb{O}} + (p_{\mathbb{O}} - p_{\mathbb{E}}) \frac{\Delta x}{2 \delta x_{\mathbb{E}}} \tag{2.114}$$

2.8.2 Wall Boundary Condition

At the control volume faces coinciding with a solid impermeable wall (B2 and B3 in Figure 2.8), the no-slip velocity condition applies, so that for face s of CV2, $\vec{u}_s = \vec{u}_{B2}$, where $\vec{u}_{B2}$ is the velocity of the wall. If the wall is fixed, then $u_s = v_s = 0$. Consequently, all the convective fluxes, involving the mass flow rate, across face s of CV2 are zero. The same conditions apply to face s of CV3. The no-slip velocity condition at a solid impermeable wall also imposes a condition on the diffusion momentum flux (or the viscous stresses) at the boundary face. For the parts of the wall aligned along the x-coordinate in Figure 2.8, the no-slip condition requires each of the velocity components to be invariant along the wall, so that for face s of CV2 and CV3, $\partial u/\partial x = \partial v/\partial x = 0$. As for the diffusion fluxes of u and v across the face, they can be approximated using one-sided interpolation schemes, the simplest of which is the first-order approximation

$$\left(\frac{\partial u}{\partial y} \right)_s \simeq \frac{u_O - u_s}{\frac{1}{2}\Delta y} \text{ for s} \in \text{CV2, CV3} \tag{2.115}$$

$$\left(\frac{\partial v}{\partial y} \right)_s \simeq \frac{v_O - v_s}{\frac{1}{2}\Delta y} \text{ for s} \in \text{CV2, CV3} \tag{2.116}$$

where $u_s = v_s = 0$ for fixed wall. For two-dimensional incompressible flow, since the velocity component is invariant along the wall, then it follows from the continuity equation that the flux across the wall of the normal component is zero, leading to $\partial v/\partial y = 0$ at face s of CV2 and CV3, which is more accurate than Eq. (2.116). Accordingly, the x and y components of Eq. (2.109) may be expressed for CV2 and CV3 as

$$\sum_{k \in \{w,n,e\}} \left(\dot{m}_k \vec{u}_k - \mu \mathcal{A}_k \left(\frac{\partial u}{\partial n} \right)_k \right) + \mu \mathcal{A}_s \frac{u_O}{\frac{1}{2}\Delta y} + \sum_{k \in \mathcal{F}_i} p_k (\hat{n}_k \cdot \hat{x}) \mathcal{A}_k = \rho (\vec{g} \cdot \hat{x}) \mathcal{V}_i \tag{2.117}$$

$$\sum_{k \in \{w,n,e\}} \left(\dot{m}_k \vec{v}_k - \mu \mathcal{A}_k \left(\frac{\partial v}{\partial n} \right)_k \right) + \sum_{k \in \mathcal{F}_i} p_k (\hat{n}_k \cdot \hat{y}) \mathcal{A}_k = \rho (\vec{g} \cdot \hat{y}) \mathcal{V}_i \tag{2.118}$$

As for the boundary conditions for the energy equation, the temperature at B2 is maintained at a constant value equal to T_w, indicating a Dirichlet boundary condition, whereas at the insulated part of the wall (B3), the heat flux across the wall is zero, indicating a Neumann boundary condition. Since the velocity at the wall is zero, it follows that the convective heat fluxes across all the control volume faces coinciding with B2 and B3 are zero. Using the Dirichlet and Neumann boundary conditions for the temperature in approximating the diffusion heat flux is discussed next.

Dirichlet Boundary Condition

The Dirichlet boundary condition at B2 states that $T = T_w$ at the control volume faces coinciding with B2. For example, $T_s = T_w$, for control volume CV2. The component of the temperature gradient normal to face s is approximated using one-sided interpolation schemes, the simplest of which is the first-order approximation

$$\left(\frac{\partial T}{\partial y} \right)_s \simeq \frac{T_O - T_s}{\frac{1}{2}\Delta y} \text{ for s} \in \text{B2} \tag{2.119}$$

where $T_s = T_w$. The discretized energy Eq. (2.110) may then be expressed for CV2 as

$$\sum_{k \in \{w,n,e\}} \left(\dot{m}_k c_v T_k - \lambda \mathcal{A}_k \left(\frac{\partial T}{\partial n} \right)_k \right) + \lambda \mathcal{A}_s \frac{T_O}{\frac{1}{2}\Delta y} = \lambda \mathcal{A}_s \frac{T_w}{\frac{1}{2}\Delta y} \tag{2.120}$$

Neumann Boundary Condition

The Neumann boundary condition (also called isoflux boundary condition) at B3 readily prescribes the component of the temperature gradient normal to face, which is zero if the wall is thermally insulated. So the control volume CV3

$$\left(\frac{\partial T}{\partial y}\right)_s = 0 \text{ for } s \in \text{B3} \qquad (2.121)$$

The temperature at the wall, which is unknown, is extrapolated by approximating $\partial T/\partial y$ at s using one-sided interpolation schemes. A first-order approximation

$$\left(\frac{\partial T}{\partial y}\right)_s \simeq \frac{T_0 - T_s}{\frac{1}{2}\Delta y} \text{ for } s \in \text{B2} \qquad (2.122)$$

so that $T_s \simeq T_0$. The discretized energy Eq. (2.110) may then be expressed for CV3 as

$$\sum_{k\in\{w,n,e\}} \left(\dot{m}_k^\rightarrow c_v T_k - \lambda A_k \left(\frac{\partial T}{\partial n}\right)_k\right) = 0 \qquad (2.123)$$

2.8.3   Symmetry Boundary Condition

When the flow is symmetric, one can take advantage of the symmetry to reduce the computational cost. For example, for the flow considered in Figure 2.8(a), the cost can be reduced by taking half of the channel as the computational domain, which is shown in Figure 2.8(b). The solution in the upper half is then a mirror image of the solution in the lower half, reflected with respect to the symmetry plane, denoted as boundary B4. The derivative of all the physical variables, except the normal velocity component, along the direction normal to the symmetry line is zero. In addition, the flow through the symmetry plane is zero so that the normal velocity component is zero. So for face n of control volume CV4,

$$\left(\frac{\partial \phi}{\partial y}\right)_n = 0 \text{ for } \phi \in \{u, p, T\}, v_n = 0 \text{ and } n \in \text{B4} \qquad (2.124)$$

So the diffusion fluxes of u, T, and p across a symmetry plane are zero. Note, however, that while the velocity component normal to the symmetry plane is zero, its flux is not, that is, $(\partial v/\partial y)_n \neq 0$. The convective fluxes across face n of control volume CV4 are also zero, which is a result of the fact that the mass flow rate across a symmetry plane is zero, that is, $\dot{m}_n^\rightarrow = 0, n \in \text{B4}$. The value of the physical property, $\phi \in \{u, p, T\}$, at face n can be related to interior nodal values using one-sided interpolation schemes of $\partial \phi/\partial y$. A first order approximation yields $\phi_n \simeq \phi_0, \phi \in \{u, p, T\}$, for CV4 $(\partial v/\partial y)_n$ can be expressed in terms of the face and interior node values using one-sided interpolation, the simplest of which is the first order approximation

$$\left(\frac{\partial v}{\partial y}\right)_n \simeq \frac{v_n - v_0}{\frac{1}{2}\Delta y} \text{ for } s \in \text{B4} \qquad (2.125)$$

where $v_n = 0$. The x and y components of the discretized momentum Eq. (2.109), and the discretized energy equation may be expressed for CV4:

$$\sum_{k \in \{w,s,e\}} \left(\dot{m}_k \vec{u}_k - \mu \mathcal{A}_k \left(\frac{\partial u}{\partial n} \right)_k \right) + \sum_{k \in \mathcal{F}_i} p_k (\hat{n}_k \cdot \hat{x}) \mathcal{A}_k = \rho(\vec{g} \cdot \hat{x}) \mathcal{V}_i \qquad (2.126)$$

$$\sum_{k \in \{w,s,e\}} \left(\dot{m}_k \vec{v}_k - \mu \mathcal{A}_k \left(\frac{\partial v}{\partial n} \right)_k \right) + \mu \mathcal{A}_n \frac{v_0}{\frac{1}{2}\Delta y} + \sum_{k \in \mathcal{F}_i} p_k (\hat{n}_k \cdot \hat{y}) \mathcal{A}_k = \rho(\vec{g} \cdot \hat{y}) \mathcal{V}_i \quad (2.127)$$

$$\sum_{k \in \{w,s,e\}} \left(\dot{m}_k c_v T_k - \lambda \mathcal{A}_k \left(\frac{\partial T}{\partial n} \right)_k \right) = 0 \qquad (2.128)$$

2.8.4   Outflow Boundary Condition

An outflow boundary should be selected to be far downstream from the flow region of interest so that the flow crossing the boundary is streamlined with no flow reversal. Under these conditions, implementation of the outflow boundary conditions for large Reynolds number flows assumes that the flow is fully developed along the streamlines crossing the boundary. If the outflow boundary B5 in Figure 2.8(b) is chosen to be sufficiently far from the step in the wall B2–B3, then the streamlines crossing the outflow boundary B5 are nearly parallel to the $x-$axis, so that the outflow boundary condition simplifies to $\partial \phi / \partial x \simeq 0$, $\phi \in \{u, v, p, T\}$. The simplest implementation of this boundary condition yields a first-order upwind approximation, which for face e of CV5 is expressed as $\phi_e = \phi_0$. The extrapolated value of ϕ at the outlet boundary faces is then corrected to satisfy the conservation of mass requirement.

Note that the accuracy of all of the one-sided first-order approximations of the normal derivatives can be improved by choosing the control volume adjacent to the boundary to have a smaller dimension along the direction of the normal. Note, however, that the aspect ratio of the control volume cannot be set to be too small. Alternatively, higher order one-sided approximations can be used by including additional interior nodes neighboring the cell containing the boundary face(s).

2.9   Solution Verification

Once the discretized system is solved, we need to examine the numerical error resulting from discretizing the domain into cells. An important source of error that can be estimated and controlled is the spatial discretization error [19]. Estimating this error allows assessment if the mesh needs further refinement. Richardson extrapolation (RE) is a method widely used to report spatial discretization errors or in other words to perform a grid-independent study. In what follows, we will briefly describe this method and its implementation.

Assume we have a grid whose global cell size is denoted by h. The discretization error is denoted by ϵ_h

$$\epsilon_h = \phi_h - \tilde{\phi} \qquad (2.129)$$

where ϕ_h is the exact solution to the discretized equations, and $\tilde{\phi}$ is the exact solution to the partial differential equations.

The discretization error is also a function of the global grid size, h, and it could be approximated [19] [20] as

$$\epsilon_h \simeq \alpha_p h^p \tag{2.130}$$

where p stands for the order of the numerical scheme employed α_p is a constant. h is assumed to be sufficiently small so that higher order terms of the Taylor series expansion of the function are neglected.

For a three-dimensional domain, the global cell size is determined as

$$h = \left(\frac{\sum_{i=1}^{N} \mathcal{V}_i}{N} \right)^{\frac{1}{3}} \tag{2.131}$$

where N is the number of grid cells and $\mathcal{V}_i$ is the volume of cell i. For a two-dimensional domain, h is computed as

$$h = \left(\frac{\sum_{i=1}^{N} \mathcal{A}_i}{N} \right)^{\frac{1}{2}} \tag{2.132}$$

where $\mathcal{A}_i$ is the area of cell i.

Combining Eqs. (2.129) and (2.130), we obtain

$$\phi_h - \tilde{\phi} \simeq \alpha_p h^p \tag{2.133}$$

When p, the order of the numerical scheme, is known we generate two solutions on two systematically refined grids and using Eq. (2.133) we produce two equations for two unknowns, $\tilde{\phi}$ and α_p:

$$\phi_{h_1} - \tilde{\phi} \simeq \alpha_p h_1^p$$
$$\phi_{h_2} - \tilde{\phi} \simeq \alpha_p h_2^p$$

When p is unknown, we generate three solutions on three different grids in order to produce three equations for three unknowns, $\tilde{\phi}$, α_p, and p

$$\phi_{h_1} - \tilde{\phi} \simeq \alpha_p h_1^p$$
$$\phi_{h_2} - \tilde{\phi} \simeq \alpha_p h_2^p$$
$$\phi_{h_3} - \tilde{\phi} \simeq \alpha_p h_3^p \tag{2.134}$$

Systematic grid refinement involves decreasing the grid spacing by a constant factor in all coordinate directions without compromising the mesh quality.

Consider two solutions, ϕ_{h_1} and ϕ_{h_2} generated using a fine grid of global sizes h_1 and a coarse grid of global sizes h_2. Let $h_2 = r_{21} h_1$, where r_{21} is the refinement factor between grids 2 and 1, and it is greater than 1.

The relative discretization error, expressed in terms of the solutions ϕ_{h_1} and ϕ_{h_2}, may be shown to be

$$\epsilon_r^1 = \frac{\frac{\phi_{h_2} - \phi_{h_1}}{\phi_{h_1}}}{r_{21}^p - 1} \tag{2.135}$$

Different interpolation schemes are used to project the coarse grid solution onto the fine grid without compromising the accuracy of either solution. For example, Roach et al. [21] uses simple two points averaging to interpolate the coarse grid solution onto the fine grid while Shyy et al. [22] uses Lagrangian interpolation for higher order of accuracy.

Another indicator of the accuracy of the solution, that we will be using in the verification, is the approximate relative error between the solutions on the two grids

$$
e_a^{21} = \left| \frac{\phi_{h_1} - \phi_{h_2}}{\phi_{h_1}} \right| \tag{2.136}
$$

In the following discussion, we will describe the Richardson extrapolation method used when the order of the scheme, p, is unknown which is the method used in all solutions verification performed in this textbook. In this case, we generate solutions on three grids: fine, medium, and coarse, of respective global sizes h_1, $h_2 = r_{21}h_1$, and $h_3 = r_{32}h_2$ where the refinement factors r_{21} and r_{32} are chosen to be larger than 1.3 as recommended by Ishmail et al. [23].

As mentioned above, the three unknowns, p, $\tilde{\phi}$, and α_p are determined by the three equations for three unknowns, Eq. (2.134)

The apparent order of the method, reported for the variable(s) of interest at the fine, medium, and coarse grids, denoted respectively by ϕ_{h_1}, ϕ_{h_2}, and ϕ_{h_3}, is computed using the following iterative procedure:

1. Set $p = p_0$, where the initial guess, p_0, is an estimate of the order of the numerical schemes used within

2. Calculate the sign function s

$$
s = \text{sign} \left(\frac{\varepsilon_{32}}{\varepsilon_{21}} \right) \tag{2.137}
$$

   where $\varepsilon_{32} = \phi_{h_3} - \phi_{h_2}$ and $\varepsilon_{21} = \phi_{h_2} - \phi_{h_1}$

3. Calculate $q(p)$

$$
q(p) = \ln \left(\frac{r_{21}^p - s}{r_{32}^p - s} \right) \tag{2.138}
$$

4. Calculate a new value for p

$$
p_{new} = \frac{1}{\ln r_{21}} \times \left| \ln \left| \frac{\varepsilon_{32}}{\varepsilon_{21}} \right| + q(p) \right| \tag{2.139}
$$

5. Calculate the error between p and p_{new}

$$
e = |p_{new} - p| \tag{2.140}
$$

6. If $e > \zeta$, a tolerance defined by the user, then
   - Update p according to $p = p_{new}$
   - Go to Step 3

7. If $e < \zeta$, then p is the value of the apparent order of the method. Note that the closer the initial guess p_0 to the effective order of the numerical method,

the faster the convergence of the iterative scheme. Using the solutions on the fine, medium, and coarse grids, two extrapolated solutions may be obtained according to

$$\phi_{ext}^{21} = \frac{r_{21}^{p} \phi_{h_1} - \phi_{h_2}}{r_{21}^{p} - 1} \qquad (2.141)$$

$$\phi_{ext}^{32} = \frac{r_{32}^{p} \phi_{h_2} - \phi_{h_3}}{r_{32}^{p} - 1} \qquad (2.142)$$

The extrapolated solution ϕ_{ext}^{21} is expected to have a higher order of accuracy than ϕ_{h_1} and ϕ_{h_2}. By implementing Richardson extrapolation, we can generate higher order solutions using lower order schemes and save on computational effort [21].

The relative error between the extrapolated solution and the fine grid solution ϕ_{h_1} is

$$e_{ext}^{21} = \left| \frac{\phi_{ext}^{21} - \phi_{h_1}}{\phi_{ext}^{21}} \right| \qquad (2.143)$$

The calculated apparent order, p, indicates the reliability of the discretization error in Eq. (2.135). When p is significantly greater than the order of the numerical schemes, the discretization error is underestimated, and the error at the fine grid solution is not necessarily bounded by it. When p is significantly smaller than the order of the numerical schemes, the error at the fine grid solution is unfairly overestimated, which could be resolved by setting upper and lower bounds to the apparent order of the scheme. Different studies have different schemes and consequently different upper and lower bounds to the apparent order [24].

By implementing the RE scheme to report the discretization error (ϵ_r^1) and approximate relative error (e_a^{21}), it is not always clear which one is optimistic and which one is conservative. In 1994, Roache et al. [25] addressed this limitation and proposed a method that allows users to calculate an error band in place of an error bound. The error band is a range of solutions that is centered around the fine grid solution. Once we calculate this error band we can assume that there is a high probability that the actual solution of the partial differential equations belongs to it. A Grid Convergence Index (GCI) is introduced

$$GCI = \frac{1.25 \, e_a^{21}}{r_{21}^{p} - 1} \qquad (2.144)$$

The absolute value of the discretization error (ϵ_r^1) is multiplied by a safety factor of 1.25 in calculating the GCI assuming three grids are used for the extrapolation. If two grids are used, then it is recommended to use a safety factor of 3.0 [26]. It is recommended to plot the error bars on the profile of the variable using the fine grid solution and the following procedure [20]:

1. Calculate the local order of accuracy (p) which is the order of accuracy at each point in our profile.

2. Calculate the global order of accuracy (p_{av}) which is the average value of the local order of accuracy.

3. Calculate the local value of the grid convergence index using Eq. 2.144. The order of accuracy used should be the average and not the local order of accuracy. When reporting error bars for profiles, the local values of the order of accuracy are only used to calculate the local values of the extrapolated solution and the global order of accuracy:

$$GCI = \frac{1.25\, e_{21}}{r_{21}^{p_{av}} - 1}$$

4. The error bar, or discretization uncertainty, on the fine grid solution is then calculated by implementing the following equation:

$$EB = \pm GCI \times \phi_{h_1}$$

It is a good practice to report the percentage occurrence of oscillatory convergence (OC) when reporting numerical uncertainty of variables by determining the number of nodes whose sign function is less than zero and calculating the percentage of this number from the total number of nodes. Oscillatory convergence is considered to be one of the limitations of the Richardson extrapolations scheme [24]. Celik et al. [27] attributed the oscillatory behavior to the employment of mixed numerical schemes and upwind schemes where the coefficients of the higher order terms do not necessarily tend to zero with a decreasing grid size h.

CHAPTER **3**

Two-Dimensional Steady State Laminar Incompressible Fluid Flow

List of Symbols	
Boundary layer thickness	$\delta(x)$
Characteristic length	L
Reynold number	Re
Density	ρ
Dynamic viscosity	μ
Entry length	L_e
Diameter of a pipe	D
Length of a pipe	$\mathcal{L}$
Inlet velocity	V_{in}
Angular coordinate	θ
Velocity vector	$\vec{u}$
Pressure	p
Mass	m
Force	F
Shear stress at the wall	τ_w
Specific total energy	e
Rate of work done by the fluid to overcome friction	$\dot{W_f}^{\rightarrow}$
Rate of heat transfer into the control surface	$\dot{Q}^{\leftarrow}$
Vector normal to the surface	$\hat{n}$
Friction head loss	h_f
Friction coefficient	f

In this chapter we will study the laminar flow of a constant density fluid inside a cylindrical pipe. The inlet flow conditions are such that the flow is axisymmetric with respect to the axis of the cylinder which reduces the problem to two spatial dimensions.

An introduction to ANSYS Fluent will be provided along with instructions for creating a model of the two-dimensional axisymmetric laminar pipe flow. Results of the simulations will be discussed and followed by a verification of the Fluent model.

3.1 Introduction

The flow of fluid in pipes is a widely studied fluid dynamics problem. Its application is commonly used in the field of engineering. Engineers are commonly interested in studying the development of the flow inside the pipe and the pressure drop along the pipe length. Calculating the energy losses in pipes is essential to the sizing of these components and to the selection of the appropriate pump. Estimation of the head loss (pressure loss) due to friction enables us to properly select the pump that establishes the desired flow rate. Although most pipe flows are turbulent, laminar pipe flows are of interest because many flows occur naturally or in engineering applications are laminar and can be solved analytically. For instance, blood flow in the cardiovascular and respiratory systems of human beings and animals is an example of laminar flow. Modeling of pipe flow and accurate estimation of the energy loss due to friction is critical to the design and sizing of endotracheal tubes, especially for infants and preterms, where the flow is laminar. As we will find, the flow rate scales as the diameter to the fourth power. This implies that for a given head provided by the pump, the air flow rate provided by the ventilation device is highly sensitive to the tube diameter, which makes proper selection of the endrotacheal tube a critical element of the treatment.

Figure 3.1 shows a typical fluid flow inside a pipe with a developing region and a fully developed region. The flow at the pipe inlet is uniform, which approximates inviscid flow exiting a pressurized reservoir and smoothly entering the pipe. The flow is stopped at the wall due to the no-slip condition at a solid impermeable boundary in the continuum regime. As the flow moves further along the positive x direction, the diffusion of momentum in the radial direction will cause this retardation of the flow by friction, which originated at the wall, to further penetrate radially inward toward the axis. The region near the wall where viscous effects are present is called the boundary layer and the outer region where this effect is negligible is called the inviscid core. The boundary layer is characterized by a thickness, $\delta(x)$, which is the radial distance from

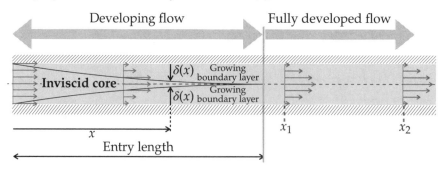

FIGURE 3.1 Entry region and fully developed flow inside the pipe.

the wall at which the axial velocity reached 99% of that in the inviscid core. In a laminar flow, the boundary layer grows as $\delta(x) \sim \sqrt{x}$ and it will eventually reach the pipe axis at some distance L_e, called the entry length. Beyond the entry length, i.e., for $x \geq L_e$, $\delta(x) = R$, where R is the radius of the pipe, and the flow is said to be fully developed, indicating that viscous effects have had their full toll and that the flow field at any cross section $x > L_e$ is similar to that at any other cross section $x' > L_e$, so that $\partial u / \partial x = 0$, i.e., the velocity is no longer a function of the distance along the pipe, x. Since friction is strongest at the wall, where the flow is stopped, the axial velocity is expected to increase as we move away from the wall radially inwards toward the axis, where the maximum speed is attained at the axis, $u_{max} = u(r = 0)$ where r is the radial coordinate. Note that to conserve mass, we must have $u_{max} > V_{in}$.

In the fully developed region, an exact solution for the laminar flow exists and is known as the Poiseuille solution [28]. The flow, driven by a constant negative pressure gradient along the $x-$direction, is characterized by the well-known parabolic velocity profile $u(r) = u_{max} \left(1 - (r/R)^2\right)$. We will use this exact solution to validate and quantify the accuracy of the numerical solution obtained in the fully developed region in Section 3.4. In the developing region, $x < L_e$, an exact solution does not exist, and we will rely on the numerical solution to describe the flow and measure the shear at the wall. We will also compare the numerically determined entry length to those measured experimentally and reported in the literature as discussed in Section 3.3.2.

The Weisback equation, proposed in 1845, enables estimation of the pressure loss due to friction as a function of the average velocity of the fluid flow and the pipe geometry. Expressions for the friction factor, which enables the estimation of the pressure loss due to friction through experimental studies by Poiseuille (1841), Hagen (1839), Darcy (1857), and later Rouse and Ince (1957), provided complete and accurate description of the head loss and the friction coefficient in laminar flow. In 1883, Reynolds presented his description of the transition from laminar to turbulent flow. Reynolds introduced Reynolds number, $Re = \rho VL / \mu$, where the characteristic length of the pipe L is its diameter D, as the key parameter that characterizes the laminar flow regime ($Re < 2000$), the turbulent flow regime ($Re > 4000$), and the transition flow regime ($2000 < Re < 4000$). Reynolds number is a measure of the ratio of the inertial forces to the viscous forces within the fluid. Correlations for the head loss and friction factor in turbulent pipe flows were experimentally investigated by Darcy (1857) and Fanning (1877). The correlation of the friction factor for turbulent flow in smooth pipes was proposed by von Karman and Prandtl (Rouse 1943, Schlichting, 1968). For turbulent flow in rough pipes, the effect of the surface roughness in the pipe wall has been incorporated, based on experimental data, into correlations for the friction factor by von Karman (1930). In addition to Moody's chart (Moody, 1944), the most widely used correlation for the friction factor in turbulent flows, relating the friction factor to Reynolds number and the surface roughness normalized by the pipe diameter, is given by Colebrook and White (1937) [29].

3.2   Problem Statement

Consider the laminar flow inside the circular pipe shown in Figure 3.2. The pipe has a length $\mathcal{L} = 10$ m and a radius $R = 0.1$ m. Liquid water with a density $\rho = 998.2$ kg/m^3 and a viscosity $\mu = 1.003$ mPa.s enters the pipe with a uniform velocity $\vec{u} = V_{in} \hat{x}$ at $x = 0$, $0 \leq r < R$; where $V_{in} = 0.0025$ m/s. Table 3.1 contains the details of the

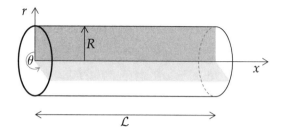

FIGURE 3.2 Axisymmetric pipe flow.

TABLE 3.1 Model Dimensions, Properties, and Flow Parameters

Parameter	Symbol	Value
Pipe dimensions		
Pipe length	$\mathcal{L}$	10 m
Pipe diameter	$D = 2R$	0.2 m
Wall roughness	ϵ	0
Fluid properties		
Density	ρ	998.2 kg/m^3
Viscosity	μ	1.003 mPa.s
Flow properties		
Average velocity	V	0.0025 m/s
Outlet (gauge) pressure	p_{out}	0 Pa
Reynolds number	Re	498

problem. The flow is considered laminar because the calculated Reynolds number is $\text{Re} = \frac{\rho V_{in} D}{\mu} = 498$ and is below the critical value of 2000.

For the setup just described, and in the absence of gravity, an axisymmetric flow field is established, for which the pressure and velocity do not depend on the angle θ. Every plane that goes through the axis of the pipe will exhibit the same streamline pattern. In addition, every plane that goes through the axis of the pipe will have a symmetry about the axis. As such, we only need to study the rectangular highlighted section of the domain shown in Figure 3.2. This planar rectangular section, $0 < r < R$ and $0 < x < \mathcal{L}$, constitutes the study domain that will be discretized into a grid. In the finite volume method, the governing equations, presented next, are discretized on the grid to yield a system of algebraic equations governing the unknown quantities at the centroids of the grid cells. Discretization of the governing equations is accomplished by applying the integral form of the conservation laws on each grid cell, taken to be a control volume.

3.3   Governing Equations and Boundary Conditions

The differential form of the conservation of mass and momentum and the boundary conditions for the steady axisymmetric laminar pipe flow are presented here. Since the flow is steady, the derivative with respect to time is zero, that is, $\partial()/\partial t = 0$. In addition, the flow is axisymmetric such that the solution cannot depend on the angle θ and therefore $\partial()/\partial\theta = 0$, so that the velocity and pressure fields, $\vec{u}(x,r)$ and $p(x,r)$, governed by the continuity and momentum equations, expressed in cylindrical coordinates in Section 1.5.2, simplify to:

$$\frac{\partial u}{\partial x} + \frac{1}{r}\frac{\partial(rv)}{\partial r} = 0 \tag{3.1}$$

$$\rho\left(u\frac{\partial u}{\partial x} + v\frac{\partial u}{\partial r}\right) = -\frac{\partial p}{\partial x} + \mu\left(\frac{\partial^2 u}{\partial x^2} + \frac{1}{r}\frac{\partial}{\partial r}\left(\frac{\partial(ru)}{\partial r}\right)\right) \tag{3.2}$$

$$\rho\left(u\frac{\partial v}{\partial x} + v\frac{\partial v}{\partial r}\right) = -\frac{\partial p}{\partial r} + \mu\left(\frac{\partial^2 v}{\partial x^2} + \frac{1}{r}\frac{\partial}{\partial r}\left(\frac{\partial(rv)}{\partial r}\right) - \frac{v}{r^2}\right) \tag{3.3}$$

where the density is taken to be a constant $\rho = 998.2$ kg/m^3 and gravity effects have been neglected. Here the velocity is expressed as $\vec{u} = u(x,r)\hat{x} + v(x,r)\hat{r}$, and Eqs. (3.1), (3.2), and (3.3) are the continuity and the momentum equations along the x and r directions. These three equations form a system of coupled nonlinear partial differential equations, where the unknowns are u, v, and p.

The boundary conditions are the no-slip velocity condition at the solid impermeable wall, $\vec{u} = \vec{0}$ at $r = R$, the uniform velocity condition at the inlet, $\vec{u} = V_{in}\hat{x}$ at $x = 0$, and the outflow pressure condition at the outlet, $p = p_{out}$ at $x = \mathcal{L}$.

3.3.1 Exact Solution in the Fully Developed Region

Recall that in the fully developed region, the axial velocity component is independent of x, that is, $\partial u/\partial x = 0$. Substituting in the continuity Eq. (3.1), we arrive at $\partial(rv)/\partial r = 0$. Integrating with respect to r, $\int_r^R \frac{\partial(rv)}{\partial r} dr = 0$ and noting that $v = 0$ at $r = R$ due to the no-slip condition at the solid wall, we conclude that the radial velocity component, v, is zero everywhere. The velocity field is then expressed as $\vec{u} = u(r)\hat{x} + 0\hat{\theta} + 0\hat{r}$. Eqs. (3.2) and (3.3) will reduce to:

$$0 = -\frac{\partial p}{\partial x} + \mu\frac{1}{r}\frac{\partial}{\partial r}\left(\frac{\partial(ru)}{\partial r}\right) \tag{3.4}$$

$$0 = -\frac{\partial p}{\partial r} \tag{3.5}$$

From Eq. (3.5), we deduce that the pressure is a function of x only, that is, $p = p(x)$. Substituting, $\partial p/\partial x = dp/dx$ into Eq. (3.4) and integrating twice in r yields

$$u = \frac{1}{\mu}\left(\frac{dP}{dx}\right)\frac{r^2}{4} + C_1\ln r + C_2 \tag{3.6}$$

The constants C_1 and C_2 can be obtained by satisfying the boundary conditions. The constant $C_1 = 0$ is calculated using the symmetry boundary condition at the axis, that is, $\frac{\partial u}{\partial r}\big|_{r=0} = 0$. The constant $C_2 = -\frac{1}{\mu}\left(\frac{dP}{dx}\right)\frac{R^2}{4}$ is calculated using the no-slip boundary condition at the wall, that is, $u(r = R) = 0$.

The velocity field in the fully developed region is then $\vec{u} = u(r)\hat{x}$, where the axial velocity component assumes the radial parabolic profile.

$$u(r) = \frac{R^2}{4\mu}\left(-\frac{dp}{dx}\right)\left(1 - \frac{r^2}{R^2}\right) \tag{3.7}$$

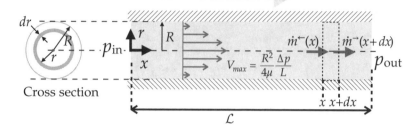

FIGURE 3.3 Computation of the mass flow rate inside the pipe.

The mass flow rate is computed by integrating the mass flux $\rho u(r)$ over the cross-section perpendicular to the axis of the pipe, that is, $\dot{m} = \int_{S_c} \rho u(r)\, dS$, where $dS = 2\pi r dr$, as depicted in Figure 3.3, so that

$$\dot{m} = \int_0^R \rho u(r)(2\pi r dr) = \frac{\rho \pi R^4}{8\mu}\left(-\frac{dp}{dx}\right) \tag{3.8}$$

Since the flow is steady, the mass flow rate is constant. This can be shown by taking a control volume between two cross sections at x and $x + dx$, as shown in Figure 3.3, and applying the integral form of the conservation of mass

$$\frac{\partial m_{CV}}{\partial t} = \dot{m}^{\leftarrow}(x) - \dot{m}^{\rightarrow}(x + dx) \tag{3.9}$$

For steady flow, the mass contained inside the control volume does not change, $\partial m_{CV}/\partial t = 0$. It follows that $d\dot{m}/dx = 0$, which upon combining with Eq. (3.8), yields $d^2p/dx^2 = 0$, implying that the pressure drops linearly from the inlet to the outlet:

$$p(x) = p_{in} - \frac{x}{\mathcal{L}}\Delta p \tag{3.10}$$

where $\Delta p = p_{in} - p_{out}$ is the pressure drop along the pipe. The axial velocity component and the mass flow rate respectively are

$$u(r) = \frac{R^2 \Delta p}{4\mu\mathcal{L}}\left(1 - \frac{r^2}{R^2}\right) \quad \text{and} \quad \dot{m} = \frac{\rho \pi R^4 \Delta p}{8\mu\mathcal{L}}. \tag{3.11}$$

The mean (or average) velocity at a given cross section of area $S_c = \pi R^2$ is

$$V = \frac{\dot{m}}{\rho S_c} = \frac{R^2 \Delta p}{8\mu\mathcal{L}} \tag{3.12}$$

so that

$$u(r) = 2V\left(1 - \frac{r^2}{R^2}\right) \tag{3.13}$$

Note the maximum velocity occurs at $r = 0$ at the centerline and is equal to twice the mean velocity.

An expression for the shear stress at the wall can be obtained by applying the conservation of linear momentum for a control volume that consists of the pipe interior. The x−component is

$$\sum F_x = \frac{\partial}{\partial t} \int_{\mathcal{V}} \rho u \, d\mathcal{V} + \int_{\mathcal{S}} \rho u \vec{u}_r \cdot \hat{n} \, d\mathcal{S} \tag{3.14}$$

where the left-hand side denotes the sum of all the forces acting on the control volume in the x−direction. These include the pressure force $F_p = p_{in} \mathcal{S}_c - p_{out} \mathcal{S}_c = \Delta p \, \mathcal{S}_c$, which drives the fluid along the direction of $-dp/dx$, and the opposing friction force, F_f imparted by the wall on the fluid at $r = R$. Note that in the fully developed region, one expects the shear stress at the wall, τ_w, to be uniform, so that $F_f = \tau_w 2\pi R \mathcal{L}$. Both terms on the right-hand side of Eq. (3.14) are zero. The first term vanishes because the momentum inside the fixed control volume does not change for a steady flow. The second term vanishes because the velocity profile at the inlet of the fully developed region is identical to the velocity profile at the outlet so that the rate of momentum crossing the control surface at the inlet is equal and opposite to that crossing the control surface at the outlet. As depicted in Figure 3.4, the friction force balances the pressure force.

$$\tau_w \, 2\pi R \mathcal{L} = -\Delta p \, \pi R^2 \Rightarrow \tau_w = -\frac{R \Delta p}{2\mathcal{L}} = -\frac{4\mu V}{R} \tag{3.15}$$

Alternatively, the shear stress at r can be obtained from Eqs. (1.41) and (1.50) as:

$$\tau(r) = \mu \frac{du}{dr} = -\frac{4\mu V}{R^2} r \tag{3.16}$$

indicating that the absolute value of the shear increases linearly from 0 at the axis to a maximum of $4\mu V/R$ at the wall, as depicted in Figure 3.5.

Fully developed flow

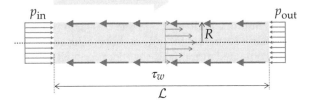

FIGURE 3.4 The balance of the friction and pressure forces.

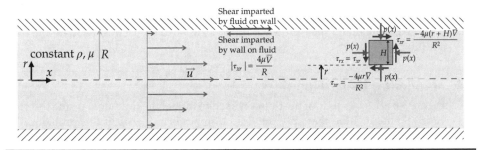

FIGURE 3.5 Shear stress.

The rate of work done by the fluid to overcome friction can be calculated by applying the first law of thermodynamics (the energy equation) for the control volume consisting of the pipe interior along the length, $\mathcal{L}$, of the fully developed region:

$$\frac{\partial}{\partial t} \int_{\mathcal{V}} \rho e \, d\mathcal{V} + \int_{\mathcal{S}} \rho \left(\mathfrak{h} + \frac{1}{2}|\vec{u}|^2 + gz \right) \vec{u}_r \cdot \hat{n} \, d\mathcal{S} = -\dot{W_f} + \dot{Q}^{\leftarrow} \tag{3.17}$$

where $e = \mathfrak{u} + \frac{1}{2}|\vec{u}|^2 + gz$ is the specific total energy, $\mathfrak{u}$ is the specific internal energy, $\mathfrak{h} = \mathfrak{u} + p/\rho$ is the specific enthalpy, $\dot{W_f}$ is the rate of work done by the fluid to overcome friction, and $\dot{Q}^{\leftarrow}$ is the rate of heat transfer into the control volume across the control surface. Since we are considering a steady flow, the first term on the left-hand side of Eq. (3.17) is zero. In addition, the rate of kinetic and potential energy transport by the inlet flow is equal and opposite to that leaving through the outlet flow, so that the second term on the left-hand side of Eq. (3.17) reduces to $\int_{\mathcal{S}} (\rho \mathfrak{u} + p) \, \vec{u}_r \cdot \hat{n} \, d\mathcal{S}$. Assuming that the pipe is insulated, then $\dot{Q}^{\leftarrow} = 0$, which implies that the only mechanism by which the internal energy may change is by viscous heating of the fluid due to its interaction with the wall, i.e., the dissipation of $\dot{W_f}$. Under typical operating conditions, the change in internal energy is negligible compared to other terms, so that

$$\dot{W_f} = \dot{m} \frac{\Delta p}{\rho} \tag{3.18}$$

Dividing by $\dot{m}g$, Eq. (3.18) may be expressed as a friction head loss, having the unit of length, as

$$h_f = \frac{\Delta p}{\rho g} = \frac{8 \mu \mathcal{L} V}{\rho g R^2} = f \frac{\mathcal{L}}{D} \frac{V^2}{2g} \tag{3.19}$$

where Eq. (3.12) has been used and f is the friction coefficient. As can be seen, for fully developed laminar flow, $f = \frac{64}{\text{Re}}$, where $D = 2R$ is the pipe diameter and $\text{Re} = \frac{\rho V D}{\mu}$ is the Reynolds number.

3.3.2   The Entry Region

In the developing or entry region, $x < L_e$, an exact solution to the Navier-Stokes equations does not exist. Information about the flow field along with quantification of key parameters can be obtained numerically or experimentally. For example, the entry length in a steady laminar pipe flow has been measured experimentally and well reported in the literature to obey the correlation

$$L_e \sim 0.06 \text{Re} D \tag{3.20}$$

where D is the pipe diameter [30] [31]. One can rely on the numerical solution to describe the flow and measure the key parameters such as the entry length and the shear at the wall. Experimentally measured parameters can also be used to validate the numerical solution.

3.4   Modeling Using Fluent

Figure 3.6 is a schematic of the domain used to model the fluid flow inside the pipe using ANSYS Fluent. Note that only the fluid domain is modeled in this problem. The solid wall is being omitted but represented by using the no-slip boundary condition at the interface between the fluid and the pipe wall.

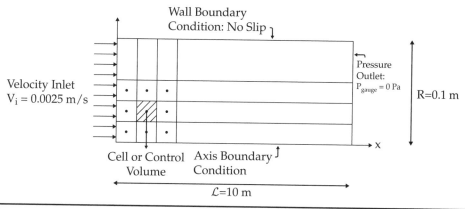

Figure 3.6 Fluent model description.

We will provide in this chapter step-by-step instructions on how to solve the laminar pipe flow problem using Fluent **ANSYS 2022 R1**. We will create the computational domain and divide it into cells (mesh the domain). We will then set up the model by assigning material properties, applying boundary conditions and specifying initial conditions before running the calculation. Using future versions of Fluent may change slightly the instructions provided.

Fluent is provided as free software for students as part of the ANSYS Academic software. ANSYS can be downloaded by visiting www.ANSYS.com/academic. The instructions to download ANSYS are included in Appendix (C).

3.4.1   Introduction to ANSYS Fluent

1. Launch ANSYS Workbench by clicking **Start → ANSYS 2022 R1 → Workbench 2022 R1**.

2. ANSYS Workbench is a simulation platform that includes different **Analysis Systems** under its **Toolbox**. To use Fluent, double-click **Fluid Flow (Fluent)** located under **Analysis Systems** in **Toolbox** as shown in Figure 3.7. This will create a new **Fluid Flow (Fluent)** analysis into the **Project Schematic**. The **Project Schematic** is the display of the different processes that can be executed in Workbench to build and run the model. Note that clicking **View**, shown in Figure 3.7, and selecting **Reset Workspace** will restore the workspace layout to its default settings. Also note that selecting **Reset Window Layout**, after clicking **View**, will restore the original window layout to when we started the Workbench application.

3. In the **Project Schematic**, rename the **Fluid Flow (Fluent)** analysis system by right-clicking the top box displaying **Fluid Flow (Fluent)** and selecting rename. We can alternatively double-click **Fluid Flow (Fluent)** at the bottom until it is highlighted and type `Chapter3` as the new name. Press **Enter**. See Figure 3.7.

4. The process for the analysis includes creating the geometry for the domain in **Geometry** application followed by meshing the domain in **Mesh** application. **Setup** application will launch Fluent where we set up the problem and run the calculations. The results of the calculations can be viewed in the **Solution** application and also in **Results** application. In this chapter, we will look at the results

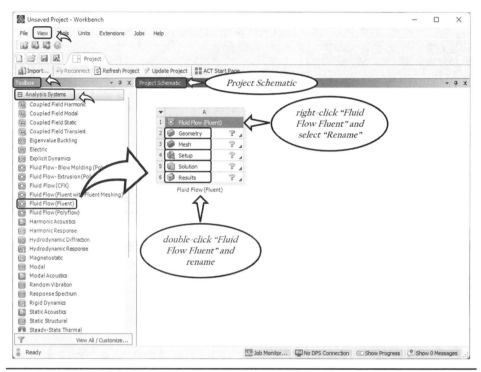

FIGURE 3.7   Toolbox, Project Schematic, and Fluid Flow (Fluent) Analysis system.

in **Solution** application. All these applications are shown in Figure 3.7, and we will go through them in the following subsections.

5. Before proceeding, we will save the Workbench project. Click **File** and select **Save As...**. In the dialog box, browse to the working folder and name the project Laminar Pipe Flow. Click **Save** to save the project. Note that ANSYS Workbench will save generated supporting files for the project in the working folder and that all supporting files are needed in order to reopen or share the project in the future.

3.4.2   Geometry

The geometry of the model can be created using ANSYS DesignModeler, or ANSYS SpaceClaim applications. We also have the option to import the geometry from an existing CAD drawing as demonstrated in Chapter 7.

In this section, the geometry of the model will be created using ANSYS DesignModeler.

Instructions to create the geometry using SpaceClaim are included in Appendix D.

Figure 3.8 is a flowchart for the process of creating the geometry using Design Modeler.

Right-click **Geometry** and select **New DesignModeler Geometry...** from the dropdown list. When ANSYS DesignModeler is launched, the screen will look like Figure 3.9.

The menu toolbar shown at the top of Figure 3.9 gives the user access to the different features and tools available in the ANSYS DesignModeler application. The **File**

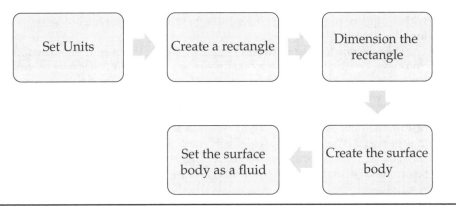

FIGURE 3.8   Flowchart of DesignModeler.

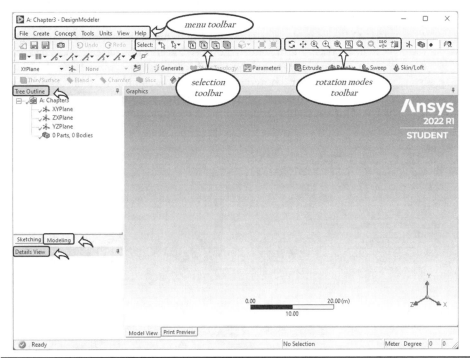

FIGURE 3.9   Modeling and sketching modes, Tree Outline, Details View, selection and rotation modes toolbars.

menu contains basic file operations. The **Create** menu contains the three-dimensional geometry creation tools. The **Concept** menu contains tools to create surfaces. The **Tools** menu has tools for modeling operations. The **View** menu is used to change the display settings. Finally, the **Help** menu provides access to online help documentation.

By default, the ANSYS DesignModeler application starts in **Modeling** mode, shown in Figure 3.9. The **Tree Outline** is displayed at the left of the screen above the **Modeling** tab. When we click the **Sketching** tab, next to the **Modeling** tab, the **Tree Outline** will be replaced with the **Sketching Toolboxes**. Below the **Tree Outline** is the **Details View** area which appears when double-clicking an object in the **Tree Outline** to make it available for editing.

The selection toolbar shown in Figure 3.9 allows the user to perform different selection tasks. The rotation modes toolbar, also shown in Figure 3.9, is very useful as it allows the user to adjust the model's view in the **Graphics** window. Take a moment to become familiar with the selection toolbar and the rotation modes toolbar.

Clicking **View** and selecting **Windows** ⇒ **Reset Layout**, shown in Figure 3.10, will restore the same window layout as when DesignModeler application is launched and may be useful if the layout is accidentally modified in the future. Note that the location of some of the icons in the figures may differ slightly from the location of icons on the user's screen as adjustment to the size of the figures was made to capture the best possible resolution.

1. **Set the Units**
   Click **Units**, shown in Figure 3.10 and select **Meter** if it is not already selected.

2. **Create a Rectangle**
   Click **XYPlane** under the **Tree Outline** to build the geometry in the XY plane.
   Click the icon representing **Look At Face/Plane/Sketch**, shown in Figure 3.10, to display the XY plane on the **Graphics** window.
   Click **Sketching**, shown in Figure 3.11, to create a sketch. A sketch is a collection of two-dimensional edges. The **Sketching Toolboxes** will appear in place of **Tree Outline** at the top left of the screen, as shown in Figure 3.11. The options

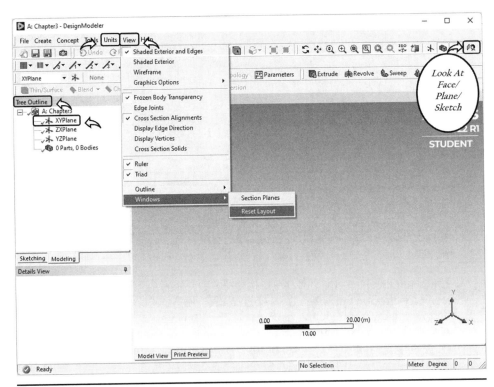

FIGURE 3.10 Reset Layout, Units, and Look at Plane.

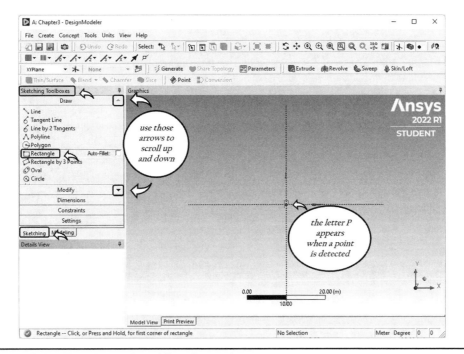

FIGURE 3.11 Sketching mode and Draw Rectangle.

available to the user in **Sketching Toolboxes** are listed and include **Draw, Modify, Dimensions, Constraints**, and **Settings**.

Click **Draw** to expand it, if it is not already expanded, and from the available list of shapes click the **Rectangle** that has a picture of a rectangle by two points. The full list of shapes under **Draw** may not be visible without scrolling up and down the list, as illustrated in Figure 3.11. The pointer of the mouse will change to reflect a pen ready to draw the rectangle.

We will select the origin as the first point of the rectangle. The pointer of the mouse will display the letter **P** when we hover over the origin indicating a point detected at the location. Notice that if we hover over the X axis or the Y axis, the pointer of the mouse will display the letter **C** indicating the detection of a curve. Make sure to see the letter **P** appearing on the screen, as shown in Figure 3.11, before selecting the origin as the first point of the rectangle. Drag the cursor to place the opposite corner of the rectangle in the first quadrant. Click anywhere in the first quadrant to select the opposite corner. The location of the opposite corner does not matter as we will dimension the rectangle in the following step.

3. **Dimension the Rectangle**
   Click **Dimensions** → Click **General** from **Sketching Toolboxes**.

   Select the upper edge of the rectangle, then drag the cursor in the positive Y axis. Click anywhere to place the dimension above the upper edge.

   Repeat the process for the right edge of the rectangle but drag the cursor in the positive X axis to place the dimension to the right side of the edge.

   Adjust the dimensions of the rectangle in **Details View**, located below the **Sketching Toolboxes**, by selecting the cell next to **H1** and typing 10, as shown

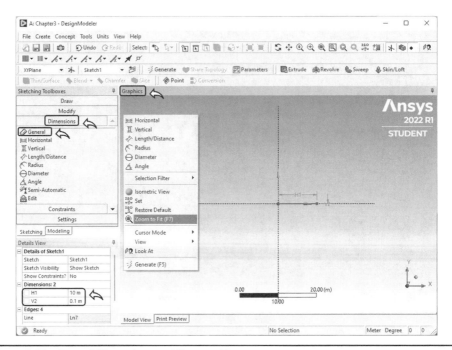

FIGURE 3.12 General Dimensions and Zoom to Fit.

in Figure 3.12. Press **Enter**. Similarly, select the cell next to **V2** and type 0.1. Press **Enter**.

Right-click anywhere inside the **Graphics** window and select **Zoom to Fit**, as shown in Figure 3.12.

Click **Dimensions** → Click **Move** to move the dimensions closer to the rectangle. Click the dimensions one at a time. Drag the cursor and click to bring the dimension closer to the rectangle. Note that the full list of options under **Dimensions** may not be displayed on the screen. Scrolling down to click **Move** may be needed.

Right-click and select **Zoom to Fit** one more time to view the rectangle created. The geometry of the pipe reflects a very long pipe chosen to capture the fully developed flow regime inside the pipe.

4. **Create the Surface Inside the Rectangle**

The area of the rectangle is the surface used as our calculation domain. To create this surface, click **Concept** shown in Figure 3.13 and select **Surfaces From Sketches**.

We will select the rectangle as the **Base Objects** in **Details View**. Select the top edge of the rectangle highlighted in Figure 3.13 and then click **Apply** next to **Base Objects** in **Details View**. All four sides of the rectangle will be highlighted. The **Base Objects** will change to **1 Sketch**. Click **Generate** shown in Figure 3.13 to create the surface. The rectangle is filled. The created surface is now added in the **Tree Outline** and is the domain for the calculation.

Expand **1 Part, 1 Body** in the **Tree Outline** and click the created surface named **Surface Body** as shown in Figure 3.14.

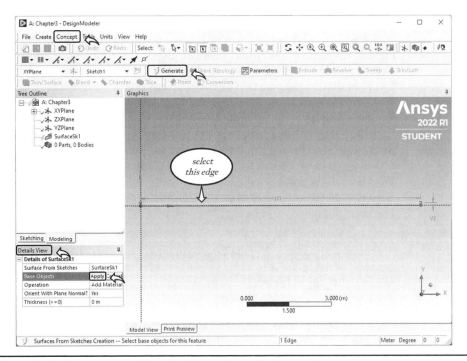

FIGURE 3.13   Concept, Base Object for Surface From Sketch, and Generate.

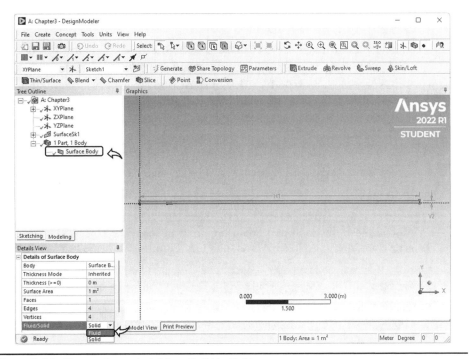

FIGURE 3.14   Change the material from Solid to Fluid.

In **Details View**, the surface created is **Solid** by default. Scroll down and click **Fluid/Solid** to activate the option for the cell. An arrow next to **Solid** will appear. Click the arrow to access the drop-down list. Change the material from **Solid** to **Fluid** as illustrated in Figure 3.14.

Note that clicking a cell to activate choices is very common in ANSYS Fluent as options using a drop-down list will be available once we click the cell.

5. **Save the Project**

Click **File** and select **Save Project**. Close the ANSYS DesignModeler application by clicking the **Close** button **X** located in the upper right corner or by clicking **File** and selecting **Close DesignModeler**. The **Geometry** in **Project Schematic** Workbench has now a green check symbol next to it indicating the application is up to date. We are ready to start meshing the model.

Geometry Summary

* Click **Units**. Select **Meter**.

* Click **XYPlane**. Click **Look At Face/Plane/Sketch**.

* Click **Sketching**. Click **Draw** → Click **Rectangle**. Draw a rectangle by two points. Select the origin as the first point. Drag the cursor and place the opposite corner in the first quadrant.

* Click **Dimensions** → Click **General**. Select the upper horizontal edge of the rectangle. Drag the cursor and click anywhere to place the dimension. Repeat for the right vertical edge. Adjust the value of **H1** to 10 and the value of **V2** to 0.1. Press **Enter**.

* Click **Concept** and select **Surfaces From Sketches**. Select the top horizontal edge of the rectangle. Click **Apply** next to **Geometry**. Click **Generate**.

* Expand **1 Part, 1 Body**. Click **Surface Body**, click **Fluid/Solid** and select **Fluid** from the drop-down list next to it and select **Fluid** next to **Fluid/Solid**.

* Click **File** and select **Save Project**.

* Click **File** and select **Close DesignModeler**.

3.4.3 Mesh

A flowchart of the steps needed to create the mesh is displayed in Figure 3.15.

We will mesh the computational domain in this section. Double-click **Mesh** under **Project Schematic** in ANSYS Workbench (see Figure 3.7). It may take a few seconds

FIGURE 3.15 Flowchart of mesh.

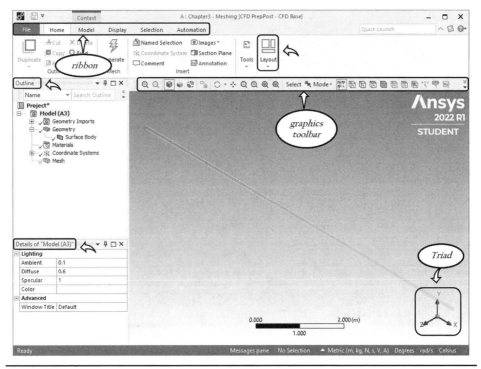

FIGURE 3.16 Outline and Details of "Model", Layout, and Triad.

to load but once Mesh application is open, the geometry will be loaded into the Mesh window as shown in Figure 3.16.

Outline on the upper left panel is the outline view of the project. The **Details View** below the **Outline** displays the details of what is established in **Outline**. The graphics toolbar allows the user to modify a selection or adjust the view in the geometry window.

The **Layout** is also shown in Figure 3.16. Clicking **Home** and selecting **Layout** ⇒ **Reset Layout** will restore the same window layout as when the Mesh application was launched.

Recall that the location of the icons on Figure 3.16 may differ slightly from the location of the icons on the user's screen as adjustment to the size of the figure was made to capture the best possible resolution.

Click the **Z** axis in the axis triad located at the bottom right corner of Figure 3.16 to orient the view to the XY plane along the Z axis.

1. **Default Mesh**

   In **Outline** view, right-click **Mesh** shown in Figure 3.17 and select **Generate Mesh**. We can also generate Mesh by clicking **Generate** in the **Home** tab, also shown in Figure 3.17. Once the mesh is generated, click **Mesh** under **Outline** to view the mesh. A default coarse mesh for the domain is created.

2. **Insert Face Meshing**

   Face Meshing will allow the user to create a structured and more uniform mesh with fewer distorted elements. Face meshing is useful when used on a rectangular face as it will create a very ordered and high quality mesh.

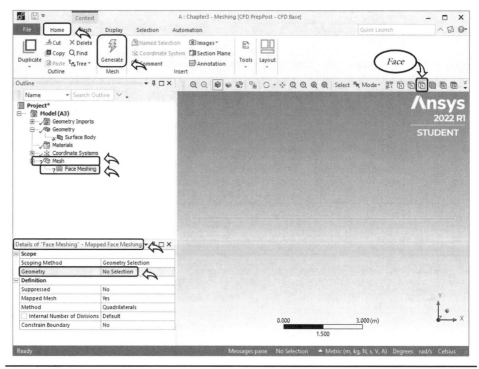

FIGURE 3.17   Generate Mesh and Face Meshing.

Right-click **Mesh** in **Outline** view and select **Insert** ⇒ **Face Meshing**. Click **No Selection** next to **Geometry** in **Details of "Face Meshing" - Mapped Face Meshing**, as shown in Figure 3.17. Activate **Face** selection filter shown in Figure 3.17 if it is not activated and select the rectangular face in the geometry window. In **Details of "Face Meshing" - Mapped Face Meshing**, click **Apply** next to **Geometry**. **Apply** will change into **1 Face**. A green check mark will appear next to **Face Meshing** in **Outline** view.

3. **Refine the Mesh**

   Click **Mesh** in **Outline** view. In **Details of "Mesh"**, select the default value of the **Element Size** under **Defaults** and change it from the default value to 0.002 as shown in Figure 3.18. The color of the mesh changes to pink indicating it is being modified. Click **Generate**.

   We can watch the progress of meshing on the lower left corner of the status bar of the Mesh window, shown in Figure 3.18.

   The element size is important and affects the number of mesh elements. A smaller element size will result in a larger number of elements suitable for fluid flow analysis.

   Once mesh is completed, a green check mark will appear next to **Mesh** if the meshing is successful. Click **Mesh** in **Outline** view to see the generated refined mesh. Zoom in to look at the grid. We can zoom in and out by scrolling the wheel of the mouse. Expand **Quality**, in **Details of "Mesh"** as shown in Figure 3.19. Click **Mesh Metric**, and from the drop-down list next to it select **Aspect Ratio**. Note the maximum value of the aspect ratio, **Max**, is **1**, an indicator there aren't

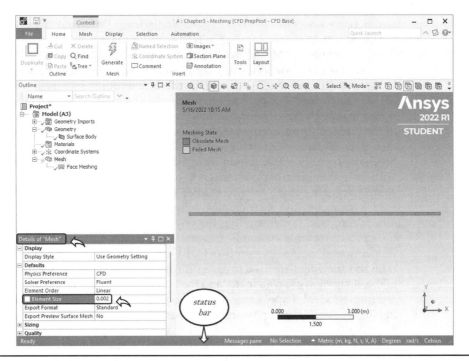

FIGURE 3.18 Element Size, status bar, and Statistics.

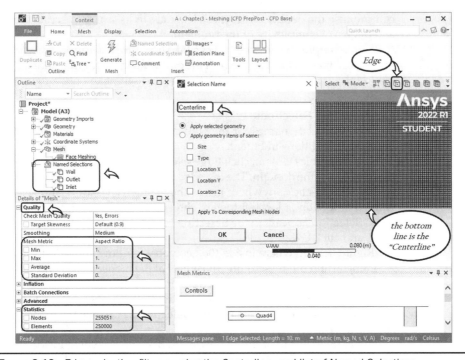

FIGURE 3.19 Edge selection filter, naming the Centerline, and list of Named Selections.

any elongated cells in our grid. Expand **Statistics** in **Details of "Mesh"**, shown in Figure 3.19, to see the number of **Nodes** and **Elements** in the mesh. Scrolling down in **Details of "Mesh"** may be necessary to expand **Statistics**.

Meshing the model is usually the most important and time consuming part of the modeling process. We divide the continuous domain into a finite number of discrete elements that constitute the mesh. The quality of the mesh is determined mainly by three factors: accuracy, efficiency, and the degree of complexity of the problem being studied [32]. Accuracy corresponds to the ability of generating a solution with the highest possible fidelity to the underlying physics. Efficiency corresponds to the ability of the mesh to generate an accurate solution with the minimum computational cost (or number of elements). The degree of complexity of the problem being studied is contingent on the results in which the user is interested and their employability. In other words, for studies that aim at modeling and understanding the physical processes of some phenomenon, the mesh constraints are more stringent and tight than those imposed on studies performed for design purposes [33]. Geometry-based metrics like the aspect ratio and skewness are used to assess the mesh quality. The former is a measure of the elongation of an element, and the latter is a measure of the deviation of a cell from its ideal shape. Large aspect ratios are inevitable in studies investigating highly complicated physical processes such as turbulent flows. Such flows necessitate clustering of elements near the walls to capture the sharp gradients (typically in one direction) of physical properties in the boundary layer, and as such the elements near the walls will be more elongated than those far away. In short, having aspect ratios (or other geometry-based metrics) that deviate, to some acceptable degree, from what is traditionally reported does not necessarily compromise the quality of the mesh if we are to look at the accuracy, efficiency, and degree of complexity of the problem.

Some complicated geometries can take hours to mesh and some complicated models can take weeks to figure out the optimum mesh for the model. The example in this chapter is a very simple model and will take only a few minutes to mesh.

4. **Create Named Selections**

We need to specify names for the boundaries of the domain in order to define boundary conditions in Fluent.

In the graphics toolbar, click **Edge** selection filter shown in Figure 3.19. Select the top edge of the domain. The top edge will change color to indicate it has been selected. Right-click anywhere inside the geometry window and select **Create Named Selection ...** located toward the end of the list of options. The default name **Selection** appears in the **Selection Name** dialog box. Change the name to Wall and click **OK**.

Right-click anywhere inside the geometry window and select **Zoom To Fit**. Zooming in and out by scrolling the wheel of the mouse may be needed in order to select and name the remaining edges of the rectangle.

Select the right side of the rectangle. Right-click anywhere inside the geometry window and select **Create Named Selection ....** Change the name to Outlet and click **OK**.

Select the left side of the rectangle. Right-click anywhere inside the geometry window and select **Create Named Selection ....** Change the name to Inlet and click **OK**.

Select the bottom side of the rectangle. Right-click anywhere inside the geometry window and select **Create Named Selection** .... Change the name to Centerline. The screen should look like Figure 3.19. Click **OK**.

The created names for the boundaries appear in the **Outline** view and are displayed when expanding **Named Selections**. See Figure 3.19.

It is important to mention that a geometrical entity cannot be selected in more than one **Named Selections** or an error message will be displayed upon updating the Mesh and translating to Fluent. This is a common mistake made by the user. **Named Selections** is used to assign boundary conditions in Fluent and having an edge or a face included in more than one **Named Selections** will create a problem as multiple boundary conditions will be assigned to the same boundary.

5. **Save the Project**
Click **File** and select **Save Project** from the drop-down list.

Click **File** and select **Close Meshing** to close the ANSYS Mesh application and return to **Project Schematic**.

In Workbench **Project Schematic**, the **Mesh** application has the lightning bolt symbol next to it indicating the application requires an update. Right-click **Mesh** and select **Update**. Once updated, a green check symbol will appear next to **Mesh** indicating the model is ready for the next step.

Mesh Summary

* Right-click **Mesh** and select **Generate Mesh**.
* Right-click **Mesh** and select **Insert** ⇒ **Face Meshing**. Activate **Face** selection and select the rectangle in the geometry window for **Geometry**.
* Click **Mesh**. Change the **Element Size** under **Defaults** to 0.002. Right-click **Mesh** and select **Generate**.
* Activate **Edge** selection filter.
* Select the top edge of the domain. Right-click and select **Create Named Selection** .... Type Wall and click **OK**.
* Repeat by selecting the right edge of the rectangle and name it Outlet.
* Repeat by selecting the left edge and name it Inlet.
* Repeat by selecting the bottom edge and name it Centerline.
* Click **File** and select **Save Project**.
* Click **File** and select **Close Meshing**.
* Right-click **Mesh** in **Project Schematic** and select **Update**.

3.4.4 Setup

A flowchart of the steps needed to set up the model in Fluent is displayed in Figure 3.20.

In ANSYS Workbench **Project Schematic**, double-click **Setup** to launch Fluent.

A dialog box shown in Figure 3.21 will appear on the screen where the different options for the user can be specified. Enable **Double Precision**. Using double precision will reduce the roundoff errors as 16 significant figures are used versus 7 significant

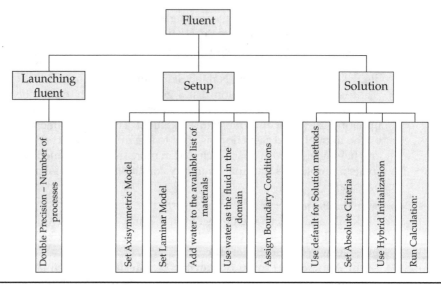

Figure 3.20   Flowchart of Fluent.

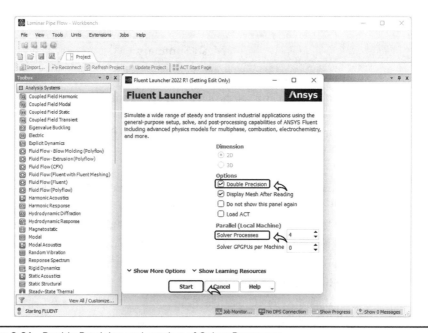

Figure 3.21   Double Precision and number of Solver Processes.

figures for single precision. Increase the number of **Solver Processes** depending on the available **Logical processors** on the machine used to perform the calculation. To find out the number of **Logical processors** on a machine, press **Ctrl** + **Shift** + **Esc** to bring up the **Task Manager** window. Click the **Performance** tab on the **Task Manager** window to display the name of the processor and the number of cores available. The number of **Logical processors** usually appears under the number of **Cores**. Note that if we enter

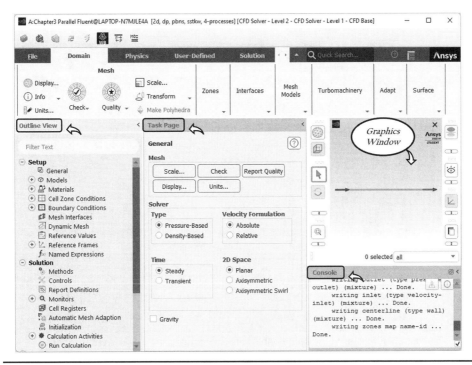

Figure 3.22 Outline View, Task Page, Graphics, and Console windows.

a number of **Solver Processes** larger than what is available on the machine, Fluent will still launch but will display a message asking the user to **Hit Return to Exit**. Click **Start**. This will launch Fluent.

A message announcing product behavioral changes from previous releases will appear. Enable the box **Don't show this message again** and click **OK**. The Fluent window is shown in Figure 3.22 and is divided into several views listed below:

> **Outline View:** We will step through each entry in this view to set up variables and properties for the simulation.

> **Task Page:** When we step through the **Outline View**, the task page changes accordingly to allow the user to make changes to properties and variables.

> **Graphics Window:** This is where we view the model, simulation progress and results.

> **Console Window:** This is the command line showing the progress of the simulation.

Figure 3.23 displays the pointer tools, view tools, and copy tools toolbars. The pointer tools toolbar allows the user to adjust the view by changing the function of the left mouse button, whereas the view tools toolbar allows the user to adjust the view by clicking any of its options. The copy tools toolbar allows the user to save images. Note the view in this figure is adjusted in size to display the full toolbars.

Arrange the workspace shown in Figure 3.23 will allow the user to change the display of the different views.

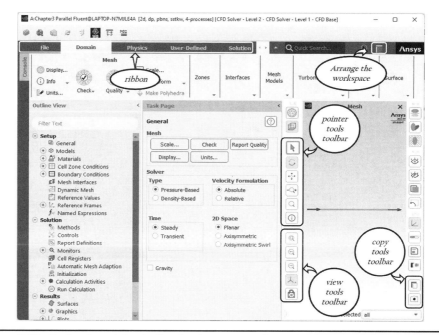

Figure 3.23 Toolbar for pointer, toolbar for view, arrange the workspace, and ribbon.

Note that all options in the **Outline View** can also be accessed from the ribbon shown in Figure 3.23.

We will go in order of what needs to be modified in the **Outline View** to setup the problem.

1. **General**

   Expand **Setup** → Double-click **General**.

   In the **Task Page**, click **Scale...** under **Mesh**. A **Scale Mesh** dialog box will display the **Domain Extents**. Make sure the dimensions are correct and close the dialog box.

   In the **Task Page** and under **Solver** options, keep the default options of **Pressure-Based** for **Type**, **Absolute** for **Velocity Formulation**, and **Steady** for **Time**. Select **Axisymmetric** for **2D Space**.

   The **Pressure-Based** solver is used for low-speed incompressible flows, while the **Density-Based** approach is used for high-speed compressible flows. In the **Pressure-Based** solver the velocity field used in the conservation of mass is calculated from the pressure equation. The pressure equation is derived from the continuity and the momentum equations in a way that the velocity field, corrected by the pressure, satisfies the continuity. However, in the **Density-Based** solver, the continuity equation is used to obtain the density field while the pressure field is determined from the equation of state [34].

2. **Models**

   Expand **Setup** → Double-click **Models**. Note that double-clicking any element will expand it. If the element is already expanded, double-clicking it will collapse it.

In the **Task Page**, double-click **Viscous - SST k-omega**. A **Viscous Model** dialog box will appear. Choose **Laminar** from the **Model** list. Click **OK**.

3. **Materials**

Expand **Setup** → Double-click **Materials**.

In the **Task Page**, the default materials in Fluent are **air** for **Fluid** and **aluminum** for **Solid**. We will add water to the **Fluid** list of materials from the **Fluent Database**.

Click **Create/Edit...** button at the bottom of the **Task Page**, shown in Figure 3.24, to open the **Create/Edit Materials** dialog box. Click **Fluent Database...**. A list titled **Fluent Fluid Materials** is displayed inside the **Fluent Database Materials** dialog box. Scroll to the bottom of the list and select **water-liquid (h2o<l>)** as shown in Figure 3.25. Click **Copy** button. Close the **Fluent Database Materials** dialog box. Close the **Create/Edit Materials** dialog box. In the **Task Page** the added material of **water-liquid** from the **Fluent Database...** is listed under **Materials** and is now available to use for our model.

4. **Cell Zone Conditions**

Expand **Setup** → Double-click **Cell Zone Conditions**.

As discussed earlier, we are modeling the fluid inside the pipe but not the solid pipe itself. Therefore, the list of **Cell Zone Conditions** in the **Outline View** contains only **Fluid**.

Expand **Fluid** under **Cell Zone Conditions** to view our domain named by default **surface_body**.

Double-click **surface_body** in the list of **Cell Zone Conditions** in the **Task Page**, shown in Figure 3.26. The **Material Name** assigned to it is **air** as shown in the **Fluid** dialog box since this is the default fluid in Fluent. Click **air** in the

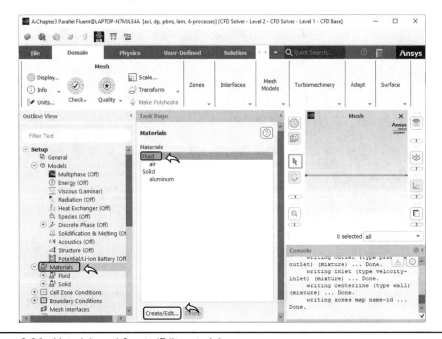

FIGURE 3.24 Materials and Create/Edit materials.

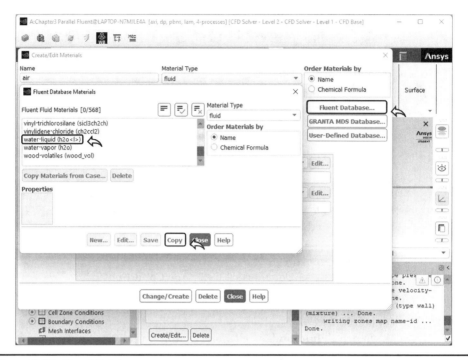

Figure 3.25   Fluent Database.

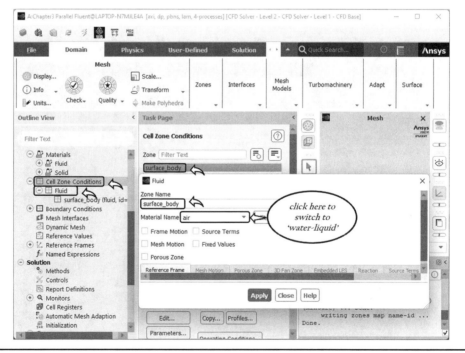

Figure 3.26   Cell Zone Conditions and change material.

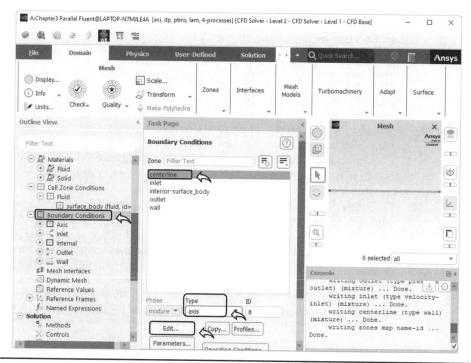

Fluid dialog box and select **water-liquid** as the **Material Name** for the **Zone Name surface_body**, as shown in Figure 3.26. Click **Apply** then click **Close** to close the **Fluid** dialog box.

5. **Boundary Conditions**

   Expand **Setup** → Double-click **Boundary Conditions**. The list of boundaries is shown in the **Task Page** in Figure 3.27.

   We will step through all the boundaries to change the types and apply boundary conditions as described earlier and shown in Figure 3.6.

   Note that if upper case letters are used in naming objects, Fluent will replace them with lower case letters. Also a space will be replaced with an underscore. Our named selections contain upper case letters, and they show as upper case in Mesh application; however, once transferred to Fluent they were changed to lower case letters as seen in the list of boundaries.

   Fluent will automatically detect the type of boundary from its name. A boundary named inlet will be assigned by default a velocity-inlet boundary type. A boundary named outlet will be assigned pressure-outlet type. A boundary named wall will be assigned stationary wall motion and so on.

   In the **Task Page**, select **centerline** from the list of **Boundary Conditions** to highlight it. At the bottom of the **Task Page**, click the arrow below **Type** to access its list of options. Scroll up the list and select **axis** as the type of boundary since we have an axisymmetric model. Click **Apply** in the **Axis** dialog box to confirm changing the **centerline** boundary type to axis. Click **Close** to close the **Axis** dialog box. The **Task Page** should display **axis** as the **Type** for **centerline Boundary Conditions** as shown in Figure 3.27.

Select **inlet** and make sure its type is **velocity-inlet**. Click **Edit...**, shown in Figure 3.27, and choose **Magnitude, Normal to Boundary** as the **Velocity Specification Method**. Set the **Velocity Magnitude [m/s]** value to 0.0025 and click **Apply**. Close the **Velocity Inlet** dialog box.

Note that **interior-surface_body** is the interior of the domain and even though it is listed under **Boundary Conditions**, we do not have the option to make changes to it when we click **Edit...**.

Select **outlet**. Make sure its type is **pressure-outlet**. Click **Edit...** and confirm the value of **Gauge Pressure [Pa]** at the outlet is 0. Click **Close** to close the **Pressure Outlet** dialog box.

Gauge pressure is used in Fluent to reduce the round-off errors caused by the way computers represent numerical values. Round-off errors can be controlled by avoiding adding numbers with very large difference in magnitude or subtracting almost equal size large numbers. By using gauge pressure, the pressure values inside the computational domain are in the same order as the pressure difference that drives the flow and therefore will reduce the round-off errors.

Select **wall**. Make sure its type is **wall**. Click **Edit...** and confirm the **Wall Motion** is set to **Stationary Wall** and the **Shear Condition** is set to **No Slip**. Click **Close** to close the **Wall** dialog box.

6. **Methods**

Expand **Solution** → Double-click **Methods**. In the **Task Page**, **Coupled** is selected as the **Scheme** for the **Pressure-Velocity Coupling**.

The default options for the remaining **Solution Methods** will be used, and those options are listed:

Pressure-Velocity Coupling → **Flux Type** → **Rhie-Chow: distance based**.

Spatial Discretization → **Gradient** → **Least Squares Cell Based**.

Spatial Discretization → **Pressure** → **Second Order**.

Spatial Discretization → **Momentum** → **Second Order Upwind**. Scroll down if needed and make sure **Pseudo Time Method** is enabled.

7. **Residuals**

Expand **Solution** → Expand **Monitors**.

Double-click **Residual** shown in Figure 3.28. In Fluent, the residual for each equation being solved is calculated at every iteration. The default residual threshold in Fluent is **0.001** for the continuity, x-velocity and y-velocity equations. We will modify the convergence criteria by changing the **Absolute Criteria** to 1e-12 as shown in Figure 3.28. **Print to Console** and **Plot** under **Options** are enabled by default so the user can view the residuals as they are calculated to monitor the convergence of the solution. Click **OK** to close the **Residual Monitors** dialog box.

8. **Initialization**

Expand **Solution** → Double-click **Initialization**.

Click **Initialize** in the **Task Page**. The **Console** will display a message when **Hybrid Initialization** is done.

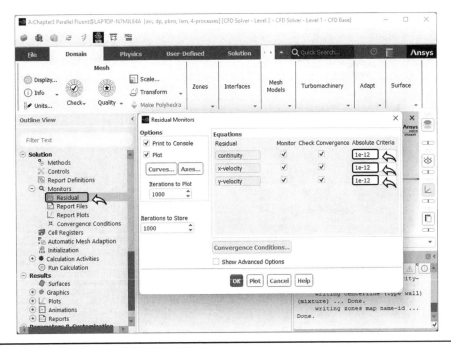

FIGURE 3.28 Residuals and Absolute Criteria.

The initial values assigned for every cell will affect the convergence speed of the calculation but will not affect the results of the calculation. However, the more reasonable the initial guessed values, the faster the solution will converge. We will use the default in Fluent for the **Initialization Methods** which is **Hybrid Initialization**, a method that solves the Laplace equation to produce a velocity field and a pressure field which smoothly connects high and low pressure values in the computational domain [34].

9. **Calculation**

Expand **Solution** → Double-click **Run Calculation**.

In **Task Page**, set the **Number of Iterations** under **Parameters** to 300. Click **Calculate** button under **Solution Advancement** located at the bottom of the **Task Page**.

A **Calculation complete** message will appear when **Absolute Criteria** are met, or when the **Number of Iterations** are completed, whichever comes first. The **Absolute Criteria** are met when the values of the residuals do not exceed the values assigned by the user.

Click **OK** to close the **Calculation complete** message.

After each iteration, Fluent calculates the imbalance of the conservation of variables for each computational cell. The imbalance is summed over all computational cells. This sum is called the unscaled residual. Fluent scales the residual using a scaling factor. The scaled residuals are useful indicators of solution convergence. The console displays the residuals after every iteration. In addition, a plot of the residuals calculated is updated and will be displayed on the graphics window while Fluent is running. The plotted residuals by default are the scaled residuals. The user has the option to change them to unscaled residuals.

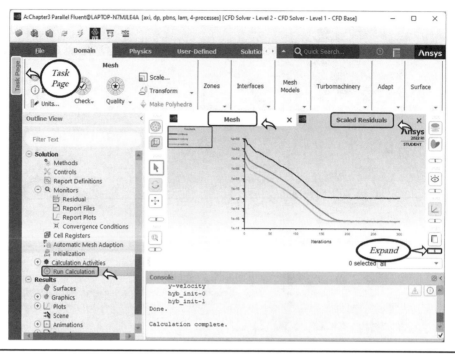

FIGURE 3.29   Scaled Residuals, Task Page, and Expand button.

Keep in mind that the number of iterations to converge will depend on the mesh and the version of Fluent used to perform the calculation.

The **Scaled Residuals** plot is shown in Figure 3.29. It is clear that the solution converged as it is reflected in the plateau residuals. However, further investigation is needed to confirm that this converged solution is the correct solution to the problem.

To maximize the graphics window, we can close the **Task Page** and/or the **Console**. To bring the **Task Page** back, click **Task Page** tab at the top left corner of the screen as shown in Figure 3.29.

Also note that some of the icons in the graphics window may be hidden, You can access hidden icons by clicking the **Expand** button shown in Figure 3.29.

10. **Flux Reports**

Expand **Results** → Double-click **Reports** to expand it.

In the **Task Page** view, double-click **Fluxes** or select **Fluxes** and click **Set Up...** under the group box of **Reports**.

In the **Flux Reports** dialog box, select **Mass Flow Rate** for **Options**, select **inlet** and the **outlet** from the list of **Boundaries** to check the total mass flux through the inlet and outlet of the model. Click **Compute**.

The **inlet** and **outlet** mass flow rates along with the **Net Results** are displayed in the **Console** as shown in Figure 3.30. Fluent uses the inward normal to the surface in calculating the mass flow rate; therefore, a positive mass flow means the mass is entering the system while a negative mass flow means that mass is leaving the system. The conserved mass is one of the indications for convergence. Click **Close** to close the **Flux Reports** dialog box.

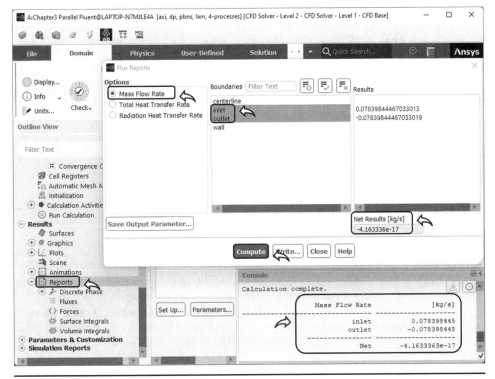

FIGURE 3.30 Flux Reports for the Mass Flow Rate.

Setup Summary

* Expand **Setup** → Double-click **General**. Select **Axisymmetric** under **2D Space**. Select **Pressure-Based** under **Type**. Select **Absolute** under **Velocity Formulation**. Select **Steady** under **Time**.

* Expand **Setup** → Double-click **Models**. Double-click **Viscous - SST k-omega**. Select **Laminar** for **Model**. Click **OK**.

* Expand **Setup** → Double-click **Materials**. Click **Create/Edit....** Click **Fluent Database...**, select **water-liquid (h2o<l>)** and click **Copy**. Close the dialog boxes.

* Expand **Setup** → Double-click **Cell Zone Conditions**. Double-click **surface_body** and select **water-liquid** next to **Material Name**. Click **Apply**, then click **Close**.

* Expand **Setup** → Double-click **Boundary Conditions**. Select **centerline** and select **axis** as the **Type**. Click **Apply** and click **Close**. Double-click **inlet**. Set the **Velocity Magnitude [m/s]** to 0.0025. Click **Apply** and click **Close**.

* Expand **Solution** → Double-click **Methods**. Use the default options for **Solution Methods**.

* Expand **Solution** → Expand **Monitors** → Double-click **Residual**. Set the **Absolute Criteria** for all equations to 1e-12. Click **OK**.

* Expand **Solution** → Double-click **Initialization**. Click **Initialize**.

* Expand **Solution** → Double-click **Run Calculation**. Set the **Number of Iterations** to 300. Click **Calculate**.

* Expand **Results** → Expand **Reports**. Double-click **Fluxes**. Select **Mass Flow Rate** for **Options**. Select **inlet** and **outlet** for **Boundaries**. Click **Compute**. Click **Close**.

* Click **File** and select **Save Project**.

* Click **File** and select **Close Fluent**.

11. **Centerline Velocity**

Expand **Results** → Double-click **Plots** to expand it.

Double-click **XY Plot** under **Plots** in **Outline View** to create the plot of the velocity of the fluid along the centerline of the pipe.

A dialog box titled **Solution XY Plot** will appear as shown in Figure 3.31.

Note that double-clicking **XY Plot** from the **Task Page** will not give the user the option to name the figure and thus the option to save the plot. We can double-click **XY Plot** from the **Task Page** if we desire to observe a certain result without saving it.

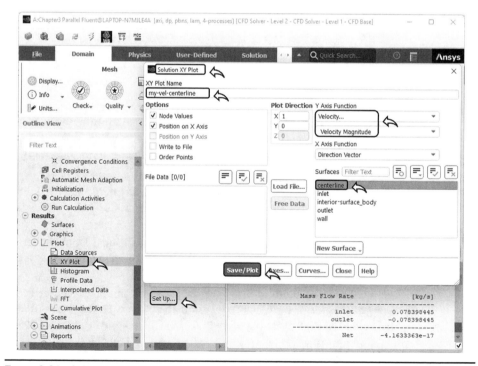

FIGURE 3.31   Solution XY Plot dialog box and Y Axis Function.

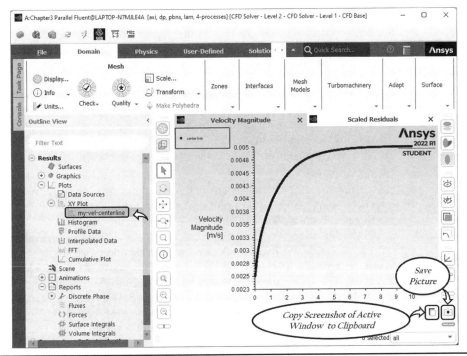

FIGURE 3.32 Centerline velocity profile, list of plots, save picture, and copy screenshot to clipboard.

A default name for the plot appears under **XY Plot Name**. Change the name to My Vel Centerline. Notice that Fluent replaces the space used in the name by a hyphen. It is a good practice not to use a space in Fluent. Also notice in Figure 3.31 that capital letters used in the name were changed to small letters.

Select **Velocity...** from the upper drop-down list and **Velocity Magnitude** from the lower drop-down list of **Y Axis Function** as shown in Figure 3.31.

Select **centerline** from the list of **Surfaces**. Click **Save/Plot** then click **Close** to close the **Solution XY Plot** dialog box.

The created plot appears under **Results → Plots → XY Plot** in **Outline View** as shown in Figure 3.32. The solution created by Fluent is at the cell centers. However, Fluent will interpolate between cells in order to create a continuous plot of the solution.

Modifications to the curve can be made by right-clicking the plot **my-vel-centerline** from the list of **XY Plot** in **Outline View** and selecting **Edit...** In the **Solution XY Plot** dialog box, click **Axes....** In the **Axes - Solution XY Plot** dialog box, select **Y** under **Axis** and change the **Type** of the **Number Format** to **general**. Click **Apply** then click **Close**. Note that if we make changes to both axes then we need to click **Apply** before switching from one axis to another in order for the changes to be saved. The changes made will not be reflected until we click **Save/Plot** in the **Solution XY Plot** dialog box. Figure 3.32 displays the profile of the centerline velocity with the applied modifications to the Y axis.

Alternatively, changes to the curve can be made by clicking **Curves...** in the **Solution XY Plot** dialog box. Click **Close** to close the **Solution XY Plot** dialog box.

To print the created plot, click the **Save Picture** camera icon shown in Figure 3.32. In case the camera icon is not visible, click **Arrange the workspace** button that was shown in Figure 3.23 and deactivate the **Console**. If the camera icon is still not visible, click the **Expand** button that was shown in Figure 3.29.

When we click the **Save Picture** camera icon, a **Save Picture** dialog box will appear. Click **Save....** In the **Select File** dialog box, the default location for all created XY Plots reside in the plot folder inside the saved project in ANSYS but the user can change the location before saving the plot and clicking **OK**.

A **Copy Screenshot of Active Window to Clipboard**, shown in Figure 3.32, can also be used to capture a screenshot of the plot created.

12. **Outlet Velocity**

We will create a plot for the velocity at the exit of the pipe (outlet) along the radial direction. Specifically, we will plot the radial direction as a function of velocity.

Expand **Results** → Expand **Plots** → Double-click **XY Plot**.

In the **Solution XY Plot** dialog box change the default name to `my-vel-outlet`. Disable **Position on X Axis** under **Options**. See Figure 3.33. Select **Mesh...** from the upper drop-down list and **Y-Coordinate** from the lower drop-down list of **Y Axis Function**. Select **Velocity...** from the upper drop-down list and **Velocity Magnitude** from the lower drop-down list of **X Axis Function**. Select **outlet** from the list of **Surfaces**. Click **Save/Plot** then click **Close**.

The created plot is now added to the list of **XY Plot** in the **Outline View**. Recall the values of the velocity in this plot are calculated at the centers of the computational cells.

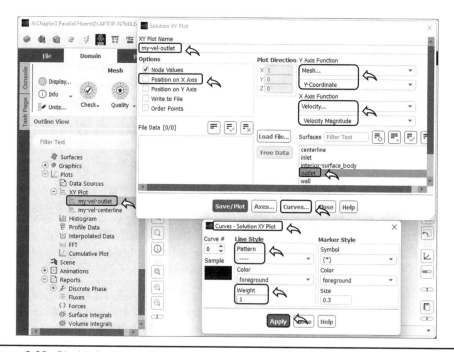

FIGURE 3.33 Disable Position on X Axis and Solution XY Plot dialog box.

Double-click **my-vel-outlet** plot from the list of **XY Plot** in the **Outline View** to edit the plot and create a smooth curve. Right-clicking **my-vel-outlet** plot and selecting **Edit...** will also allow the user to edit the plot. In the **Solution XY Plot** dialog box, click **Axes....** In the **Axes - Solution XY Plot** dialog box, select **Y** under **Axis** and change the **Type** of the **Number Format** to **general**. Click **Apply** then click **Close**. Click **Curves...** in **Solution XY Plot** dialog box. Select the dashed line option which is the second option under **Pattern** and set the value of **Weight** to 1 in the **Curves - Solution XY Plot** dialog box, as shown in Figure 3.33. Click **Apply** then click **Close** to close the **Curves - Solution XY Plot**. Click **Save/Plot** to save and plot the changes. Click **Close** to close the **Solution XY Plot** dialog box.

Figure 3.34 displays the radial velocity at the outlet of the pipe. For a better display of the outlet velocity, the **Outline View** window is collapsed in addition to the **Console** window and **Task Page** window. The no-slip boundary condition is reflected in this figure with the velocity being 0 at the wall of the pipe where R = 0.1 m. The velocity increases to reach its maximun value at the center of the pipe.

Finally, we will export the data for the outlet velocity as it will be used in Section 3.5 for verification purposes. Click **Arrange the workspace** button, and activate **Outline View**. To export the data, double-click **my-vel-outlet** plot from the list of **XY Plot** in **Outline View**. Enable **Write to File** under **Options** in the **Solution XY Plot** dialog box. Click **Write...** at the bottom of the dialog box. Name the file my-vel-outlet and change the location to save the file to the **Desktop**

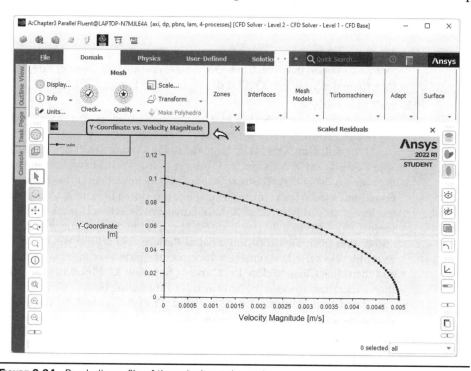

FIGURE 3.34 Parabolic profile of the velocity at the outlet.

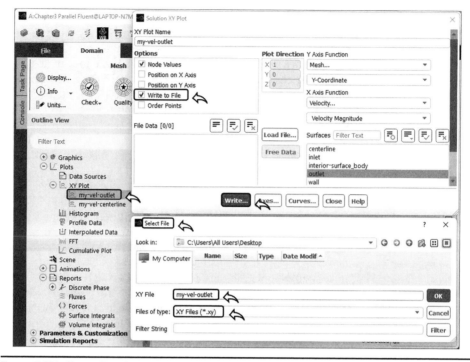

FIGURE 3.35 Export Data by using Write to File.

in the **Select File** dialog box. Note the output file is a data file with **.xy** extension that we can open in Excel. Click **OK** to save and close the **Select File** dialog box, then close the **Solution XY Plot** dialog box. See Figure 3.35.

13. **Wall Shear Stress**

In this section, we will create a plot for the shear stress along the wall of the pipe.

Expand **Results** → Expand **Plots** → Double-click **XY Plot**.

In the **Solution XY Plot**, change the default name to my-shear-wall. Disable **Position on X Axis** under **Options**. Select **Wall Fluxes...** from the upper drop-down list and **Wall Shear Stress** from the lower drop-down list of **Y Axis Function**. Select **Mesh...** from the upper drop-down list and **X-Coordinate** from the lower drop-down list of **X Axis Function**. Select **wall** from the list of **Surfaces**. Click **Curves....** Under **Line Style** select the dashed line which is the second option from the drop-down list of **Pattern** and set the weight to 1. Under **Symbol,** select the first option, which corresponds to no symbol. Click **Apply,** and then click **Close** to close the **Curves - Solution XY Plot**. Click **Save/Plot** and then click **Close** to close the **Solution XY Plot** dialog box.

The created plot is added to the list of **XY Plot**. Figure 3.36 displays the wall shear stress along the wall of the pipe.

14. **Velocity Contours**

Expand **Results** → Expand **Graphics**.

Double-click **Contours** from the **Outline View** or from the **Task Page** to create the velocity contours inside the pipe.

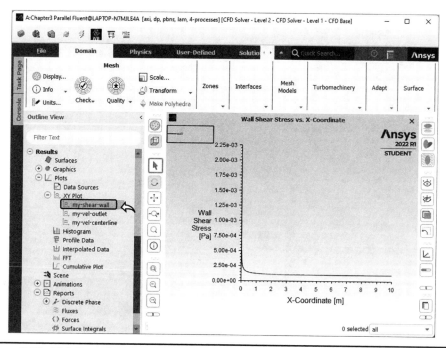

FIGURE 3.36 Wall Shear Stress as a function of axial distance.

Unlike the **XY Plot**, double-clicking from the **Task Page** will allow saving the created contour.

In the dialog box titled **Contours**, change the name from contour-1 to my-vel. Select **Velocity...** from the upper drop-down list and **Velocity Magnitude** from the lower drop-down list for **Contours of** as shown in Figure 3.37. Select **interior-surface_body** from the list of **Surfaces**.

Note that if we do not select a surface, then Fluent will plot using the whole domain for a two-dimensional model. For a three-dimensional model, the reader must specify the surface or no plot will be created. The **Colormap Options...** shown in Figure 3.37 will allow making changes to the color map location, font, number format and more.

Click **Save/Display** then click **Close** to close the **Contours** dialog box. The created contour appears under **Results** → **Graphics** → **Contours** in **Outline View** and is shown in Figure 3.38.

Figure 3.38 displays the velocity contours in the domain.

A screenshot of the results can be made by clicking the **Copy Screenshot of Active Window to Clipboard** icon shown in Figure 3.32.

15. **Pressure Contours**
    Expand **Results** → Expand **Graphics** → Double-click **Contours** to create the pressure contours inside the pipe. **Contours** can also be accessed from the **Task Page**.

    In the dialog box titled **Contours**, change the name to my-pres. Select **Pressure...** from the upper drop-down list and **Static Pressure** from the lower drop-down list for **Contours of**. Select **interior-surface_body** from the list of **Surfaces**.

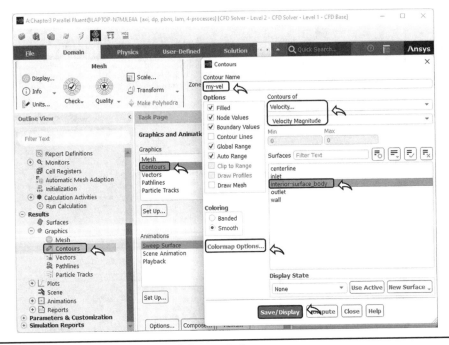

Figure 3.37   Contours dialog box and Colormap Options.

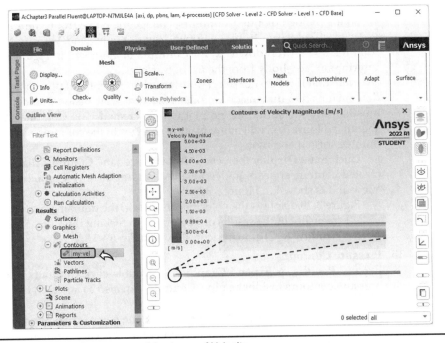

Figure 3.38   List of contours and Contours of Velocity.

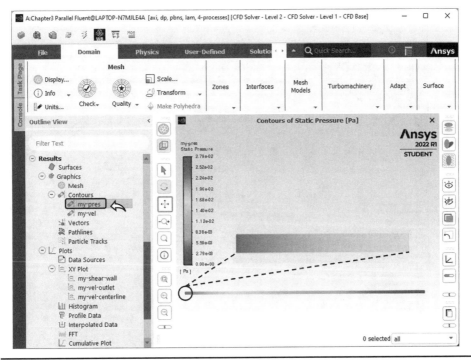

FIGURE 3.39 Contours of Static Pressure.

Click **Colormap Options...** to make the desired changes to the color map location, font, number format and more. Click **Apply** after making changes to the **Colormap** dialog box and click **Close**.

Click **Save/Display** then click **Close** to close the **Contours** dialog box.

Figure 3.39 displays the pressure contours in the domain.

16. **Save the Project**

Click **File** and select **Save Project** from the drop-down list.

Settings have changed! dialog box appears and displays a warning that settings have changed with the following options: **Use settings changes for current calculations only** and **Use settings changes for current and future calculations**. The default option is the first option. Keep the default option and click **OK**. Click **File** again and select **Close Fluent**.

A warning message asking if the reader wants to quit the application appears. Click **OK**.

ANSYS case and data files are automatically saved when we exit Fluent and return to Workbench. The case files contain the mesh, boundary conditions, cell zone conditions, and solution parameters for our problem. The data files contain the values of the flow field quantities in each mesh element and the convergence history for the flow field. If the reader has no need to make changes to the mesh, run design optimization, or postprocessing in CFD post, then running Fluent using the case and data files can be very useful. The reader can export at any time the case and data files by clicking **File** and selecting **Export** ⇒ **Case and Data ...** from the drop-down list.

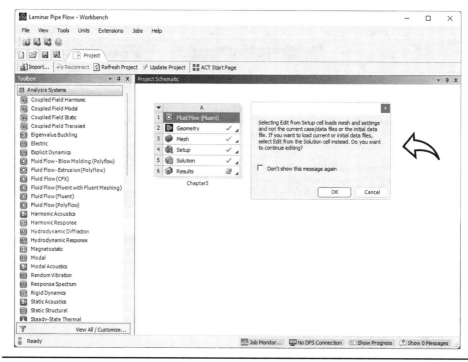

FIGURE 3.40   Warning message when Setup is launched.

17. **Sharing Files and Closing Workbench**
To share the Workbench project, click **File** and select **Archive....** Click **Save** and select from the **Archive Options** the option to share the results and solution or the option to share the project and settings without the solution. Note that an archived Workbench file can be opened with the same or newer release of ANSYS, but not with an older version of the software.

Click **File** and select **Exit** from the drop-down list to close Workbench.

If we need to return to the project at a later time, launch ANSYS Workbench by clicking **Start → ANSYS 2022 R1 → Workbench 2022 R1**. From the directory where the project was saved, double-click **Laminar Pipe Flow.wbpj**.

In ANSYS Workbench **Project Schematic**, double-click **Solution**, or right-click **Solution** and select **Edit...** to launch the **Solution** application. If we launch **Setup**, a warning message that the current mesh and settings will be loaded and not the current data files where the solution is stored will be displayed as shown in Figure 3.40. By clicking **OK** and launching **Setup**, the previously generated solutions will be lost.

3.5   Verification

As part of the verification of the solution, we will perform a grid independent study using Richardson Extrapolation (RE) scheme and following the procedure described in Section 2.9. We will also compare the Fluent numerical results to the analytical results presented in Section 3.3.1.

TABLE 3.2 Mesh Statistics

Mesh # j	Quality	N_j	h_j (mm)	$r_{j,j-1}$
1	Fine	40,000	5	
2	Medium	10,000	10	2
3	Coarse	2,500	20	2

3.5.1 Grid Independent Study

In this section, we implement the RE scheme covered in Section 2.9 to assess the convergence of the solution. We use three meshes: fine, medium, and coarse mesh to perform the grid independent study. For each of the three meshes, the number of elements N_j; global cell size h_j; and the refinement factor r, are listed in Table 3.2.

The global cell sizes are calculated according to Eq. (2.132):

$$h_j = \left(\frac{A_{dom}}{N_j} \right)^{\frac{1}{2}}, \quad \text{for } j = 1, 2, 3 \tag{3.21}$$

where the area of the computational domain is $A_{dom} = 10 \text{ m} \times 0.1 \text{ m} = 1 \text{ m}^2$. The corresponding refinement factors are $r_{j,j-1} = h_j/h_{j-1}, j > 1$. Note that grids 1, 2, and 3 are chosen such that r_{21} and r_{32} are > 1.3, such that they can be used to generate three solutions required by the RE scheme. The numerical uncertainty is reported for key variables that are important to the objectives of the study. For the steady laminar pipe flow studied in this chapter, we report the discretization error for the outlet velocity and for the wall shear stress.

Outlet Velocity Profile

The velocity, ϕ_{ij}, at 10 equidistant points, $i = 1, \ldots, npts$ where $npts = 10$, distributed along the outlet of the domain is listed in Table 3.3 for grids $j = 1, 2,$ and 3.

TABLE 3.3 Convergence Metrics Using RE for the Outlet Velocity

i	y_i (cm)	ϕ_{i_1} (mm/s)	ϕ_{i_2} (mm/s)	ϕ_{i_3} (mm/s)	s_i	$(e_a^{21})_i$ %	$(e_a^{32})_i$ %	$\not{p}_i$	$(GCI^{21})_i$ $\times 100$	EB_i $\times 10^6$
1	0	4.9845	4.9473	4.8032	1	0.75	2.91	1.95	0.28	14.2
2	0.1	4.9346	4.8977	4.7313	1	0.75	3.40	2.17	0.28	14.1
3	0.2	4.7848	4.7490	4.6108	1	0.75	2.91	1.95	0.28	13.6
4	0.3	4.5351	4.5012	4.3516	1	0.75	3.32	2.14	0.28	12.9
5	0.4	4.1852	4.1539	4.0333	1	0.75	2.90	1.95	0.28	11.9
6	0.5	3.7350	3.7071	3.5824	1	0.75	3.36	2.16	0.28	10.6
7	0.6	3.1845	3.1607	3.0691	1	0.75	2.90	1.94	0.29	9.1
8	0.7	2.5336	2.5145	2.4254	1	0.75	3.54	2.23	0.29	7.3
9	0.8	1.7822	1.7687	1.7172	1	0.76	2.91	1.93	0.29	5.1
10	0.9	0.9303	0.9231	0.8806	1	0.78	4.61	2.56	0.30	2.8

TABLE 3.4   Extrapolated Solution of the Outlet Velocity Profile Using RE Scheme

i	$(\phi_{ext}^{21})_i$ (mm/s)	$(\phi_{ext}^{32})_i$ (mm/s)	$(e_{ext}^{21})_i$ %	$(e(e_{ext}^{32})_i$ %
1	4.9975	4.9975	0.26	1.00
2	4.9451	4.9451	0.21	0.96
3	4.7973	4.7973	0.26	1.01
4	4.5450	4.5450	0.22	0.96
5	4.1961	4.1961	0.26	1.01
6	3.7431	3.7431	0.22	0.96
7	3.1929	3.1929	0.26	1.01
8	2.5387	2.5387	0.20	0.95
9	1.7870	1.7870	0.27	1.02
10	0.9318	0.9318	0.16	0.93

Executing the RE scheme yields the values listed in Table 3.3, of the sign function s_i, approximate percent relative errors $(e_a^{21})_i$, $(e_a^{32})_i$, local order of accuracy ρ_i, grid convergence index $(GCI^{21})_i$ and error bar EB_i, for points $i = 1, \ldots, npts$. The extrapolated values $(\phi_{ext}^{21})_i$, $(\phi_{ext}^{32})_i$ and extrapolated relative errors $(e_{ext}^{21})_i$, $(e_{ext}^{32})_i$ at points $i = 1, \ldots, npts$ are listed in Table 3.4. Upon inspecting the convergence parameters for the outlet velocity profile listed in Table 3.3, it can be seen that the local order of accuracy ρ ranges between 1.93 and 2.56 with a global average $\rho_{av} = 2.098$, which is close to the overall order of the numerical scheme. Since $s_i > 0$ for all points, the Oscillatory Convergence (OC) did not occur at any of the 10 points, that is, $OC = 0$. OC is considered to be one of the limitations of the RE scheme and is discussed in Section 2.9. Figure 3.41 shows the velocity profile

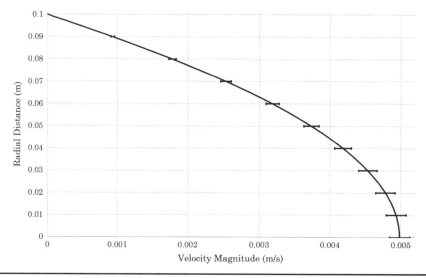

FIGURE 3.41   Error bars amplified by a factor of 10 for the velocity at the outlet.

TABLE 3.5 Convergence Metrics Using RE for the Wall Shear Stress

i	x_i (m)	ϕ_{l_1} (mPa)	ϕ_{l_2} (mPa)	ϕ_{l_3} (mPa)	s_i	$(e_a^{21})_i$ %	$(e_a^{32})_i$ %	p_i	$(GCI^{21})_i$ ×100	EB_i ×10^6
1	0.1	0.2303	0.2591	0.2191	−1	12.51	15.45	0.47	7.02	16.2
2	1	0.1220	0.1204	0.1165	1	1.32	3.21	1.26	0.74	0.9
3	2	0.1092	0.1082	0.1049	1	0.95	3.05	1.66	0.54	0.6
4	3	0.1044	0.1036	0.1006	1	0.82	2.87	1.80	0.46	0.5
5	4	0.1022	0.1014	0.0986	1	0.76	2.81	1.87	0.43	0.4
6	5	0.1011	0.1003	0.0975	1	0.74	2.80	1.91	0.42	0.4
7	6	0.1005	0.0997	0.0969	1	0.73	2.81	1.93	0.41	0.4
8	7	0.1002	0.0994	0.0966	1	0.73	2.83	1.93	0.41	0.4
9	8	0.1000	0.0993	0.0964	1	0.74	2.85	1.94	0.41	0.4
10	9	0.0999	0.0992	0.0964	1	0.74	2.86	1.94	0.41	0.4
11	10	0.0998	0.0990	0.0962	1	0.78	2.84	1.86	0.44	0.4

at the outlet as predicted on the fine grid. Also shown in the figure are the error bars at the 10 points of Table 3.3. The bars are amplified by a factor of 10 for visibility purposes.

The magnitude of the error bar assumes a maximum value of $\pm 1.42 \times 10^{-5}$ m/s at $i = 1$. The maximum discretization uncertainty, which is twice the $(GCI^{21})_1$ corresponding to the maximum error bar magnitude, is 0.56%.

Wall Shear Stress
To report the uncertainty of the wall shear stress, we export the values of the shear stress at $npts = 10$ points on the wall. The values, ϕ_{ij}, $i = 1, \ldots, npts$, generated at the fine ($j = 1$), medium ($j = 2$), and coarse ($j = 3$) grids, are summarized in Table 3.5.

Figure 3.42 shows the variations of the shear stress at the wall with the axial distance from the inlet, as predicted by the numerical solver using the fine grid. The uncertainty indicated by the error bars (increased by a factor of 10) is a maximum at the inlet and monotonically decreases with the axial distance from the inlet. There are two observations to be made. The first observation is that the percent of relative error associated with the extrapolated wall shear stress at $x = 0.1$ m is larger than 25%, which is too high. If the objective is to accurately estimate the wall shear stress near the entrance, then further grid refinement is needed, especially near the wall in the entrance region. The second observation is that despite the large error in the wall shear stress in the entry region, the error decreases quickly to reach acceptable values for $x \geq 1$ m, as can be seen in Table 3.6. This is a manifestation of the fact that the fully developed flow velocity profile does not depend on the detailed inlet velocity profile, but rather on the inlet mass flow rate, the fluid density and viscosity, and the pipe diameter. In contrast, the entry length, which is the distance from the inlet at which a fully developed flow is established, depends on the velocity profile at the inlet.

3.5.2 Comparison with Exact Solution and/or Empirical Correlations
Entry Length
As discussed in Section 3.3.2, the entry length for laminar flow inside the pipe is related to Reynolds number and to the pipe diameter according to Eq. (3.20). For the laminar

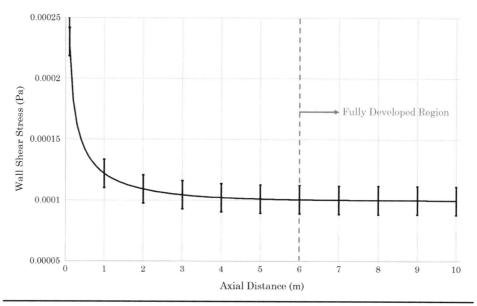

FIGURE 3.42 Amplified error bars for the wall shear stress.

TABLE 3.6 Extrapolated Solution of the Wall Shear Stress Using RE Scheme

i	$(\phi_{ext}^{21})_i$ (mPa)	$(\phi_{ext}^{32})_i$ (mPa)	$(e_{ext}^{21})_i$ %	$(e_{ext}^{32})_i$ %
1	0.1564	0.3618	47.26	28.39
2	0.1231	0.1231	0.93	2.24
3	0.1097	0.1097	0.44	1.39
4	0.1048	0.1048	0.33	1.14
5	0.1025	0.1025	0.29	1.04
6	0.1013	0.1013	0.27	1.01
7	0.1007	0.1007	0.26	0.99
8	0.1004	0.1004	0.26	0.99
9	0.1003	0.1003	0.26	0.99
10	0.1002	0.1002	0.26	1.00
11	0.1000	0.1000	0.30	1.07

pipe flow considered, $\mathrm{Re} = \frac{\rho V D}{\mu} = 498$ and $D = 0.2$ m, so that the calculated entry length is $L_e = 6$ m.

Figure 3.32 is the solution obtained from Fluent for the velocity along the centerline of the pipe. The estimated entry length in this figure is 6 m. The velocity along the centerline increases in the entry region of the pipe and becomes constant at $x = 6$ m. This is in accordance with Figure 3.42 where the shear stress at the wall becomes constant at $x = 6$ m.

The estimated entry length is in agreement with the value predicted using the empirically determined Eq. (3.20).

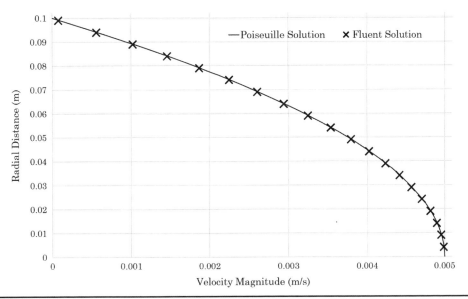

FIGURE 3.43 Comparison between numerical and analytical results for the radial velocity.

Velocity Profile
The exact solution for the velocity profile inside the pipe in the fully developed region was derived in Section 3.3.1 and is given by Eq. (3.13). Figure 3.43 compares the exact value for the velocity at the outlet of the pipe to the numerical solution obtained from Fluent. The figure shows good agreement between the analytical Poiseuille solution and the Fluent solution.

Wall Shear
The wall shear in the fully developed region is derived in Section 3.3.1 and is given by Eq. (3.15). Since the fluid density is constant and the cross-sectional area of the pipe is fixed, conservation of mass implies that the average velocity is equal to the inlet velocity $V_{av} = V_{in} = 0.0025$ m/s. For $\mu = 0.001$ kg/m.s and $R = 0.1$ m, the shear at the wall, calculated using Eq. (3.15), equals 0.0001 N/m^2.

The calculated shear at the wall is in good agreement with the solution obtained from Fluent and presented in Figure 3.36.

CHAPTER 4

Three-Dimensional Steady State Turbulent Incompressible Fluid Flow

List of Symbols	
Pressure	p
Dynamic viscosity	μ
Length of the pipe	$\mathcal{L}$
Reynolds number	Re
Characteristic length	L
Local temporal mean velocity vector	$\bar{\vec{u}}$
Local temporal fluctuating velocity vector	$\vec{u}'$
Time	t
Diameter of the pipe	D
Time scale	τ
Density	ρ
Gravitational acceleration	$\vec{g}$
Filtering function	G
Sub-grid scale stress	τ_{ij}
Turbulent kinetic energy	KE_{turb}
Kinetic energy of the fluctuating component of the velocity per unit mass	k
Wave number	κ
Spectral turbulent kinetic energy density	$E(\kappa)$
Eddy length scale	l_{eddy}
Dissipation rate of turbulent kinetic energy	ε
Generation rate	G
Source term	S

Continued

List of Symbols	
Wall roughness	ϵ
Wall shear	τ_w
Kinematic viscosity	ν
Friction head loss	h_f
Friction factor	f
Entry length	L_e

This chapter will cover the turbulent fluid flow in a three-dimensional pipe. We will build on the geometry created in Chapter 3 by revolving the two-dimensional model around the centerline to create a three-dimensional pipe. In addition, the parameters for the flow are chosen such that the flow inside the pipe is turbulent.

The problem presented in this chapter could be simplified and solved as a two-dimensional problem due to the symmetries of the geometry and boundary conditions. However, we will build a three-dimensional model to introduce the reader to three-dimensional modeling. The results of the simulations will be discussed and followed by a verification of the Fluent model.

4.1 Introduction to Turbulence

The Navier-Stokes (NS) existence and smoothness problem remains one of the few remaining unsolved Millenium prize problems.[1] Mathematicians are yet to prove that smooth solution of the NS equations exists for the three-dimensional case subject to given initial conditions. This has not prevented the development of increasingly advanced methods for numerically modeling fluid flow. These numerical methods are commonly validated against data from laboratory experiments and then applied in a variety of engineering applications.

The challenge stems from the absence of a complete theoretical understanding of turbulent flow. Scientists rely on statistical descriptions of fluid properties characterized by a sufficiently large value of Reynolds number, reflecting the degree of instability as the ratio of the destabilizing inertia effects to the stabilizing viscous effects. To shed light on the underlying challenge, we consider a fully developed incompressible axisymmetric flow in a horizontal pipe of circular cross section. Under these conditions, the solution of the NS equations, assuming steady flow, yields the famous parabolic velocity profile of Poiseuille flow, $u(r) = \frac{1}{4\mu} \frac{\Delta p}{\mathcal{L}} \left(R^2 - r^2 \right)$. Experimental measurements show that the mathematically predicted velocity profile matches the experimentally observed profile for values of Reynolds number less than 2000. For larger values of Reynolds number, the experimentally observed velocity profile departs significantly from the laminar parabolic profile. The shear stress at the wall in a turbulent flow is typically larger than that in a laminar flow, implying that a larger pressure loss is incurred for a given flow rate.

In a landmark experiment, Osbourne Reynolds (1842–1912) observed the pipe flow behavior as the inlet velocity is increased. To do so, he injected a thin dye streak at the pipe inlet, as seen in Figure 4.1. Reynolds observed that when the average velocity is

---

[1]The millenium problems are hosted by the CMI, https://www.claymath.org/millennium-problems

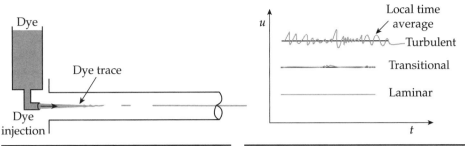

FIGURE 4.1 A Schematic depicting injection of dye streak at the inlet center of a pipe.

FIGURE 4.2 Typical time variation for the laminar, transition, and turbulent flow regimes.

small, the dye propagates along the pipe axis as a nearly well-defined line. In this *laminar* flow regime, the dye streak undergoes little smearing due to molecular diffusion of the dye in the surrounding water. As the flow rate is increased, irregular intermittent dye bursts were observed. In this *transitional* flow regime, the dye bursts are characterized by considerable spatio-temporal fluctuations. Further increase in the average velocity brings about transition to yet another flow regime (*turbulent*), where intense mixing rapidly spreads the dye all over the pipe.

Figure 4.2 shows a plot of typical time variation of the velocity at a point along the pipe axis for the laminar, transition, and turbulent flow regimes. In the transitional and turbulent flow regimes, the flow is intrinsically unsteady, irrespective of the care one takes to establish a steady flow by maintaining a constant volume flow rate and time-independent flow conditions. This is in contrast with the laminar flow, where a steady flow can be established under these conditions. Further inspection of Figure 4.2 reveals that, in the turbulent and transitional flow regimes, the velocity fluctuates about a mean velocity which is stationary in time. Thus, while the actual local, instantaneous velocity field is unsteady, the mean velocity field can often be taken as steady under the aforementioned conditions. One can reasonably assume that the fluctuations in the velocity are responsible for the strong and rapid mixing of the die. We note here that in this case, the angular and radial velocity components fluctuate about a mean of 0. These fluctuations in all of the velocity components enhance mixing of the die with the surrounding water beyond mixing expected by molecular diffusion. Fluctuations across many spatio-temporal scales are fundamental to turbulence. For this reason statistical and spectral methods (e.g., Fourier transforms) are often used to characterize turbulent flow.

Figure 4.3 is a simplified representation of a turbulent flow showing an example of eddies of different sizes. According to Fay [2], "Eddies are regions of swirling flow that,

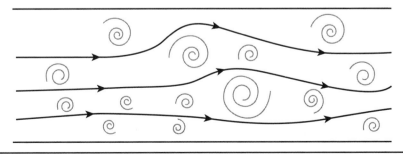

FIGURE 4.3 Eddies in a turbulent flow.

for a time, retain their identities as they drift with the flow but which ultimately break up into smaller eddies." The velocity field in a turbulent flow can be reconstructed by the superposition of a large number of eddies that span different time and length scales. The size of the largest eddy, L_1, defines the characteristic length scale of the problem. Due to the instability of the flow at high Re, these eddies break into smaller eddies of characteristic size $L_2 < L_1$. The smaller eddies, in turn, break into even smaller eddies, $L_3 < L_2$. This process of eddies repeatedly breaking up into smaller and smaller eddies is known as a turbulent "cascade" which terminates when the eddies reach a scale, L_s, where viscous damping dissipates their energy. As an eddy breaks down into smaller eddies, some of the kinetic energy transferred from the larger eddy to the smaller one is lost due to the fact that work is done by the flow to overcome friction. This lost work is dissipated as local heat, which is then converted to a rise in internal energy. For large eddies, the work lost in the breakdown process is small compared to the eddies' kinetic energy. The relative loss of energy to viscous effects increases in a nonlinear fashion as smaller eddies are generated until all the kinetic energy of the smallest eddies is converted into internal energy by viscous dissipation.

4.2   Turbulence Modeling

4.2.1   The Turbulence Energy Spectrum

The turbulent kinetic energy, KE_{turb}, of the fluid in a given volume, $\mathcal{V}$, is the integral of the kinetic energy of the fluctuating component of the velocity per unit mass, k, over the mass,

$$KE_{turb} \equiv \int_{\mathcal{V}} \rho k \, d\mathcal{V} \tag{4.1}$$

where the turbulent kinetic energy per unit mass is the time average of the square of the fluctuating component of the velocity field

$$k \equiv \frac{1}{2}\overline{|\vec{u}'|^2} = \frac{1}{2}\left(\overline{u'^2} + \overline{v'^2} + \overline{w'^2}\right) \tag{4.2}$$

In the limit of continuum, the sum becomes the integral

$$k \equiv \frac{1}{2}\int_0^\infty E(\kappa) \, d\kappa \tag{4.3}$$

where κ is the wave number and $E(\kappa)$ is the spectral turbulent kinetic energy density (turbulent kinetic energy per unit wave number). Important information is revealed by showing how k is distributed among the different wave numbers. Such information is represented in terms of the turbulent energy spectrum, which shows how the spectral energy density varies over the entire range of wave-numbers. A log-log plot of ξ, the spectral turbulent kinetic energy density, normalized by $2kD$, versus the wavelength, normalized by the diameter, is presented in Figure 4.4 for a pipe flow of Re $= 1E6$. The energy spectrum of Figure 4.4 shows how turbulent kinetic energy is distributed over structures of various scale-lengths. It can be seen that eddies of size equal to the diameter $(\kappa D = 1)$ carry the most kinetic energy, which is consistent with the observation that the most rapidly growing disturbances are those with length scale of the order of D, leading to large eddies of velocity amplitude of about 10% of the average velocity.

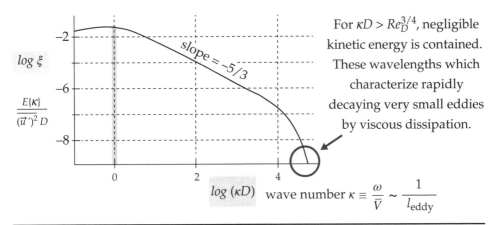

For $\kappa D > Re_D^{3/4}$, negligible kinetic energy is contained. These wavelengths which characterize rapidly decaying very small eddies by viscous dissipation.

FIGURE 4.4   Energy spectrum of a turbulent flow.

It can also be seen that the eddies of size less than $D/31263$ (that is, $\log(\kappa D) > 4.5$), contain negligible turbulent kinetic energy. Another important observation is that for eddies of sizes $1/\kappa$ in the range $0 < \log(\kappa D) < 4$, the log of the normalized spectral kinetic energy, $\log \xi$, decreases linearly (slope $= -5/3$) with the log of the normalized wave number, $\log(\kappa D)$, indicating a power-law decay of the spectral turbulent kinetic energy for smaller eddies. The fact that smaller eddies contain smaller turbulent kinetic energy is a manifestation of the fact the viscous dissipation is dominant at this scale. For very small eddies characterized by $\log(\kappa D) > 4$, we can see that the decay in spectral kinetic energy due to viscous dissipation is so large that no kinetic energy is contained in eddies characterized by $\log(\kappa D) > 4.5$.

4.2.2   Reynolds Averaging

The spatio-temporal fluctuations in the velocity in a turbulent flow have a huge impact on the apparent diffusivity due to the nonlinear term $\vec{u} \cdot \nabla \vec{u}$. We start by expressing the velocity vector field as the sum of a mean flow and a fluctuating flow

$$\vec{u}(\vec{x}, t) = \overline{\vec{u}}(\vec{x}, t) + \vec{u}'(\vec{x}, t) \tag{4.4}$$

where the $\overline{\vec{u}}(\vec{x}, t)$ is local temporal mean, defined as

$$\overline{\vec{u}}(\vec{x}, t) = \frac{1}{T} \int_t^{t+T} \vec{u}(\vec{x}, t') \, dt' \tag{4.5}$$

and $\vec{u}'$ is the fluctuating velocity. The local time averaging is carried out over a time interval T that satisfies the condition $\tau_s \ll T \ll \tau_p$, where τ_p is the characteristic time scale of the mean field (for the pipe flow $\tau_p = D/V$) and τ_s is the time scale of the slowest fluctuation.[2] Similarly, we express

$$\rho(\vec{x}, t) = \overline{\rho}(\vec{x}, t) + \rho'(\vec{x}, t) \tag{4.6}$$

where $\overline{\rho}$ is the local mean density and ρ' is the fluctuating density.

---

[2]Note the similarity with defining, in the context of the continuum hypothesis, the macroscopic properties over a length scale that is too small compared to the problem length scale, and too large compared to the molecular length scale.

Starting with the local time average of the NS equations,

$$\overline{\frac{\partial \rho}{\partial t} + \nabla \cdot (\rho \vec{u})} = 0 \tag{4.7}$$

$$\overline{\rho \left(\frac{\partial \vec{u}}{\partial t} + \nabla \cdot (\vec{u}\vec{u}) \right)} = \overline{-\nabla p + \mu \nabla^2 \vec{u} + \frac{\mu}{3} \nabla (\nabla \cdot \vec{u}) + \rho \vec{g}} \tag{4.8}$$

where we will assume that the dynamic molecular viscosity μ is constant. Substituting Eqs. (4.4) and (4.6) in the NS equations and substituting the pressure term with its fluctuating and average terms, assuming constant density flow ρ and noting that $\overline{A+B} = \overline{A} + \overline{B}$, $\overline{\vec{u}'}(\vec{x}, t) = \vec{0}$, and $\overline{\rho'}(\vec{x}, t) = 0$

$$\frac{\partial \overline{\rho}}{\partial t} + \nabla \cdot (\overline{\rho \vec{u}}) = 0 \tag{4.9}$$

$$\frac{\partial \overline{\rho \vec{u}}}{\partial t} + \nabla \cdot (\overline{\rho} \overline{\vec{u}} \, \overline{\vec{u}}) = -\nabla \overline{p} + \mu \nabla^2 \overline{\vec{u}} + \frac{\mu}{3} \nabla \left(\nabla \cdot \overline{\vec{u}} \right) - \nabla \cdot (\overline{\rho} \, \overline{\vec{u}'\vec{u}'}) + \overline{\rho} \vec{g} \tag{4.10}$$

Equations (4.9) and (4.10) are referred to as the Reynolds-averaged Navier-Stokes (RANS) equations. These velocity fluctuations bring about additional viscous stresses, $\overline{\rho} \, \overline{\vec{u}'\vec{u}'}$ (called the Reynolds stresses), that often dominate the viscous stresses associated with molecular diffusion in the mean field. Solving Eqs. (4.9) and (4.10) for the local mean fields requires modeling of the Reynolds stresses, the closure problem. We would need a separate equation for this higher order moment of the velocity, but that equation itself would include an even higher moment of velocity. This chain would continue on until we truncate at a higher moment and "close" the problem.

4.2.3 Turbulence Closure Models

Turbulence closure models aim at solving Eq. (4.10) by solving transport equations that govern key turbulence parameters, instead of direct handling of the Reynolds stress term. A common method that relates the Reynolds stresses in Eq. (4.10) to the mean field is the Boussinesq hypothesis

$$-\rho \overline{u_i' u_j'} = \mu_t \left(\frac{\partial \overline{u_i}}{\partial x_j} + \frac{\partial \overline{u_j}}{\partial x_i} \right) - \frac{2}{3} \left(\rho k + \mu_t \frac{\partial \overline{u_k}}{\partial x_k} \right) \delta_{ij} \tag{4.11}$$

where μ_t is the turbulent (or eddy) viscosity and k is the turbulence kinetic energy defined in Eq. (4.2). Many turbulence closure models are based on the Boussinesq hypothesis. The hypothesis, which assumes that the turbulent viscosity is isotropic, is a good approximation in many engineering flows.

The $k - \varepsilon$ Turbulence Closure Model

In the $k-\varepsilon$ turbulence closure model, the turbulence kinetic energy, k, defined in Eq. (4.2), and its rate of dissipation, ε, are evolved by solving the associated transport equations

$$\frac{d(\rho k)}{dt} = \nabla \cdot \left[(\mu + \frac{\mu_t}{\sigma_k}) \nabla k \right] + G_k + G_b - \rho \varepsilon - Y_M + S_k \tag{4.12}$$

$$\frac{d(\rho \varepsilon)}{dt} = \nabla \cdot \left[(\mu + \frac{\mu_t}{\sigma_\varepsilon}) \nabla \varepsilon \right] + C_{1\varepsilon} \frac{\varepsilon}{k} (G_k + C_{3\varepsilon} G_b) - C_{2\varepsilon} \rho \frac{\varepsilon^2}{k} + S_\varepsilon \tag{4.13}$$

where G_k and G_b are respectively the rate of generation of turbulent kinetic energy per unit volume due to the mean velocity gradients and buoyancy. Default values for the constants $C_{1\varepsilon}, C_{2\varepsilon}, C_{3\varepsilon}$ and for the k and ε Prandtl numbers are $C_{1\varepsilon} = 1.44, C_{2\varepsilon} = 1.92, C_{3\varepsilon} = 0, \sigma_k = 1.0,$ and $\sigma_\varepsilon = 1.3$. In compressible flows, the fluctuations in the dilation contribute to the overall dissipation ratio. This contribution is captured in Y_M. The parameters S_k and S_ε are source terms. The turbulent viscosity is expressed in term of k and ε as

$$\mu_t = \rho C_\mu \frac{k^2}{\varepsilon} \tag{4.14}$$

where $C_\mu = 0.09$.

4.2.4 Filtering

Modeling turbulence using Large Eddy Simulation involves filtering the fields (density, velocity, temperature) in space according to

$$\bar{\phi}(\vec{x}, t) = \int_\Omega \phi(\vec{x}', t)\, G(\vec{x}, \vec{x}')\, d\vec{x}' \tag{4.15}$$

where $G(\vec{x}, \vec{x}')$ is the filtering function, whose role is to filter eddies of size smaller than a specified value, usually taken to be the grid size. The function $G(\vec{x}, \vec{x}')$ transforms ϕ from the physical domain $(\vec{x}', t)$ into a spectral or frequency space. Note that in the frequency domain, G is a low-pass filter. The filtered NS equations for incompressible flows, with constant μ, are

$$\frac{\partial \bar{\rho}}{\partial t} + \nabla \cdot (\bar{\rho}\bar{\vec{u}}) = 0 \tag{4.16}$$

$$\frac{\partial \bar{\rho}\bar{\vec{u}}}{\partial t} + \nabla \cdot (\bar{\rho}\bar{\vec{u}}\,\bar{\vec{u}}) = -\nabla\bar{p} + \mu\nabla^2\bar{\vec{u}} + \frac{\mu}{3}\nabla\left(\nabla \cdot \bar{\vec{u}}\right) + \nabla \cdot \tau + \bar{\rho}\vec{g} \tag{4.17}$$

where the sub-grid scale stress, τ_{ij}, is

$$\tau_{ij} = \rho\left(\overline{u_i u_j} - \bar{u}_i\,\bar{u}_j\right) \tag{4.18}$$

4.3 Problem Statement

We will now address the modeling in Fluent of turbulent fluid flow in a three-dimensional cylindrical pipe as shown in Figure 4.5. The working fluid (water) and the pipe geometry are identical to those used to study laminar pipe flow in Chapter 3. Table 4.1 provides

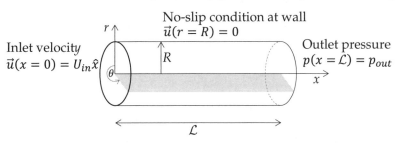

FIGURE 4.5 The domain and boundary conditions.

TABLE 4.1 Model Dimensions, Properties, and Flow Parameters

Parameter	Symbol	Value
Pipe dimensions		
Pipe length	$\mathcal{L}$	10 m
Pipe diameter	$D = 2R$	0.2 m
Wall roughness	ϵ	0
Fluid properties		
Density	ρ	998.2 kg/m^3
Viscosity	μ	1.003 mPa.s
Flow properties		
Average velocity	V	0.05 m/s
Outlet (gauge) pressure	p_{out}	0 Pa
Reynolds number	Re	9952

the pipe dimensions, the fluid properties and the flow parameters for the problem. In contrast to the laminar flow studied in Chapter 3, the velocity at the inlet is chosen such that Reynolds number Re is well above 2000 so that the flow is in the turbulent regime. The three-dimensional NS equations will be solved for the turbulent water flow in the three-dimensional pipe model shown in Figure 4.5. The velocity and pressure distributions inside the pipe will be compared to empirical data reported by experimental studies.

4.4 Governing Equations and Boundary Conditions

For a steady, incompressible fluid in the absence of body forces, the continuity and momentum equations reduce to,

$$\nabla \cdot \vec{u} = 0 \tag{4.19}$$

$$\rho \vec{u} \cdot \nabla \vec{u} = -\nabla p + \mu \nabla^2 \vec{u} \tag{4.20}$$

The boundary conditions, shown in Figure 4.5, are the no-slip condition at the wall, $\vec{u}(r = R) = \vec{0}$, a given inlet velocity profile $\vec{u}(x = 0) = U_{in}\hat{x}$, and a specified outlet pressure, $p(x = \mathcal{L}) = p_{out}$. In Section 4.5, numerical simulations are carried out using Fluent to study the turbulent flow in a pipe of length $\mathcal{L} = 10$ m and diameter $D = 2R = 0.2$ m. The pipe wall is taken to be smooth so that the wall roughness $\epsilon = 0$. The inlet axial speed is $U_{in} = 0.05$ m/s and the outlet gauge pressure is $p_{out} = 0$ Pa. Since we are studying a fully turbulent flow, we proceed to express the governing equation in terms of mean variables,

$$\nabla \cdot \bar{\vec{u}} = 0 \tag{4.21}$$

$$\rho \nabla \cdot (\bar{\vec{u}}\bar{\vec{u}}) = -\nabla \bar{p} + \mu \nabla^2 \bar{\vec{u}} - \rho \nabla \cdot (\overline{\vec{u}'\vec{u}'}) \tag{4.22}$$

The Reynolds stresses are expressed in terms of the mean velocity gradients according to Eq. (4.11), where μ_t is given by Eq. (4.14) and k and ε are evolved by solving Eqs. (4.12) and (4.13).

4.4.1 Flow in the Fully Developed Region

This section amounts to giving various expressions we will use to verify the Fluent simulation.

Turbulent flow along a wall consists of four regions, or layers, characterized by the distance from the wall, y. These layers are the viscous sublayer, the buffer layer, the overlap layer, and the turbulent layer. The viscous sublayer is the region closest to the wall and is highly impacted by the flow retardation due to friction at the wall, resulting in small values of local Reynolds number, Re_δ. In this viscous sublayer, of thickness δ_{sub}, viscous effects dominate inertial effects. Since δ_{sub} is very small, one can approximate $\frac{\partial u}{\partial y} \simeq \frac{u}{y}$, so that the shear at the wall is expressed as $\tau_w = \mu\frac{u}{y}$ or $\frac{\tau_w}{\rho} = \frac{\nu u}{y}$. The term $\sqrt{\tau_w/\rho}$, which has the unit m/s, is referred to as the friction velocity, u^*,

$$u^* = \sqrt{\frac{\tau_w}{\rho}} \tag{4.23}$$

With Eq. (4.23), we arrive at the law of the wall characterizing the flow in the viscous sublayer:

$$\text{Viscous sublayer } u^+ = y^+, \text{ where } u^+ = \frac{u}{u^*} \text{ and } y^+ = \frac{y}{\nu/u^*} \tag{4.24}$$

where y^+ is the normalized distance from the solid wall. In the viscous sublayer, experimental measurements show that for smooth surfaces, $0 \leq y^+ \leq 5$, so that $0 \leq \delta_{sub} \leq \frac{25\nu}{u_\delta}$ or $\mathrm{Re}_\delta = \frac{u_\delta \delta_{sub}}{\nu} \leq 25$, where u_δ is the velocity at the edge of the viscous sublayer. For a given fluid, a larger average fluid velocity results in a larger velocity at the edge of the viscous sublayer, and the viscous sublayer thickness gets smaller. In the overlap (or inertial) layer, experimental results reveal the logarithmic dependence

$$\text{Overlap layer } u^+ = \frac{1}{C} \ln y^+ + B \tag{4.25}$$

where the approximate values $C \simeq 0.41$ and $B \simeq 5.0$ were found to be valid over the full range of turbulent flows in pipes with smooth walls. Noting that the friction head loss, h_f, is related to the friction factor by Eq. (3.19), and that the pressure drop is $\Delta p = \rho g h_f$, the conservation of linear momentum applied to the control volume occupied by the fluid in the pipe, shows that the pressure force $\Delta p \pi R^2$ balances the opposing shear force at the wall, $\bar{\tau}_w 2\pi R\mathcal{L}$, so that the average shear at the wall is

$$\bar{\tau}_w = \frac{\Delta p \pi R^2}{2\pi R\mathcal{L}} = \frac{\rho f V^2}{8} \tag{4.26}$$

It follows from Eq. (4.23) that u^* is related to the Darcy friction factor as

$$u^* = \sqrt{\frac{f V^2}{8}} \tag{4.27}$$

Setting $y = R - r$ in Eq. (4.25),

$$\frac{u(r)}{u^*} = \frac{1}{C} \ln \frac{(R - r)u^*}{\nu} + B \tag{4.28}$$

The average velocity is obtained by integrating $V = \int_0^R u(r)2\pi r dr$.

$$\frac{V}{u^*} \simeq 2.44 \ln \frac{Ru^*}{\nu} + 1.34 \tag{4.29}$$

Upon replacing u^* with its expression from Eq. (4.27), we arrive at a correlation between Reynolds number and the Darcy friction factor for fully developed turbulent flow in a pipe with a smooth wall

$$\frac{1}{\sqrt{f}} \simeq 2\log(\text{Re}\sqrt{f}) - 1.02 \tag{4.30}$$

The pressure drop in a pipe flow may be expressed in terms of the Darcy friction factor f according to Eq. (3.18). For turbulent flow in a pipe with a rough wall, the friction factor is a function of the fluid properties (density ρ and viscosity μ), pipe diameter D and wall roughness ϵ, and flow velocity $V = U_{in}$. In dimensionless form, this dependence assumes the form $f = \mathcal{F}\left(\text{Re}, \frac{\epsilon}{D}\right)$. An accurate expression of this dependence was proposed by Colebrook for fluids flows in pipes of different wall roughness values [31].

$$\frac{1}{\sqrt{f}} = -2\log\left(\frac{\epsilon/D}{3.7} + \frac{2.51}{\text{Re}\sqrt{f}}\right) \tag{4.31}$$

For a smooth pipe wall, which is the case considered for the numerical study in Section 4.5.3, $\epsilon = 0$ and the friction factor depends on the Reynolds number only.

The maximum velocity, which occurs at the centerline, can be related to the average velocity and the friction factor by setting $r = 0$ in Eq. (4.28), using Eq. (4.29) to eliminate the log term, and then replacing u^* with its expression in Eq. (4.27) to yield,

$$u_{max} \simeq V(1 + 1.3\sqrt{f}) \tag{4.32}$$

An expression for the velocity profile as a function of the friction factor, Reynolds number, the average velocity can be obtained by replacing u^* in Eq. (4.28) with its expression in Eq. (4.27) to yield

$$\bar{u}(r) = V\frac{\sqrt{2f}}{4}\left(\frac{1}{C}\ln\frac{\left(1-\frac{r}{R}\right)\text{Re}\sqrt{2f}}{8} + B\right) \tag{4.33}$$

The velocity profile in the fully developed region is commonly approximated by the power law profile

$$\bar{u}(r) = u_{max}\left(1 - \frac{r}{R}\right)^{1/n} \tag{4.34}$$

Using the expression in Eq. (4.32) for the maximum velocity, and noting that $\int_0^R u(r)2\pi r dr = V\pi R^2$, we arrive at the following expression for the cross-sectional velocity profile in the fully developed region

$$\bar{u}(r) = V(1 + 1.3\sqrt{f})\left(1 - \frac{r}{R}\right)^{\frac{3}{2}\left(-1+\sqrt{1+\frac{52}{45}\sqrt{f}}\right)} \tag{4.35}$$

A popular power law profile is the one-seventh power law profile

$$\bar{u}(r) = V\frac{60}{49}\left(1 - \frac{r}{R}\right)^{1/7} \tag{4.36}$$

which corresponds for the profile in Eq. (4.35) for $f = 0.02982$. When compared to the parabolic profile of the fully developed laminar pipe flow, the turbulent velocity profile of Eq. (4.36) is more uniform away from the wall and experiences a sharper drop across the boundary layer to reach 0 at the wall ($r = R$).

Since the wall shear is a function of the radial component of the velocity gradient at the wall, it follows that the shear at the wall is larger when the flow is turbulent.

Due to intense mixing of the eddies of different sizes, the boundary layer along the length of the pipe in a turbulent pipe flow grows much faster with the distance from the inlet, when compared to the laminar pipe flow boundary layer. This results in a shorter entry length, which depends on the pipe diameter and Reynolds number according to the relation reported by Bhatti and Shah (1987) and Zhi-qing (1982) [35]:

$$\frac{L_e}{D} \simeq 1.359 \, \text{Re}^{1/4} \tag{4.37}$$

4.5 Modeling Using Fluent

Figure 4.5 is a schematic of the turbulent fluid flow problem to be modeled and solved in Fluent. Similar to Chapter 3, we will only model the fluid inside the pipe. The solid wall effect is modeled using the no-slip boundary condition at the wall.

We will create the computational domain for the calculation, then we will mesh it. We will then set up the model by assigning material properties, applying boundary conditions, and specifying initial conditions. Finally, we will choose the solver and run the calculation.

This section will provide step-by-step instructions on how to solve the three-dimensional turbulent pipe flow problem using Fluent. Note that this tutorial is written with the assumption that the reader has completed the tutorial in Chapter 3 and is comfortable in using all the tools introduced in Chapter 3. The instructions covered in the previous chapter will not be repeated in this chapter.

Launch ANSYS Workbench. Double-click **Fluid Flow (Fluent)** to create a new **Fluid Flow (Fluent)** analysis into the **Project Schematic** and rename it `Chapter4`. Refer to Section 3.4.1 if directions are needed.

Save the workbench project in your working directory and name it `Turbulent Pipe Flow`.

4.5.1 Geometry

In this section, the geometry of the model will be created using ANSYS **DesignModeler**. Figure 4.6 is a flowchart for the process of creating the geometry using **DesignModeler**.

Instructions to build the geometry using SpaceClaim are included in Appendix D.

To set **DesignModeler Geometry...** as the default **Geometry** application where double-clicking **Geometry** will launch it, click **Tools** and select **Options...** in Workbench. Click **Geometry Import** and select **DesignModeler** from the drop-down list of **Preferred Geometry Editor**. Click **OK** to close the **Options** dialog box. See Figure 4.7.

Double-click **Geometry** to launch DesignModeler.

1. **Set the Units**
   Click **Units** and select **Meter**.

2. **Create and Dimension a Rectangle**
   Click **XYPlane** under the **Tree Outline**.

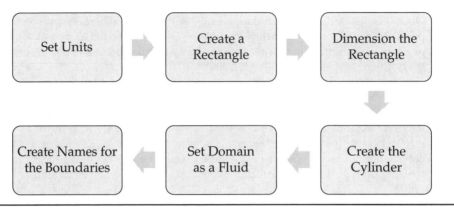

FIGURE 4.6   Flowchart of DesignModeler.

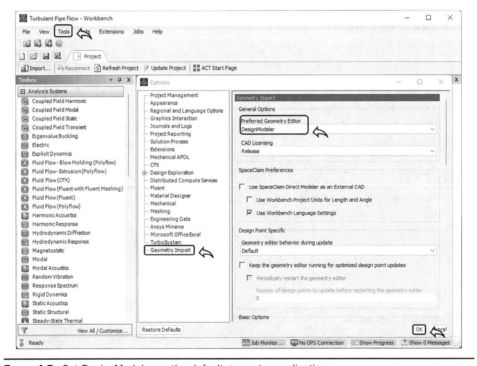

FIGURE 4.7   Set DesignModeler as the default geometry application.

Click **Look at Face/Plane/Sketch** to display the XY plane on the screen. Click **Sketching** to switch to the sketching mode.

Click **Draw** → Click **Rectangle** by two points. Click the origin to select it as the first point. Make sure the pointer of the mouse displays the letter **P** before selecting the origin. Drag the cursor and click anywhere in the first quadrant to place the opposite corner of the rectangle.

Click **Dimensions** → Click **Horizontal**. The **Horizontal** option works by dimensioning a horizontal distance between two points or two edges. Select the left vertical edge of the rectangle, select the right vertical edge of the rectangle,

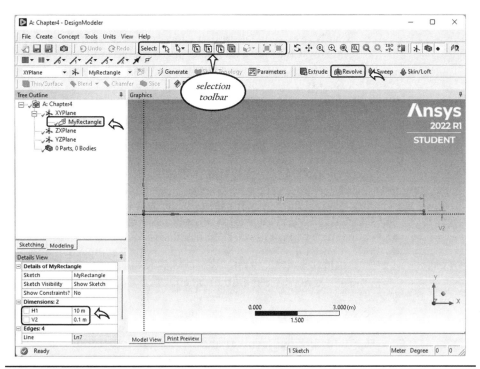

FIGURE 4.8   Dimensions and Revolve feature.

then drag the cursor and click anywhere to place the horizontal dimension in the **Graphics** window.

Click **Dimensions** → Click **Vertical**. The **Vertical** option works by dimensioning the vertical distance between two points or two edges. Select the top horizontal edge of the rectangle. Select the bottom horizontal edge of the rectangle. Drag the cursor and click anywhere to place the vertical dimension in the **Graphics** window.

The **General** option for **Dimensions**, covered in Chapter 3, can also be used and will produce the same result as **Horizontal** and **Vertical** options.

Adjust the dimensions of the rectangle in **Details View** by changing the value of **H1** to 10 and the value of **V2** to 0.1. See Figure 4.8.

Right-click anywhere inside the **Graphics** window and select **Zoom to Fit**.

Click **Dimensions** → click **Move** to move the dimensions closer to the rectangle. Click the dimensions one at a time and move to the desired location.

Right-click and select **Zoom to Fit** one more time to view the rectangle created.

Click **Modeling** button located next to **Sketching** and expand **XYPlane** in **Tree Outline** view. **Sketch1** is the rectangle created. Right-click **Sketch1** and select **Rename**. Type MyRectangle as the new name of the sketch. Press **Enter**. The screen should look like Figure 4.8.

3. **Revolve the Rectangle**
We will create the cylinder by revolving the rectangle around its bottom horizontal edge.

Click **Revolve** shown in Figure 4.8.

The **Revolve** is a 3D feature that can also be accessed through the **Create** menu by clicking **Create** and then selecting **Revolve** from its drop-down list.

Revolve1 appears in the **Tree Outline** view with its details in **Details View**. Note that **Revolve1** in the **Tree Outline** view has a lightning bolt symbol next to it indicating it is not completed. Also note the message appearing at the bottom of the screen intending to help the user complete **Revolve1**. See Figure 4.9. DesignModeler and Mesh applications will display a message at the bottom left of the screen to provide guidance to the user. We need to select the **Geometry** to be revolved and the **Axis** to revolve around in **Details View**.

Select **MyRectangle** in **Tree Outline** view and click **Apply** next to **Geometry** in **Details View** to select the rectangle as the geometry to be revolved.

Click the highlighted **Not selected** cell next to **Axis** in **Details View**. The **Not selected** cell will change into two options: **Apply** and **Cancel**. Make sure the **Selection Filter: Edges** button, shown at the top of Figure 4.9, is activated. Select the bottom horizontal edge of the rectangle in the **Graphics** window. Zooming out, by scrolling the wheel of the mouse, may be needed if the red arrow, shown in Figure 4.9, is not visible on the screen. Click **Apply** next to **Axis** in **Details View**.

Click **Generate** to create the cylinder.

It is possible to make changes to the cylinder after generating it. This can be done by right-clicking **Revolve1** and selecting **Edit Selections**.

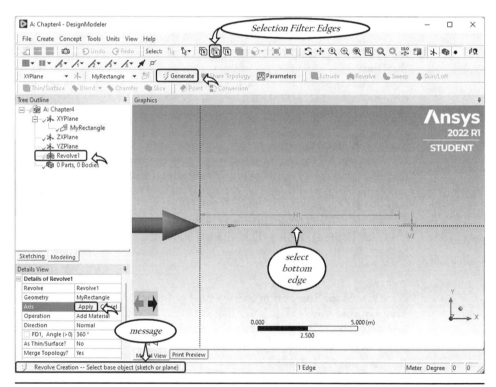

FIGURE 4.9 Details View of Revolve1.

Expand **1 Part, 1 Body** in **Tree Outline** view to see the cylinder created. Its default name is **Solid**. Right-click **Solid** and select **Rename** to rename it `my-domain`.

In **Details View** of **my-domain**, click **Fluid/Solid** to activate the drop-down list and change the domain from **Solid** to **Fluid**.

Note that most Fluent cells require clicking the cell in order to activate the drop-down list and allow the user to make changes. The drop-down list inside the cell is not visible to the user until it is activated.

4. **Display and Views**

Click the **Rotate** icon in the rotation modes toolbar to activate it. See Figure 4.10. The mouse pointer changes into the rotate symbol. Press and hold the left mouse button and drag the mouse to rotate the cylinder to the desired view. Dragging side to side will rotate the view about the vertical axis while dragging up and down will rotate the view about the horizontal axis. Deactivate the **Rotate** icon by clicking it again.

Hover the mouse over the rotation modes icons shown in Figure 4.10 and become familiar with the **Pan**, **Zoom**, **Box Zoom**, and **Zoom to Fit** icons as they are very useful while creating the geometry. Note that in addition to the **Zoom** icon, scrolling the wheel of the mouse allows the user to zoom in and out while viewing the geometry. Clicking an icon will activate it. Clicking it again will deactivate it.

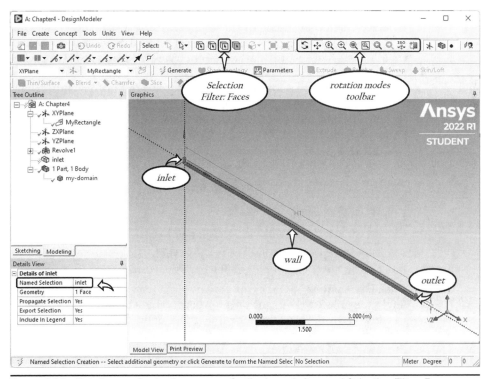

FIGURE 4.10 Rotation modes toolbar, names for the boundaries, and Selection Filter: Faces.

Click the **Pan** icon. The **Pan** icon allows the user to translate the cylinder. Press and hold the left mouse button to drag the cylinder with the mouse to a different location.

Click the cyan **ISO** ball in the triad to display an isometric view of the model when finished exploring the different icons in the rotation modes toolbar.

5. **Create Named Selections**

We will perform naming the boundaries of the model in DesignModeler, as opposed to performing it in the Mesh application in the previous chapter. Naming the boundaries in DesignModeler can make it easier to mesh the model and apply boundary conditions. Figure 4.10 displays the boundaries of the model along with their names.

Activate the **Selection Filter: Faces** icon shown in Figure 4.10. The user may need to click **my-domain** in **Details View** first, if the **Selection Filter: Faces** icon is greyed out.

Select the face on the left side of the cylinder. Rotating the model and zooming in will make it easy to select the face. The face will change color when selected. Right-click anywhere in the **Graphics** window and select **Named Selection** from the list. In **Details View**, click **Apply** next to **Geometry** to select the face for **Geometry**. The cell next to **Geometry** will display **1 Face** indicating one face is selected. Select **NamedSel1** next to **Named Selection** and type inlet, as shown in Figure 4.10. Press **Enter** and click **Generate**.

Select the curved surface of the cylinder. Right-click and select **Named Selection** from the list. In **Details View**, click **Apply** next to **Geometry**. Select **NamedSel2** next to **Named Selection** and rename it wall. Click **Generate**.

Select the face on the right side of the cylinder. Rotate the model and zoom in to select the face. Right-click and select **Named Selection** from the list. In **Details View**, click **Apply** next to **Geometry**. Select **NamedSel3** next to **Named Selection** and rename it outlet. Click **Generate**.

The created names for the boundaries appear in the **Tree Outline** view.

Click **File** and select **Save Project**. Close the DesignModeler application by clicking **File** and selecting **Close DesignModeler**. The **Geometry** in Workbench **Project Schematic** has now a green check symbol next to it indicating the application is up to date. We are ready to start meshing the model.

Geometry Summary

* Click **Units**. Select **Meter**.

* Click **XYPlane**. Click **Look At Face/Plane/Sketch**.

* Click **Sketching**. Click **Draw** → Click **Rectangle**. Select the origin as the first point. Drag the cursor and place the opposite corner in the first quadrant.

* Click **Dimensions** → Click **Horizontal**. Select the vertical edges of the rectangle, then place the dimension anywhere on the screen. Click **Dimensions** → Click **Vertical**. Select the horizontal edges of the rectangle, then place the dimension anywhere on the screen. Set the value of **H1** to 10 and the value of **V2** to 0.1. Press **Enter**.

* Click **Modeling**. Expand **XYPlane** and rename **Sketch1** MyRectangle.

∗ Click **Revolve**. Select **MyRectangle** for **Geometry**. Activate **Selection Filter: Edges** and select the bottom horizontal edge of the rectangle for **Axis**. Click **Generate**.

∗ Expand **1 Part, 1 Body** and rename the created body `my-domain`. Select **Fluid** next to **Fluid/Solid**.

∗ Activate **Selection Filter: Faces**. Select the face on the left side of the pipe. Right-click and select **Named Selection**. Click **Apply** next to **Geometry** and type `inlet` next to **Named Selection**. Click **Generate**.

∗ Repeat and create a **Named Selection** on the curved face of the pipe. Name it `wall`.

∗ Repeat and create a **Named Selection** on the right face on the right side of the pipe. Name it `outlet`.

∗ Click **File** and select **Save Project**.

∗ Click **File** and select **Close DesignModeler**.

4.5.2   Mesh

A flowchart of the steps needed to create the mesh is displayed in Figure 4.11.

Open the Mesh application by double-clicking **Mesh** in ANSYS Workbench **Project Schematic**. Right-clicking the **Mesh** cell and selecting **Edit...** from the list of options will also open the Mesh application.

Once **Mesh** is open, the geometry will be loaded into it and the screen looks like Figure 4.12. The **Named Selections** created in **Geometry** are also loaded. Expand **Named Selections** in **Outline** view to see the list of **Named Selections**.

1. **Default Mesh**

   To mesh the model, click **Home** tab, then click **Generate**, shown in Figure 4.12. We can also right-click **Mesh** in **Outline** view and select **Generate Mesh**.

   The progress of meshing will be displayed at the bottom left of the screen in the status bar, shown in Figure 4.12. Note that Mesh application uses one core to mesh the model so it may take long time to complete.

   Once completed, click **Mesh** in **Outline** view to display the mesh created.

   Click **Mesh** in the **Outline** view and use the **Rotate**, **Pan**, and **Zoom** icons in the graphics toolbar, shown in Figure 4.12, to examine the details of the mesh created. Activate one of the icons. Press and hold the left mouse button then drag the mouse side by side or up and down to change the view. To deactivate the icon, click **Mode**, shown in Figure 4.12, and select **Single Select** from the drop-down list.

Figure 4.11 Flowchart of mesh.

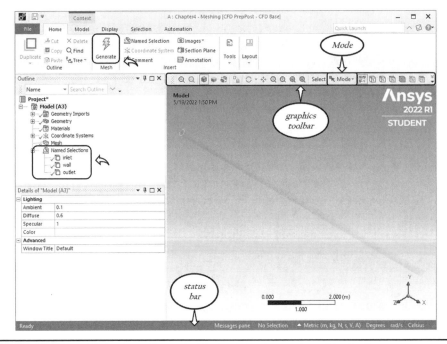

FIGURE 4.12   Named Selections, Generate Mesh, status bar, and graphics toolbar.

Click **Mesh** in **Outline** view. Expand **Statistics** located at the bottom of the **Details of "Mesh"** to display the number of **Elements** and **Nodes** in the mesh.

2. **Insert Sizing**
Right-click **Mesh** in **Outline** view and select **Insert** ⇒ **Sizing** to increase the number of cells on the round edges of the inlet and outlet.

In **Details of "Sizing"-Sizing**, the highlighted regions require the user's input. Click **No Selection** next to **Geometry** to change it to **Apply** and **Cancel**. Click **Edge** selection filter and while holding the **Ctrl** button, select the circular edges of the inlet and outlet shown in Figure 4.13. Click **Apply** next to **Geometry**. Notice that once we click **Apply**, **Sizing** in the **Outline** view changes to **Edge Sizing**. Click **Type** and from the drop-down list next to it, change the type from **Element Size** to **Number of Divisions**. Note that clicking **Element Size** directly will also allow the user to change its type. Change the **Number of Divisions** by selecting **1** in the cell next to it and typing 38. Select **Default (1.2)** in the cell next to **Growth Rate** and type 1. Press **Enter**. Right-click **Mesh** in **Outline** view and select **Generate Mesh**.

3. **Insert Inflation**
In order to capture the boundary layers that will form at the interface between the fluid and the wall of the cylinder, we will refine the mesh near the wall.

Right-click **Mesh** in **Outline** view and select **Insert** ⇒ **Inflation**. **Inflation** will refine the mesh of the domain near the wall.

Click **Face** selection filter. While holding the **Ctrl** button, select the inlet and outlet faces of the cylinder. Click **Apply** next to **Geometry** in **Details of "Inflation"-Inflation**.

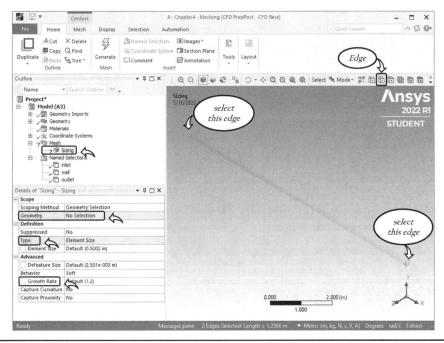

Figure 4.13   Insert Sizing, select circular edges, and Type in Details of "Sizing".

Click **No Selection** next to **Boundary** and make sure **Edge** selection filter is activated. While holding the **Ctrl** button, select the round edges of the inlet and outlet of the cylinder. Click **Apply** next to **Boundary** in **Details of "Inflation"-Inflation**.

Click **Inflation Option** and from the drop-down list next to it select **First Layer Thickness**. Click **Please Define** next to **First Layer Height** and type 0.001. Press **Enter**. Select the default value in the cell next to **Maximum Layers** and type 16. Press **Enter**. Select the default value in the cell next to **Growth Rate** and type 1.1. Press **Enter**. See Figure 4.14.

Right-click **Mesh** in **Outline** view and select **Generate Mesh**. Expand **Quality** in **Details of "Mesh"** and select **Aspect Ratio** next to **Mesh Metric**. Note that the maximum value of the aspect ratio, **Max**, is **16.59**. The cells having the highest aspect ratios are located in the region where inflation layers are added. Refining the mesh by adding inflation layers near the walls results in a one-directional grid refinement yielding more elongated cells near the walls and thus higher aspect ratios. The value of the aspect ratio can be reduced by reducing the element size of the body where inflation layers are added. However, this can increase the computational effort. Another method to decrease the aspect ratio is to increase the **First Layer Height**, the distance between the walls and the first inflation layer, which might affect the accuracy of the solution.

4. **Close Mesh Application**
   Click **File** and select **Save Project** from the drop-down list.
      Click **File** and select **Close Meshing** to close the ANSYS Mesh application and return to **Project Schematic**. Right-click **Mesh** and select **Update**. The green

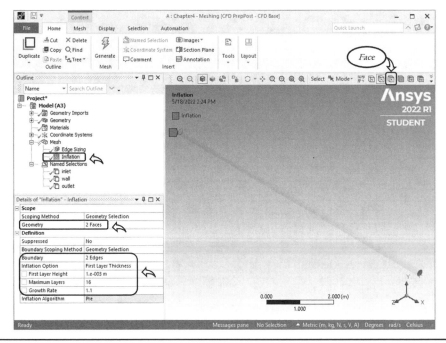

FIGURE 4.14 Details of "Inflation".

check symbol next to **Mesh** indicates the model is now ready for the ANSYS **Setup** application.

Mesh Summary

* Right-click **Mesh** and select **Generate Mesh**.
* Right-click **Mesh** and select **Insert** ⇒ **Sizing**. Activate **Edge** selection filter and control-select the inlet and outlet edges for **Geometry**. Select **Number of Divisions** next to **Type**. Set the **Number of Divisions** to 38 and the **Growth Rate** to 1.1. Right-click **Mesh** and select **Generate Mesh**.
* Right-click **Mesh** and select **Insert** ⇒ **Inflation**. Activate **Face** selection filter and control-select the inlet and outlet faces in the geometry window for **Geometry**. Activate **Edge** selection filter and control-select the circular edges of the inlet and outlet for **Boundary**. Select **First Layer Thickness** next to **Inflation Option**. Set the **First Layer Height** to 0.001, the **Maximum Layers** to 16, and the **Growth Rate** to 1.1. Right-click **Mesh** and select **Generate Mesh**.
* Click **Mesh**, expand **Quality** and select **Aspect Ratio** next to **Mesh Metric**.
* Click **File** and select **Save Project**.
* Click **File** and select **Close Meshing**.
* Right-click **Mesh** in **Project Schematic** and select **Update**.

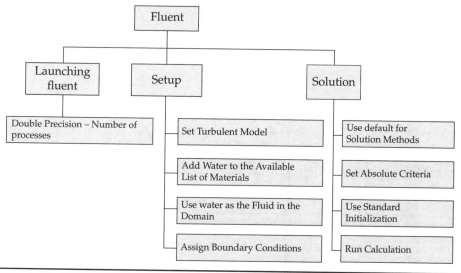

FIGURE 4.15 Flowchart of Fluent.

4.5.3 Setup

A flowchart of the steps needed to set up the model in Fluent is displayed in Figure 4.15.

In ANSYS Workbench **Project Schematic**, double-click **Setup** to launch Fluent. Note that the **Dimension** is automatically set to **3D** since Fluent detects a three-dimensional model. Enable **Double Precision** in the dialog box. Increase the number of **Solver Processes** depending on the available processors for usage. Click **Start**.

This will launch Fluent. The screen looks like Figure 4.16.

The grid in the graphics window is ON by default when displaying 3D objects. The grid can be disabled by clicking **File**, selecting **Preferences...** and disabling **Ground Plane Grid** under **Graphics** in the **Preferences** dialog box. See Figure 4.17.

Scroll up the **Console** window to review information like the number of processes used and the number of elements in the model.

Rotate the view in the graphics window using the **Rotate View**. Use the **Zoom In/Out** and the other icons in the pointer tools toolbar of Fluent, shown in Figure 4.16, to check the model. Recall that once an icon is activated, we need to press and hold the left mouse button while dragging the mouse side by side or up and down to adjust the view. An icon activated will stay activated until the user clicks another icon. Clicking an active icon will not deactivate it.

We will go in order of what needs to be modified in the **Outline View** to setup the problem.

1. **General**

     Expand **Setup** → Double-click **General**.

        In the **Task Page** view, click **Scale...** under **Mesh**.

        Make sure the dimensions are correct in the **Scale Mesh** dialog box. The **Scale Mesh** dialog box allows the conversion of the mesh to SI units. It also allows the scaling of the model by applying custom scale factors to the individual coordinates of the mesh.

        Close the **Scale Mesh** dialog box.

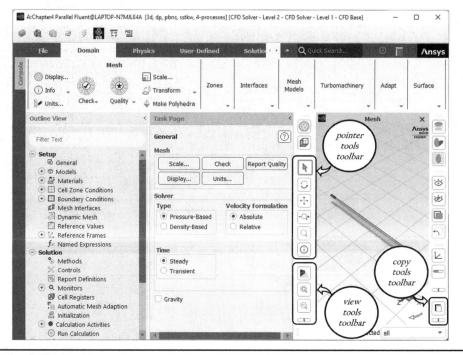

FIGURE 4.16 Graphics toolbar.

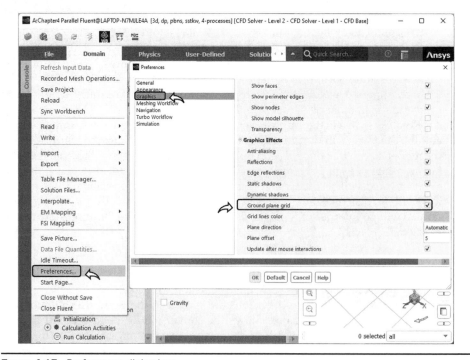

FIGURE 4.17 Preferences dialog box.

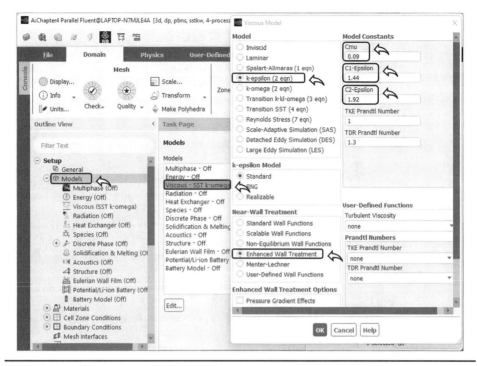

FIGURE 4.18 Turbulence model.

For the **Solver** options, keep the default options of **Pressure-Based** for **Type**, **Absolute** for **Velocity Formulation**, and **Steady** for **Time**.

2. **Models**
Expand **Setup** → Double-click **Models**.
In the **Task Page** view, double-click **Viscous - SST k-omega**. Choose **k-epsilon (2 eqn)** from the **Model** list. We will use **Standard** under **k-epsilon Model**.
Select **Enhanced Wall Treatment** under **Near-Wall Treatment** as shown in Figure 4.18. Click **OK**.
A turbulent boundary layer has a very thin viscous sub-layer next to the wall and requires a very fine mesh resolution. The use of **Enhanced Wall Treatment** option in Fluent will capture the viscous sub-layer inside the turbulent boundary layer while using large mesh near the wall.
The default values for **Model Constants** are used.

3. **Materials**
Expand **Setup** → Double-click **Materials**.
In the **Task Page** view, click **Create/Edit...** button, located at the bottom of the page, to open the **Create/Edit Materials** dialog box. Click **Fluent Database...**. Scroll to the bottom of the **Fluent Fluid Materials** list and select **water-liquid (h2o<l>)**. Click the **Copy** button. Close the **Fluent Database Materials** dialog box. Close the **Create/Edit Materials** dialog box.
The added material of **water-liquid** from the **Fluent Database...** appears in the list of **Materials** in the **Task Page**.

4. **Cell Zone Conditions**

Expand **Setup** → Double-click **Cell Zone Conditions**.

The list of **Cell Zone Conditions** in the **Outline View** contains only **Fluid** because the solid pipe is not being modeled.

Expand **Fluid** under **Cell Zone Conditions**. The domain for the model named **my-domain** in ANSYS DesignModeler is listed under **Fluid**.

In the **Task Page** view, double-click **my-domain** under the list of **Cell Zone Conditions**. The default **Material Name** of **my-domain** is **air**. Change it to **water-liquid**. Click **Apply** then click **Close** to close the **Fluid** dialog box.

5. **Boundary Conditions**

Expand **Setup** → Double-click **Boundary Conditions**.

We will step through all the boundaries listed in the **Task Page** view and make changes where necessary.

Select **inlet** and make sure its **Type** is **velocity-inlet**. Click **Edit...** and choose **Magnitude, Normal to Boundary** as the **Velocity Specification Method**. Set the **Velocity Magnitude [m/s]** value to 0.05 and click **Apply**. Close the **Velocity Inlet** dialog box.

Note that **interior-my-domain** is the interior of the domain and we do not have the option to make changes to it.

Select **outlet** to make sure its **Type** is **pressure-outlet**. Click **Edit...** and confirm the value of **Gauge Pressure [Pa]** at the outlet is 0. Use the default values for backflow conditions in the pressure outlet dialog box. These values are used by ANSYS Fluent in the case where fluid is flowing through the outlet which might occur during the solution procedure. Close the **Pressure Outlet** dialog box.

Select **wall** to make sure its **Type** is **wall**. Click **Edit...** and confirm the **Wall Motion** is set to **Stationary Wall** and the **Shear Condition** is set to **No Slip**. Close the **Wall** dialog box.

6. **Methods**

Expand **Solution** → Double-click **Methods**. Keep the default options for the **Solution Methods**:

Pressure-Velocity Coupling → **Scheme** → **Coupled**.

Pressure-Velocity Coupling → **Flux Type** → **Rhie-Chow: distance based**.

Spatial Discretization → **Gradient** → **Least Squares Cell Based**.

Spatial Discretization → **Pressure** → **Second Order**.

Spatial Discretization → **Momentum** → **Second Order Upwind**.

Spatial Discretization → **Turbulent Kinetic Energy** → **First Order Upwind**.

Spatial Discretization → **Turbulent Dissipation Rate** → **First Order Upwind**.

Scroll down if needed and make sure **Pseudo Time Method** is enabled.

7. **Residuals**

Expand **Solution** → Expand **Monitors** → Double-click **Residual**.

Note that in addition to the residuals for the continuity, x-velocity, y-velocity, and z-velocity, we have two additional residuals corresponding to the two additional equations for k and epsilon.

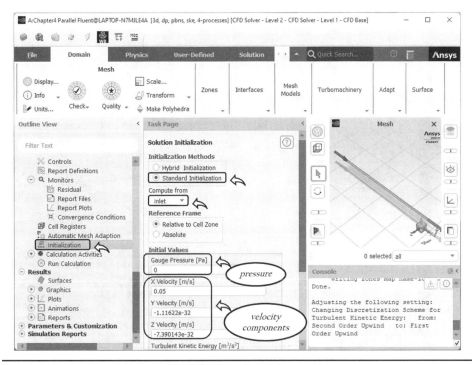

FIGURE 4.19 Standard Initialization and Compute from inlet.

Change the **Absolute Criteria** to 1e-06 for all six equations to be solved . Click **OK** to close the **Residual Monitors** dialog box.

8. **Initialization**

Expand **Solution** → Double-click **Initialization**.

Instead of using **Hybrid Initialization** as was done in the previous chapter, we will use an initial guess for all cells that is equal to the values at the inlet.

Select **Standard Initialization** for the **Initialization Methods**, as shown in Figure 4.19. Select **inlet** for **Compute from**.

The initial value for the **Gauge Pressure [Pa]** is 0. The initial values for velocities are **0.05** m/s in the x direction and negligible in the y and z directions. The initial values for the **Turbulent Kinetic Energy [m²/s²]** and **Turbulent Dissipation Rate [m²/s³]** are calculated based on the prescribed turbulence intensity and the diameter of the pipe.

Click **Initialize**. When the initialization is completed, the pointer of the mouse changes from **Busy** to **Normal Select**.

9. **Run Calculation**

Expand **Solution** → Double-click **Run Calculation**. In the **Task Page** view, set the **Number of Iterations** under **Parameters** to 500. Click the **Calculate** button under **Solution Advancement** located at the bottom of the **Task Page** view.

A **Calculation complete** message will appear when the calculation is completed. Click **OK**. The message appears when the convergence criteria are met or when the **Number of Iterations** requested by the user are completed, whichever comes first.

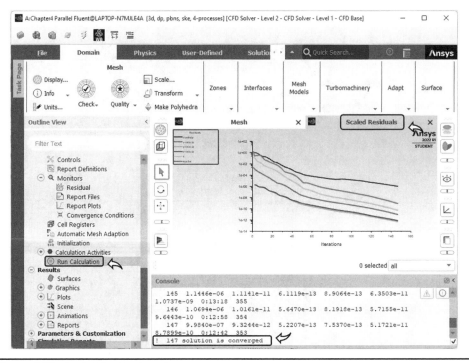

FIGURE 4.20   Scaled Residuals and convergence message.

The **Console** window will display the number of iterations at which the calculation was completed.

Click **Scaled Residuals** in the graphics window to display the plot of the residuals as a function of iterations. The screen should look like Figure 4.20.

In order to continue running the calculation until all requested number of iterations are completed, we need to disable the convergence criteria.

Expand **Solution** → Expand **Monitors** → Double-click **Residual**. Disable **Check Convergence** for all six equations. Click **OK** to close the **Residual Monitors** dialog box.

Expand **Solution** → Double-click **Run Calculation**.

The number of iterations is **500** since this is the number used earlier. Click **Calculate** to run the simulation for an additional 500 iterations. A message will appear to warn the user that settings have changed with options for the calculation. Keep the default option which is **Use settings changes for current calculation only** and click **OK**.

A **Calculation complete** message will appear when the additional 500 iterations are completed. Click **OK**.

10. **Flux Reports**

Expand **Results** → Double-click **Reports** to check the total mass flux through the inlet and outlet of the model.

In the **Task Page** view, double-click **Fluxes**.

In the **Flux Reports** dialog box, select **Mass Flow Rate** for **Options**, select **inlet** and **outlet** from the list of **Boundaries** to check the mass balance for the

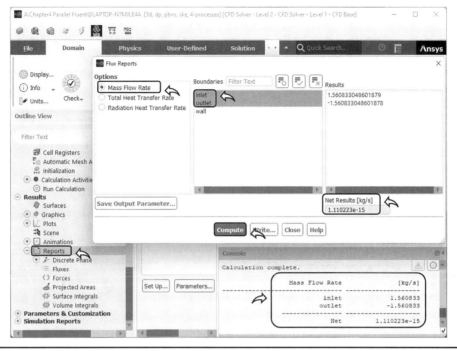

FIGURE 4.21 Flux Reports for the Mass Flow Rate.

model. Click **Compute**. The **Net Results** is displayed in the dialog box, and also it is displayed in the **Console** window as shown in Figure 4.21. The net mass imbalance is less than 0.2% of the total mass flux and is an indication of convergence.

Click **Close** to close the **Flux Reports** dialog box.

11. **Save the Project**

Save the project by clicking **File** and selecting **Save Project** from the drop-down list. Click **File** and select **Close Fluent**.

The ANSYS Fluent case and data files are automatically saved upon exiting ANSYS Fluent application.

Setup Summary

* Expand **Setup** → Double-click **General**. Make sure **Pressure-Based** is selected for **Type**, **Absolute** is selected for **Velocity Formulation** and **Steady** is selected for **Time**.

* Expand **Setup** → Double-click **Models**. Double-click **Viscous - SST k-omega**. Select **k-epsilon (2 eqn)** for **Model** and **Enhanced Wall Treatment** for **Near-Wall Treatment**. Click **OK**.

* Expand **Setup** → Double-click **Materials**. Click **Create/Edit...**. Click **Fluent Database...**, select **water-liquid (h2o<l>)** and click **Copy**. Close the dialog boxes.

∗ Expand **Setup** → Double-click **Cell Zone Conditions**. Double-click **my-domain** and select **water-liquid** next to **Material Name**. Click **Apply**, then click **Close**.

∗ Expand **Setup** →Double-click **Boundary Conditions**. Double-click **inlet** and set the **Velocity Magnitude [m/s]** to 0.05. Click **Apply**. Click **Close**.

∗ Expand **Solution**→Double-click **Methods**. Use default options for **Solution Methods**.

∗ Expand **Solution** → Expand **Monitors** → Double-click **Residual**. Set the **Absolute Criteria** for all equations to 1e-06. Click **OK**.

∗ Expand **Solution** → Double-click **Initialization**. Select **Standard Initialization** and select **inlet** for **Compute from**. Click **Initialize**.

∗ Expand **Solution** → Double-click **Run Calculation**. Set the **Number of Iterations** to 500. Click **Calculate**.

∗ Click **OK** in the **Calculation complete** dialog box. Expand **Solution** → Expand **Monitors** → Double-click **Residual**. Disable **Check Convergence** for all six equations and click **OK**. Expand **Solution** → Double-click **Run Calculation**. Click **Calculate** and then click **OK** for **Use settings changes for current calculation only**.

∗ Expand **Results** → Double-click **Reports**. Double-click **Fluxes**. Select **Mass Flow Rate** for **Options**. Select **inlet** and **outlet** for **Boundaries**. Click **Compute**. Click **Close**.

∗ Click **File** and select **Save Project**.

∗ Click **File** and select **Close Fluent**.

4.5.4 Solution

In Chapter 3, we looked at the results in **Solution** application. This is a limited but a fast way to look at the solution while the user is debugging the model. In this chapter, we will launch **Results** application from the ANSYS Workbench **Project Schematic** which is a better application to display the final results of the simulation in the form of plots and movies.

Double-click **Results** in ANSYS Workbench **Project Schematic** to launch CFD-Post. Note that **DesignModeler**, **Mesh**, and **Fluent** applications should be closed in order to open **Results** application or an error may occur.

The results of ANSYS Fluent are automatically loaded into CFD-Post application. Figure 4.22 displays the CFD-Post application.

In Figure 4.22, the **3D Viewer** area shows that the geometry is on the right side. The view of the geometry can be manipulated using the 3D viewer toolbar located above the **3D Viewer** area.

The **Outline** is displayed on the left side and is divided into top and bottom areas. The top area contains the tree view listing the objects in the model, as well as the objects created by the user. The bottom area is the details view which appears when we double-click an object in the tree view to make it available for editing.

Results in CFD-Post can be plotted on existing surfaces in the model or on new surfaces created by the user.

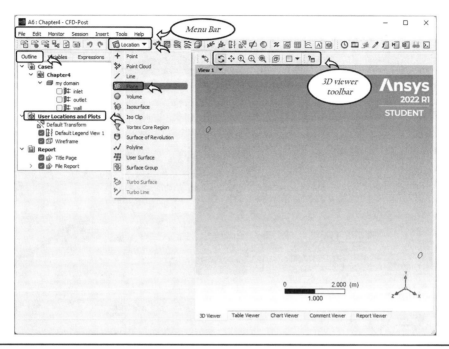

FIGURE 4.22   Viewer toolbar and create Plane.

In this section, we will create velocity and pressure contours on a plane that goes through the axis of the cylinder. We will also generate an XY plot for the velocity along the centerline of the cylinder.

1. **Velocity Contours**

   Since our model is a three-dimensional model, we need to create a two-dimensional surface within the model in order to display the velocity contours on a surface.

   We will create a plane inside the pipe that cuts through its axis. The velocity contours will be the same for any plane going through the axis of the cylinder since the model is axisymmetric. We will create a vertical plane using the following three points: (0,0,0), (10,0,0), and (0,0.1,0).

   To create the plane, click **Location** shown in Figure 4.22 and select **Plane** from the list of options. Type My Plane for **Name** and click **OK**. In **Details of My Plane** view, select **Three Points** next to **Method** (see Figure 4.23) and enter the coordinates of the three points: (0,0,0), (10,0,0), and (0,0.1,0). Click **Apply**. A plane cutting through the pipe is created as shown in Figure 4.23. **My Plane** appears under **User Locations and Plots** in the **Outline** view.

   Click the **Contour** icon shown in Figure 4.23. You can also access **Contour** from the **Insert** menu. Type My Vel next to **Name**. Click **OK**. Note that names created by the user need to be unique and different from the names of Fluent variables, or an error message appears stating that the name already exists, and the user is asked to use a different name.

   In **Details of My Vel** view, select **My Plane** next to **Locations**. Select **Velocity** next to **Variable**. Click **Apply**. The screen looks like Figure 4.24.

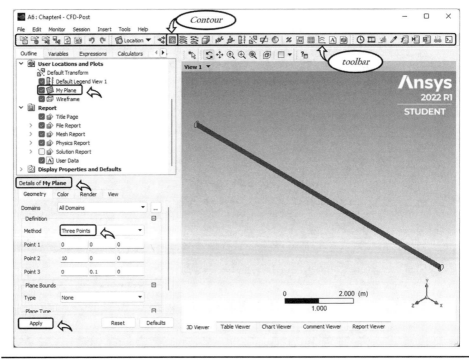

FIGURE 4.23   Create a Plane using Three Points and Contour.

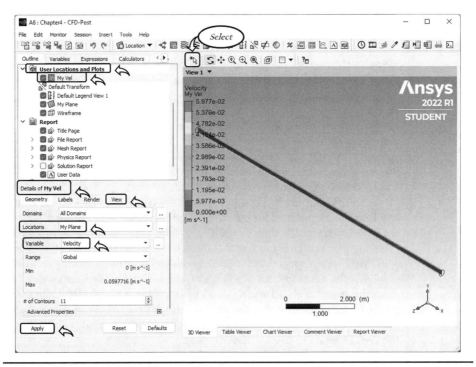

FIGURE 4.24   View in Details of Contour and Select icon.

The plot **My Vel** appears under **User Locations and Plots**.

Double-clicking any object in **User Locations and Plots** will allow making changes to it in **Details View**. We will improve the look of the velocity contours. First, we will scale it about the origin so we can have a better view of the velocity contours.

To scale our view, click **View** shown in Figure 4.24. Scroll down and enable **Apply Scale**. The three cells next to **Scale** correspond to the three different directions, X, Y, and Z, and they are set by default to **1**. Change the scale in the Y direction in the second cell to `10`. Click **Apply**. The plane displayed is now scaled in the Y direction by a factor of 10. This scaling is performed for viewing purposes only. Disable **My Vel** from the list of **User Locations and Plots**.

In addition to the **Rotate**, **Pan**, **Zoom**, etc., icons in the 3D viewer toolbar, we have the **Select** icon shown in Figure 4.24. Click the **Select** icon and hover over the wireframe in the **3D Viewer** area. Notice the cursor will change into a box. Right-click the wireframe and the available options specific to the wireframe. These options are different from the options available when you right-click away from the wireframe in the **3D Viewer** area, which are specific to the **3D Viewer** area. To deactivate the **Select** icon, click any icon in the 3D viewer toolbar.

Click any arrow of the triad in the **3D Viewer**. Note that the triad can be used only when the viewing mode is on so make sure the **Select** icon is deactivated. Click the **Z** axis in the triad located at the bottom right of the screen to redirect the view to the XY plane.

Enable **My Vel** in the **Outline** view. Click **Fit View** icon shown in Figure 4.25 to re-center and re-scale the view. Click **Zoom Box** and drag a rectangular box

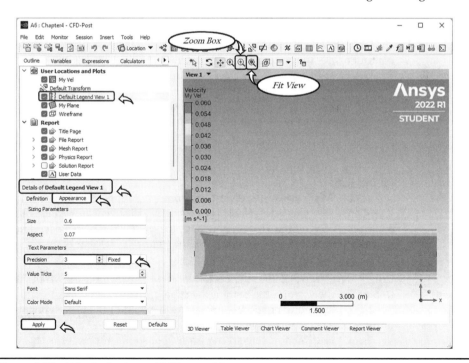

FIGURE 4.25 Velocity Contours, Fit View, and Appearance.

over your selection then release to zoom in. Clicking the cyan **ISO** ball in the triad will allow the isometric viewing of the full model. Click the **Z** axis in the triad. Click the **Pan** icon in the 3D viewer toolbar. Press and hold the left mouse button and drag the velocity contours to the bottom of the screen and away from the legend.

Next, we will improve the legend. Double-click **Default Legend View 1** under **User Locations and Plots** as shown in Figure 4.25. In **Details of Default Legend View 1**, click **Appearance** shown in Figure 4.25, and change **Precision** under the **Text Parameters** to **Fixed**. Click **Apply**. The screen looks like Figure 4.25.

2. **Pressure Contours**

   We will create the pressure contours in **My Plane** by duplicating the velocity contours. Duplicating the velocity contours will create a new object of the same type (contours) and with the exact same settings (scaling) of velocity contours.

   Right-click **My Vel** under **User Locations and Plots** in the **Outline** view and select **Duplicate** from the list of options. Type My Pres next to **Name** and click **OK**.

   My Pres contour appears in the list of objects created by the user under **User Locations and Plots**. Double-click **My Pres** to edit it.

   In **Details of My Pres** view, change the **Variable** from **Velocity** to **Pressure**. Click **Apply**. The two contours of velocity and pressure are displayed on top of each other. Disable **My Vel** under **User Locations and Plots** in the **Outline** view to view **My Pres**.

   Figure 4.26 shows the pressure contours inside the pipe. Increase the # **of Contours** to 20 to capture the variation of pressure inside the pipe. Remember to click **Apply** after changing the # **of Contours**.

   Note that the scaling and the changes to the legend appear on the pressure contours since the settings were duplicated.

   Disable **My Vel** and **My Pres** contours from the list of **User Locations and Plots**.

   Click the **Probe** icon in the toolbar shown in Figure 4.26 and click the left mouse button at any random location inside **My Plane**. The coordinates for the selected location are displayed at the bottom of the page of the 3D viewer area. Click the cell next to the coordinates to activate the list, and select **Pressure**. The value of the pressure at the selected location will be displayed as shown in Figure 4.26.

   Enable **My Pres** under **User Locations and Plots**. Right-click anywhere in the **3D Viewer** area and select **Save Picture...** to save a picture of the pressure contours. A **Save Picture** dialog box appears. Click the **Browse** icon next to **File** to choose the directory where to save the picture. Click **Save**.

3. **XY Plot for the Shear Stress**

   Click **Chart Viewer** tab at the bottom of the viewers area shown in Figure 4.27, to plot the shear along the wall of the pipe.

   First, we need to create the line along the wall of the pipe where the values of the wall shear are sampled.

   Click **Location** and select **Line** from the list of options. Type My Wall next to **Name** and click **OK**. In **Details of My Wall** view, enter (0,0.1,0) for **Point**

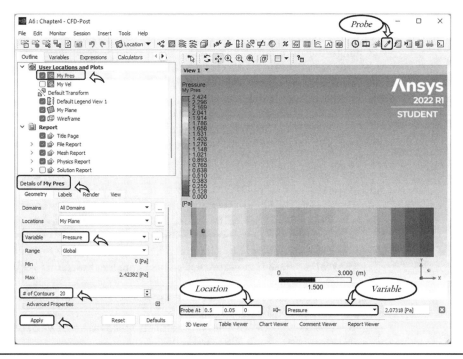

FIGURE 4.26   Pressure Contours, Probe icon, Probe Location, and Probe Variable.

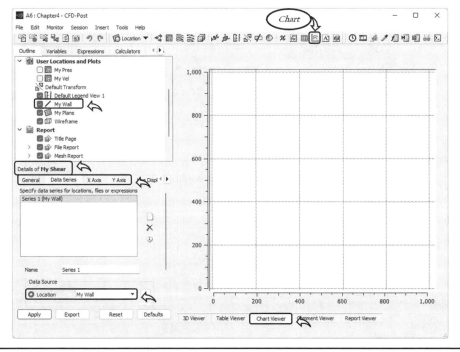

FIGURE 4.27   Insert a Chart and Details of the chart.

1 and (10,0.1,0) for **Point 2**. **Point 1** and **Point 2** are the two endpoints of the line we are creating.

Change the number of **Samples** to 100. A large number of samples will create a smooth curve.

Click **Apply**. The line, **My Wall**, is created and appears under **User Locations and Plots** in **Outline** view.

To view the created line, click **3D Viewer** tab at the bottom of the viewers, area. Disable all contours and make sure **Wireframe** is enabled so it is used as a reference for the created line. Note the created line and the wireframe are not scaled.

Click **Chart** in the toolbar, shown in Figure 4.27, or click **Insert** and select **Chart**. Type My Shear next to **Name** and click **OK**.

In **Details of My Shear** view and under **General**, the user can change the **Title** as it will be displayed on the chart. The user has the option to disable the **Display Title** option in **Details of My Shear**.

Under **Data Series** next to **General**, click the arrow next to **Location** and select **My Wall** as shown in Figure 4.27. This is the location along which we are plotting the variable.

Under **X Axis**, select **X** from the drop-down list next to **Variable**. **X** is located toward the bottom of the drop-down list.

Under **Y Axis**, select **Wall Shear** from the drop-down list of **Variable** to specify we are plotting the wall shear variable.

Click **Apply**. The plot created appears under **Report** and is displayed in the **Chart Viewer** area as shown in Figure 4.28.

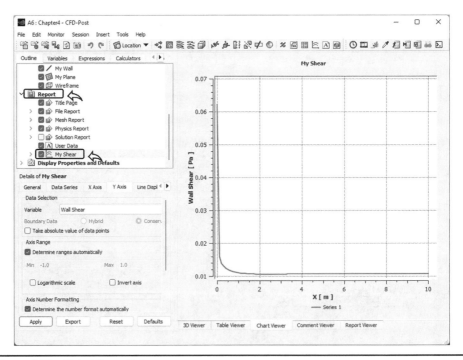

FIGURE 4.28 Plot of the wall shear.

The user can switch between viewing the contours of velocity and pressure or the charts created by clicking the **3D Viewer** tab or the **Chart Viewer** tab at the bottom of the viewers, area.

4. **XY Plot for the Velocity**

We will plot the velocity profile in the radial direction at different locations inside the pipe. We will start by creating the lines along which the values of the velocity are sampled.

Click **Location** and select **Line**. Type `My Line1` next to **Name** and click **OK**. In **Details of My Line1** view, enter (0.5,-0.1,0) for **Point1** and (0.5, 0.1,0) for **Point2**, the two endpoints of the radial (vertical) line at x = 0.5 m. Change the number of **Samples** to 100 and click **Apply**. **My Line1** is created and appears under **User Locations and Plots** in **Outline** view.

Return to the **3D Viewer** area by clicking it, at the bottom of the viewers, area, and visually check the location of the line created. Enable the velocity contours **My Vel** and disable the pressure contours **My Pres** and notice the line is in the entrance region of the flow.

We will repeat by creating two additional radial lines in the fully developed region, at x = 5 m and x = 8 m.

Click **Location** and select **Line**. Type `My Line2` next to **Name** and click **OK**. In **Details of My Line2** view, enter (5,-0.1,0) next to **Point 1** and (5,0.1,0) next to **Point 2**. Change the number of **Samples** to 100 and click **Apply**.

Click **Location** and select **Line**. Type `My Line3` next to **Name** and click **OK**. In **Details of My Line3** view, enter (8,-0.1,0) for **Point 1** and (8,0.1,0) for **Point 2**. Change the number of **Samples** to 100 and click **Apply**.

Check the location of the lines created in **3D viewer**.

Click **Chart** in the toolbar. Type `My Vel Radial` next to **Name** and click **OK**.

In **Details of My Vel Radial** view, disable **Title** under **General**. Under **Data Series**, click the drop-down list next to **Location** and select **My Line1**. Select **Series1** and type x=0.5 m next to **Name**, shown in Figure 4.29, to change the name of the **Data Series** from **Series 1** to x=0.5 m. Press **Enter**.

Under **X Axis**, select **Velocity** from the drop-down list next to **Variable**.

Under **Y Axis**, select **Y** from the drop-down list of **Variable** as we will plot the velocity along the radial direction, called Y in Fluent.

Click **Apply**. Figure 4.29 displays the velocity profile at x = 0.5 m inside the pipe.

We will add the velocity along **My Line2** and **My Line3** to this chart.

Click **Data Series** under **Details of My Vel Radial**. Click **New** shown in Figure 4.29. Change the name of the new data series from **Series 2** to x = 5 m. Click the drop-down list next to **Location** and select **My Line2**. The **X Axis** and **Y Axis** variables are copied from the previous **Data Series** and are **Velocity** and **Y** respectively. Click **Apply**. The radial velocity profile at x = 5 m is added to the chart.

Repeat the process by adding one more **Data Series** for the radial velocity at x = 8 m. Name the new data series x = 8 m and select **My Line3** for its **Location**. Click **Apply**.

Figure 4.30 displays the velocity profile along the three different locations, **x = 0.5 m, x = 5 m**, and **x = 8 m** inside the pipe.

The plot created **My Vel Radial** appears under **Report** in **Outline** view.

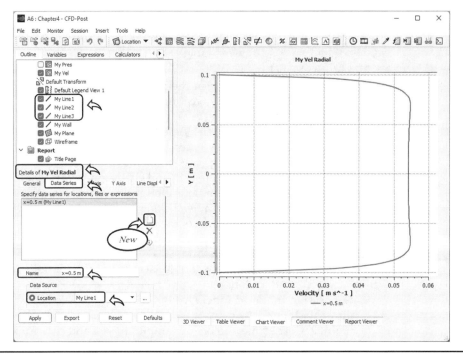

FIGURE 4.29 Velocity as a function of radial distance, at x = 0.5 m.

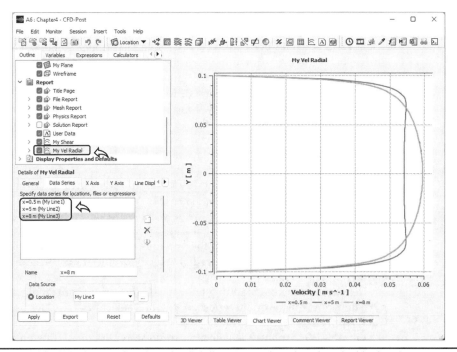

FIGURE 4.30 Velocity as a function of radial distance, at three locations inside the pipe.

5. Save the project and close Fluent
   Click **File** and select **Save Project** to save the project. Close **Results**. ANSYS
   CFD-Post files created are automatically saved when exiting the application.

4.6 Verification

A convergence study using the Richardson Extrapolation scheme is presented in this
section. The design parameters of interest are the distributions, along the distance from
the inlet, of the wall shear stress and maximum speed. In addition, comparisons with
empirical correlations are also presented in this section.

4.6.1 Grid Independent Study

We will report the uncertainty estimators of the wall shear stress and the velocity magnitude along the centerline. By following the procedure described in Section 2.9, we will
generate three systematically refined grids. Table 4.2 displays the number of elements
of the generated grids 1, 2, and 3. Because the number of elements in the Fluent student
license is capped at 512,000 elements, we used the Fluent academic license when using
grid 1. Since the model is three-dimensional, the global cell sizes, listed in Table 4.2, are
calculated according to the Eq. (2.131):

$$h_j = \left(\frac{V_{dom}}{N_j}\right)^{\frac{1}{3}} = \left(\frac{\pi R^2 \mathcal{L}}{N_j}\right)^{\frac{1}{3}}, \quad \text{for } j = 1, 2, 3 \tag{4.38}$$

where N_j is the number of cells and V_{dom} is the volume of the computational domain.
The corresponding refinement factors, $r_{j,j-1} = h_j/h_{j-1}, j > 1$, are also listed in Table 4.2.
The refinement factors were selected to be greater than 1.3, so that the solutions on the
generated grids are used to implement the grid independent study. Near solid boundaries, the grid should be sufficiently fine to resolve the large gradients in the velocity
that arise in the boundary layer, especially in turbulent flows. A key parameter is y^+,
discussed in Section 4.4.1, that characterizes the flow in the viscous sublayer. For smooth
surfaces, experimental measurements show that $y^+ \leq 5$ in the viscous sublayer. As such,
the grid refinement near the wall should be such that the maximum of y^+, referred to
as max(Wall y^+), reported by Fluent meets the same condition. For grids 1, 2, and 3,
max(Wall y^+) reported by Fluent is listed in the last column of Table 4.2. Since y^+ is part
of the solution, the user usually has to go through various iterations of successively refining the grid near the wall. For each iteration, the user inspects at the completion of the
simulation the value of max(Wall y^+) and then accordingly decides on whether or not to
further refine the grid near the wall. Note that using too large of a refinement factor will
significantly increase the number of elements, which will in turn increase the computational effort it takes to generate the solution. For instance, if the user sets $r_{21} = 3$ where

TABLE 4.2 Mesh Statistics

Mesh # j	Quality	N_j	h_j(m)	$r_{j,j-1}$	max(Wall y^+)
1	Fine	2,655,950	0.00491		2.42
2	Medium	505,780	0.00853	1.74	3.98
3	Coarse	188,784	0.01185	1.39	5.31

TABLE 4.3 Convergence Metrics Using Richardson Extrapolation for the Wall Shear Stress

i	x_I (m)	ϕ_{I_1} (mPa)	ϕ_{I_2} (mPa)	ϕ_{I_3} (mPa)	s_i	$(e_a^{21})_I$ %	$(e_a^{32})_I$ %	p_i	$(GCI^{21})_I$ ×100	EB_I ×10^6
1	0.1	16.1551	16.0578	16.4935	−1	0.60	2.71	4.16	0.08	12.4
2	0.25	13.6683	13.4593	13.5828	−1	1.53	0.92	0.77	0.19	26.6
3	0.5	12.0946	12.0097	12.0882	−1	0.70	0.65	0.12	0.09	10.8
4	0.75	11.3983	11.3724	11.4652	−1	0.23	0.82	3.46	0.03	3.3
5	1	11.0299	11.0335	11.1399	1	0.03	0.96	10.41	0.00	0.5
6	1.25	10.8184	10.8370	10.9518	1	0.17	1.06	5.89	0.02	2.4
7	1.5	10.6940	10.7214	10.8430	1	0.26	1.13	4.99	0.03	3.5
8	1.75	10.6225	10.6562	10.7839	1	0.32	1.20	4.58	0.04	4.3
9	2	10.5850	10.6238	10.7572	1	0.37	1.26	4.31	0.05	4.9
10	2.25	10.5705	10.6136	10.7519	1	0.41	1.30	4.13	0.05	5.5
11	2.5	10.5715	10.6174	10.7587	1	0.43	1.33	4.02	0.06	5.8
12	2.75	10.5815	10.6282	10.7701	1	0.44	1.34	3.99	0.06	5.9
13	3	10.5944	10.6403	10.7816	1	0.43	1.33	4.02	0.06	5.8
14	4	10.6316	10.6715	10.8102	1	0.38	1.30	4.34	0.05	5.1
15	5	10.6419	10.6799	10.8187	1	0.36	1.30	4.47	0.05	4.8
16	6	10.6435	10.6813	10.8204	1	0.36	1.30	4.49	0.05	4.8
17	7	10.6435	10.6814	10.8204	1	0.36	1.30	4.48	0.05	4.8
18	8	10.6434	10.6813	10.8205	1	0.36	1.30	4.48	0.05	4.8
19	9	10.6434	10.6813	10.8204	1	0.36	1.30	4.48	0.05	4.8
20	10	10.6384	10.6748	10.8130	1	0.34	1.29	4.58	0.04	4.6

$h_2 = 0.00853$ m, the number of elements associated with a cell size $h_1 = \frac{h_2}{3} = 0.002835$ m is around 14 million elements and will significantly increase the computational effort.

Wall Shear Stress

The values of the average wall shear stress at 20 cross sections distributed along the pipe axis, at locations $x_i, i = 1, \ldots, 20$, are reported in Table 4.3 for the fine ($j = 1$), medium ($j = 2$), and coarse ($j = 3$) grids. The wall shear stress, averaged of the angle θ, is computed as

$$\phi_{ij} = \left(\frac{1}{2\pi} \int_0^{2\pi} \tau_{xr}(x_i, R, \theta) \, d\theta \right)_j \tag{4.39}$$

where τ_{xr} is the axial shear stress acting on a surface normal to the radial direction. Table 4.3 also summarizes the results of implementing the Richardson Extrapolation scheme. In this table, we observe that in the fully developed regions, $x_i > 2.75$ m, the local order of accuracy ranges between 4.02 and 4.58 with an average value of 4.42. We note here that this value exceeds the theoretical values of the order of accuracy of the schemes employed which could lead to underestimating the error bars. In the developing flow region, $x_i < 2.75$ m, oscillatory convergence is observed for $x_i \leq 0.75$ m. According to the results listed in Table 4.3, the maximum value of the error bar,

TABLE 4.4 Extrapolated Solution of the Wall Shear Stress with Relative Errors

i	$(\phi_{ext}^{21})_i$ (mPa)	$(\phi_{ext}^{32})_i$ (mPa)	$(e_{ext}^{21})_i$ %	$(e_{ext}^{32})_i$ %
1	16.1659	15.9089	0.07	0.94
2	14.0623	13.0302	2.80	3.29
3	13.3551	10.0228	9.44	19.82
4	11.4028	11.3285	0.04	0.39
5	11.0299	11.0299	0.00	0.03
6	10.8176	10.8176	0.01	0.18
7	10.6922	10.6922	0.02	0.27
8	10.6196	10.6196	0.03	0.34
9	10.5811	10.5811	0.04	0.40
10	10.5656	10.5656	0.05	0.45
11	10.5659	10.5659	0.05	0.49
12	10.5757	10.5757	0.05	0.50
13	10.5888	10.5888	0.05	0.49
14	10.6276	10.6276	0.04	0.41
15	10.6384	10.6384	0.03	0.39
16	10.6400	10.6400	0.03	0.39
17	10.6400	10.6400	0.03	0.39
18	10.6399	10.6399	0.03	0.39
19	10.6399	10.6399	0.03	0.39
20	10.6353	10.6353	0.03	0.37

$\pm 2.66 \times 10^{-5}$ Pa, occurs at $x_i = 0.25$ m which corresponds to a maximum discretization uncertainty of $2 \times (GCI^{21})_2 = 0.38\%$. In Table 4.4, we report the extrapolated solutions along with their relative errors. Figure 4.31 displays the variation of the shear stress at the wall. The error bars displayed in this figure are amplified by a factor of 20 for visibility purposes only. As can be seen in Figure 4.31 and Table 4.3, the error bars stabilize in the fully developed region.

Centerline Velocity

The profile of the centerline velocity is of interest because it can be validated against existing empirical relations for the entry length. For the purpose of reporting its uncertainty estimators for each of the three grids, we extracted the velocity magnitudes at 11 points on the centerline. The convergence metrics calculated using the Richardson Extrapolation scheme on those 11 points are summarized in Table 4.5. In this table, we observe that the local order of accuracy ranges between 0.46 and 2.14 with an average of 1.031. This value is close to the theoretical values employed in our numerical schemes. However, upon inspection, we notice that there is a significant difference between the values of the local order of accuracy which stabilizes at 0.46 in the fully developed region. When compared to the order of the numerical schemes employed, $p = 0.46$ could be

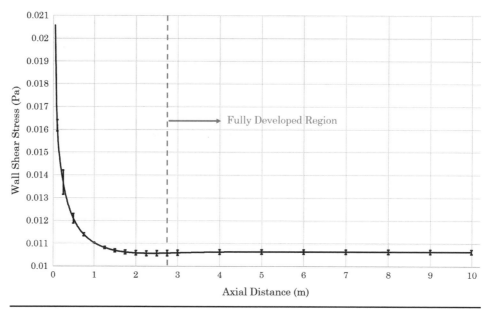

FIGURE 4.31 Average wall shear stress (ϕ_{i_1}) as a function of x. The error bars are presented at locations x_i, $i = 1, \ldots, 20$.

TABLE 4.5 Convergence Metrics Using Richardson Extrapolation for the Centerline Velocity

i	y_i (m)	ϕ_{i_1} (mm/s)	ϕ_{i_2} (mm/s)	ϕ_{i_3} (mm/s)	s_i	$(e_a^{21})_i$ %	$(e_a^{32})_i$ %	p_i	$(GCI^{21})_i$ ×100	EB_i ×10^6
1	0.1	50.9839	51.1347	51.3331	1	0.30	0.39	1.87	0.48	245.4
2	1	56.5431	56.7239	56.9905	1	0.32	0.47	2.15	0.52	294.1
3	2	59.4527	59.5219	59.6203	1	0.12	0.17	2.06	0.19	112.7
4	3	59.7287	59.8007	59.8842	1	0.12	0.14	1.57	0.20	117.0
5	4	59.6343	59.7344	59.8207	1	0.17	0.14	0.86	0.27	162.9
6	5	59.5914	59.7000	59.7818	1	0.18	0.14	0.54	0.30	176.8
7	6	59.5819	59.6918	59.7717	1	0.18	0.13	0.46	0.30	178.7
8	7	59.5809	59.6906	59.7702	1	0.18	0.13	0.46	0.30	178.6
9	8	59.5810	59.6906	59.7703	1	0.18	0.13	0.46	0.30	178.4
10	9	59.5810	59.6907	59.7703	1	0.18	0.13	0.46	0.30	178.4
11	10	59.5871	59.6952	59.7739	1	0.18	0.13	0.46	0.30	176.0

considered small, and this will lead to enlarging the error bars in the fully developed region. The maximum value for the error bar is found to be $\pm 2.94 \times 10^{-4}$ m/s which corresponds to a maximum discretization uncertainty of 1.04%. In Table 4.6, we report the extrapolated solutions along with their relative errors. Figure 4.32 displays the errors bars for the centerline velocity profile. The error bars in this figure are magnified by a factor of 10 for visibility purposes.

TABLE 4.6 Extrapolated Solution of the Centerline Velocity with Relative Errors

i	$(\phi_{ext}^{21})_i$ (mm/s)	$(\phi_{ext}^{32})_i$ (mm/s)	$(e_{ext}^{21})_i$ %	$(e_{ext}^{32})_i$ %
1	50.9005	50.9005	0.16	0.46
2	56.4638	56.4638	0.14	0.46
3	59.4199	59.4199	0.06	0.17
4	59.6766	59.6766	0.09	0.21
5	59.4692	59.4692	0.28	0.45
6	59.2809	59.2809	0.52	0.71
7	59.2056	59.2056	0.64	0.82
8	59.1976	59.1976	0.65	0.83
9	59.2015	59.2015	0.64	0.83
10	59.2024	59.2024	0.64	0.82
11	59.2165	59.2165	0.63	0.81

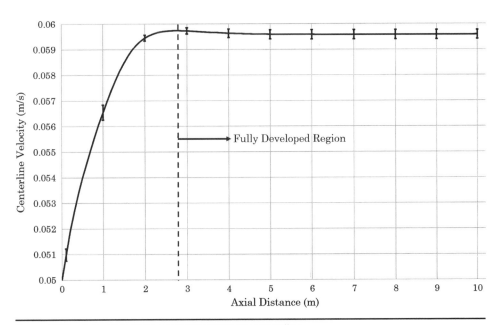

FIGURE 4.32 Error bars for the velocity along the centerline.

4.6.2 Comparison with Empirical Correlations

Entry Length

An expression for the entry length as function of Reynolds number and pipe diameter, proposed by Bhatti and Shah (1987) and Zhi-qing (1982), is presented in Eq. (4.37) [35]. In the model studied in this chapter, $\mathrm{Re} = \frac{\rho v D}{\mu} = 9952$ and $D = 0.2\,\mathrm{m}$, which yields an estimated entry length of $L_e = 2.715\,\mathrm{m}$.

Using Fluent, the estimated entry length in Figures 4.31 and 4.32 corresponding to the wall shear stress and the centerline axial profiles, respectively, is $L_e = 2.7$ m, and is in agreement with the value predicted empirically by Eq. (4.37).

Darcy Friction Factor

The value for the Darcy friction factor in the fully developed region may be calculated using the correlation by Colebrooke in Eq. (4.31) or using Moody's chart. In our Fluent model, we used the default value for the wall roughness which indicates a perfectly smooth wall. For $Re = 9952$ and $\epsilon = 0$, Eq. (4.31) yields $f = 0.031$. The values of the wall shear stress and the pressure drop in the fully developed region are then estimated to be 9.6 mPa and 1.4 Pa respectively, using Eq. (4.26).

Based on the conservation of linear momentum, Eq. (4.26) provides a relation between the pressure drop and the wall shear stress in the fully developed region. To calculate the pressure drop in Fluent, we create a cross-sectional plane at $x = 3$ m. We calculate the area-weighted average of pressure on this plane, which was found to be 1.49 Pa. Since we set the gauge pressure at the outlet to 0 Pa, the pressure drop per unit length in the fully developed region is $(1.49 - 0)/7 = 0.213$ Pa/m. The average wall shear stress in the fully developed region is then calculated from Eq. (4.26) to be around 10.6 mPa, also displayed in Figure 4.31. The friction factor, estimated using Eq. (4.26), is $f = 0.034$, which is in agreement with the value predicted empirically. The relative percent errors in the friction factor, the wall shear stress, and the pressure drop between Fluent and empirical solutions are around 9.68%, 10.42%, and 6.43% respectively.

Velocity Profile

For a turbulent flow, an exact expression for the time-averaged velocity profile in a cross section in the fully developed region does not exist. However, an approximation led to an

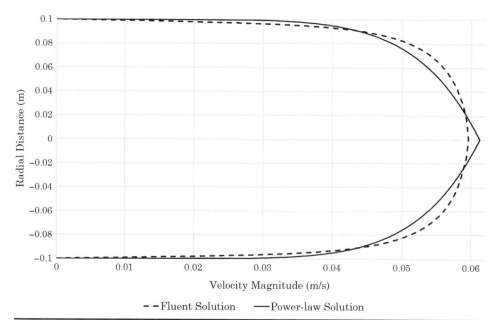

FIGURE 4.33 Comparison between Fluent solution and Power-law solution for the velocity at x = 8 m.

expression for the velocity profile and it is given by Eq. (4.36). To compare the solution obtained by Fluent to the approximate solution, we extracted the velocity profile at a cross section at $x = 8$ m. The comparison is provided in Figure 4.33 where we observe that there is a good agreement in the region near the walls. However, we notice that the power-law profile fails to exactly predict the velocity profile in the core region of the pipe. The empirical value for the maximum velocity is $V \times \frac{60}{49} = 0.05 \times \frac{60}{49} = 0.0612 \frac{m}{s}$. Using Eq. (4.32), which includes dependence on the friction factor, the centerline velocity in the fully developed region is 0.062 m/s. The centerline velocity in the fully developed region predicted by Fluent is shown in Figure 4.32 and is equal to 0.0595 m/s. The relative percent error between the solution obtained by Fluent and the empirical solution is less than 3%.

CHAPTER 5

Convection Heat Transfer for Two-Dimensional Steady State Incompressible Flow

List of Symbols	
Total energy	E
Internal energy	U
Kinetic energy	KE
Potential energy	PE
Work	W
Heat	Q
Pressure	p
Volume	$\mathcal{V}$
Torque	$\mathcal{T}$
Temperature	T
Thermal conductivity	k
Heat flux	$\vec{q}$
Area	$\mathcal{A}$
Length	$\mathcal{L}$
Density	ρ
Dynamic viscosity	μ
Specific heat at constant pressure	c_p
Convection heat transfer coefficient	h
Nusselt number	Nu
Reynolds number	Re
Prandtl number	Pr
Kinematic viscosity	ν
Characteristic length	L
Thermal diffusivity	α

Continued

List of Symbols	
Wall roughness	ϵ
Hydrodynamic boundary layer thickness	δ
Thermal boundary layer thickness	δ_T
Radius	R
Diameter	D
Average velocity	V
Specific heat at constant volume	c_v
Specific internal energy	u
Temperature	T
Specific enthalpy	$\hbar$
Thermal entry length	$L_{e,T}$
Hydrodynamic entry length	L_e

This chapter will cover the problem of convection heat transfer for the case of one phase fluid flowing inside a heated pipe. In addition to the hydrodynamic effects of the flow that were covered in the Chapters 3 and 4, the thermal effect in the heated pipe will be covered in this chapter. This is a two-dimensional axisymmetric problem, similar to the problem presented in Chapter 3. We will use the same model geometry created in Chapter 3 and will add the solid pipe to the model.

Instructions to build the two-dimensional model of the heat transfer and the fluid inside the pipe will be presented. Results of the simulations will be discussed and followed by a verification of the Fluent model.

5.1   Introduction to Heat Transfer

Heat transfer plays an important role in various engineering fields and for different applications (medical, HVAC, etc.), and therefore it is very important to understand it. A condensed review covering the three modes of heat transfer is presented in this section.

The first law of thermodynamics states that the change in total energy stored in a control mass undergoing a thermodynamic process from equilibrium state 1 to equilibrium state 2 is balanced by work and heat energy transfers across the boundary

$$E_2 - E_1 = W^{\leftarrow}_{1-2} + Q^{\leftarrow}_{1-2} \tag{5.1}$$

where the total energy is the sum of the internal energy, kinetic, and potential energy, $E = U + KE + PE$, and $W^{\leftarrow}_{1-2}$ and $Q^{\leftarrow}_{1-2}$ are respectively the work done on the system and the heat added during the process. For work to take place, a force is needed along a displacement. This force could be due to pressure and the displacement could be due to expansion (or contraction), so that the work done by the system is $W^{\rightarrow}_{1-2} = \int_1^2 p d\mathcal{V}$. Work can also be exchanged across the system boundary through a shaft, in which case $W^{\rightarrow}_{1-2} = \int_1^2 \mathcal{T} d\theta$, where $\mathcal{T}$ is the torque and θ is the angle.

Heat transfer across the boundary is caused by a temperature difference, and it flows from the high temperature region to the low temperature region. In general, heat transfer takes place in two distinct modes, by molecular collisions and by transport of waves. These two modes, referred to as conduction and radiation, in addition to convection heat transfer, which is heat conduction compound with fluid motion effects, are discussed next.

5.1.1   Conduction

Conduction heat transfer in matter takes place due to a temperature difference. Noting that the temperature is a local statistical average of the kinetic energy associated with the molecular velocity, the energy of the molecules in the high temperature region is larger than that in the low temperature region. Because motion of the molecules has a random component, which is proportional to the local temperature, molecules will exchange energy by collisions. During the collision, the energy lost by the high temperature molecule is gained by the low temperature molecule. Conduction heat transfer is therefore a diffusion process, where energy is transferred by virtue of random molecular collisions in matter. Other diffusion processes include diffusion mass transfer, which takes place due to a concentration difference, and diffusion momentum transfer, which takes place due to a difference in momentum.

Since heat flows from the high temperature region to the low temperature region, the direction of this flow is along the unit vector, $\hat{n}$, that defines the (spatial) gradient of the temperature, ∇T. Noting that ∇T points from the low to the high temperature, then $\hat{n} = -\nabla T/|\nabla T|$. One may also think of $\hat{n}$ as the local unit vector normal to the isothermal surfaces, pointing from the surface of the lower temperature to that of the higher temperature. The conduction heat flux may then be expressed as $\vec{q} = q\,\hat{n}$, where the magnitude of the heat flux is proportional to the magnitude of the temperature gradient, $q = k|\nabla T|$. The thermal conductivity, k, is a property of the material which measures how well the material conducts heat by virtue of its molecular composition and the thermodynamic conditions. The conduction heat flux is then expressed by Fourier's law as

$$\vec{q} = -k\,\nabla \tag{5.2}$$

The rate of conduction heat transferred across a surface out of a control mass (or a control volume) is then the integral over the surface area

$$\dot{Q}^{\rightarrow} = \int_{A} \vec{q} \cdot \hat{n}\, d\mathcal{A} \tag{5.3}$$

where $\hat{n}$ is the unit normal vector to $d\mathcal{A}$ pointing away from the control mass. As an example, consider steady heat transfer across a wall separating two rooms at temperatures T_H and T_C, as depicted in Figure 5.1. The wall surface area is such that its length, $\mathcal{L}$, and width, $\mathcal{W}$, are much larger than the wall thickness, Δx. Due to the temperature difference, $T_H - T_C$, across the wall, the conduction heat flux in the wall is predominantly along the thickness of the wall, since $\Delta x \ll \mathcal{L}$ and $\Delta x \ll \mathcal{W}$. This may also be demonstrated mathematically by expressing $\vec{q} = -k\left(\frac{\partial T}{\partial x}\hat{x} + \frac{\partial T}{\partial y}\hat{y} + \frac{\partial T}{\partial z}\hat{z}\right)$. Noting that the wall dimensions along the x, y, and z directions are respectively Δx, $\mathcal{L}$, and $\mathcal{W}$, then

$$\frac{\frac{\partial T}{\partial y}}{\frac{\partial T}{\partial x}} \sim \frac{\Delta x}{\mathcal{W}} \ll 1$$

$$\frac{\frac{\partial T}{\partial z}}{\frac{\partial T}{\partial x}} \sim \frac{\Delta x}{\mathcal{L}} \ll 1$$

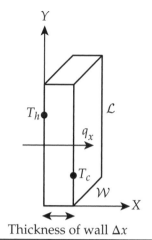

FIGURE 5.1 Conduction heat transfer across a wall.

so that the heat flux is approximated as $\vec{q} \simeq -k\frac{\partial T}{\partial x}\hat{x}$, which is perpendicular to the wall surface area. At steady state, the rate of heat transfer in the positive x direction across any $y - z$ planar section in the wall is equal to that across any other planar section,

$$\frac{d\dot{Q}}{dx} = 0 \tag{5.4}$$

where

$$\dot{Q} = \int_{A} -k\frac{\partial T}{\partial x}\,d\mathcal{A} \tag{5.5}$$

so that $d^2T/dx^2 = 0$, yielding a linear temperature distribution in the wall, $T = T_H + \frac{x}{\Delta x}(T_C - T_H)$. The rate of heat transferred into the wall across the left side $(\dot{Q}^{\leftarrow})$ is equal to the rate of heat transferred out of the wall across the right side $(\dot{Q}^{\rightarrow})$,

$$\dot{Q} = k\mathcal{A}\frac{T_H - T_C}{\Delta x} \tag{5.6}$$

where $\mathcal{A} = \mathcal{L}\mathcal{W}$.

5.1.2 Convection

Convection heat transfer is conduction heat transfer aided by fluid flow. In the absence of a flow, heat transfer between the fluid and a solid is purely conductive. In this case, the rate at which heat is transferred is limited by the thermal conductivity of the fluid and by the temperature gradient that arises. When the fluid is moving, larger temperature gradients can be created, which significantly enhance the heat transferred between the solid and the fluid. This is because, in a solid cooling application (e.g., cooling of an electronic component), the flow continuously replaces the fluid particles heated by transfer from the wall by colder particles from the upstream. The opposite happens in a heating application (e.g., hair blow drying). When the fluid flow is caused by buoyancy (gravity acting on a varying density), the convection is called natural (or free) convection. When the flow is driven by other forces, such as a pressure gradient, it is called forced convection. For the example shown in Figure 5.2, one can intuitively observe that

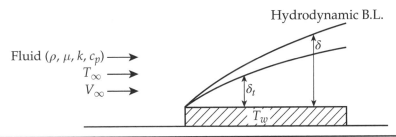

FIGURE 5.2 Cooling of an electronic component by forced convection.

cooling of the hot electronic component is dependent on the fluid flow velocity V_∞, fluid properties (density ρ, dynamic viscosity μ, thermal conductivity k, specific heat c_p), the interfacial area, $\mathcal{A}$, between the solid and the fluid, in addition to the temperature difference, $T_w - T_\infty$, between the wall and the fluid bulk. This dependence is commonly expressed as Newton's law of cooling

$$\dot{Q} = h\,\mathcal{A}\,(T_w - T_\infty) \tag{5.7}$$

where the convection heat transfer coefficient, h, is a function of the flow velocity V and the fluid properties ρ, μ, k, and c_p. Noting that the flow velocity at the wall is zero due to the no-slip condition, the heat transfer between the fluid and the solid in the region very close to the wall is conductive. As such, the rate of heat transfer depends on the area, $\mathcal{A}$, the fluid conductivity, k, and the temperature gradient at the wall, which depends on the fluid velocity and the fluid properties, in addition to the wall and fluid temperatures. The dependence of h on the flow conditions and fluid properties is commonly expressed as

$$h = f(\rho, V, \mathcal{L}, \mu, k, c_p) \tag{5.8}$$

which, in dimensionless form, is expressed as

$$\mathtt{Nu} = f(\mathtt{Re}, \mathtt{Pr}) \tag{5.9}$$

where the Nusselt number, $\mathtt{Nu} = \frac{hL}{k}$, measures the ratio of the convective to the conductive heat transfer. Dependence on the flow is explicitly captured through the Reynolds number, $\mathtt{Re} = \frac{\rho VL}{\mu}$, which measures the ratio of inertia to viscous effects. Prandtl number, $\mathtt{Pr} = \frac{\nu}{\alpha}$, where $\nu = \mu/\rho$ and $\alpha = k/(\rho c_p)$, which is the ratio of the kinematic diffusivity to the thermal diffusivity, is the key parameter for characterizing the ratio of the thermal boundary layer thickness to the hydrodynamic boundary layer thickness. The thermal and hydrodynamic boundary layers that form at the surface of the hot plate, shown in Figure 5.2, play an important role in heat transfer as they affect the rate at which the fluid is moving and subsequently affect the temperature gradient at the wall. Except for simple flows, such as laminar flow over a flat plate, the dependence of $\mathtt{Nu}$ on $\mathtt{Re}$ and $\mathtt{Pr}$, and possibly other dimensionless numbers, is empirically established from experimental measurements.

5.1.3 Radiation

The convection and conduction modes of heat transfer are called transport phenomena as they require the presence of a material or medium to happen. Radiation heat transfer does not require a medium. It is a surface phenomenon and happens between surfaces.

Radiation heat transfer is electromagnetic radiation emitted by a body due to its temperature. Stefan Boltzman's law for radiation is

$$q = \varepsilon \cdot \mathcal{A} \cdot \sigma \cdot T_s^4 \tag{5.10}$$

where ε is the emissivity which is also the capability of a material to radiate ($\varepsilon = 1$ for a black body), $\mathcal{A}$ is the surface of the body, σ is Stefan Boltzman's constant ($\sigma = 5.67 \times 10^{-8}$ W/m^2/K^4), and T_s is the surface temperature.

5.2   Problem Statement

Consider a system for heating fluid by passing it through a solid pipe of a thick wall, as shown in Figure 5.3. The solid pipe is electrically heated and its outer surface is insulated such that all the heat generated inside the solid pipe is transferred to the fluid passing through it. The temperature of the fluid will vary along the pipe to conserve the energy of the system.

Table 5.1 provides the pipe dimensions and properties, the fluid properties and the flow parameters for the problem. The velocity at the inlet is chosen such that Reynolds number $\text{Re} = \frac{\rho V D}{\mu}$ is well above 2,000, the transition Reynolds number, so that the flow is in the turbulent regime.

The flow is axisymmetric and therefore the shaded area in Figure 5.3 will be used as the domain for the Fluent model. We will build a two-dimensional model to solve the fluid flow and heat transfer inside the pipe.

The heat generated inside the pipe wall will be conducted within the pipe wall and the biggest part will be transferred to the fluid via forced convection heat transfer mode. The entering flow will reach fully developed hydrodynamic and thermal conditions once the hydrodynamic and thermal boundary layer thicknesses ($\delta(x)$ and $\delta_T(x)$) respectively reach the pipe inner radius, as depicted in Figure 5.4.

Convection heat transfer exchanged between the moving fluid and the inner surface of the pipe wall is controlled by the radial profile of the temperature inside the thermal boundary layer. Cooling of the pipe wall (or heating of the fluid) is dependent on the

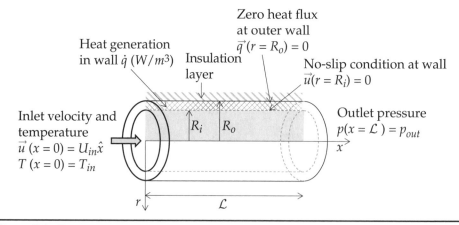

FIGURE 5.3 The domain and boundary conditions.

TABLE 5.1   Model Dimensions, Properties, and Flow Variables

Variable	Symbol	Value
Pipe dimensions		
Pipe length	$\mathcal{L}$	10 m
Pipe inner diameter	$D_i = 2R_i$	0.2 m
Pipe outer diameter	$D_o = 2R_o$	0.4 m
Wall roughness	ϵ	0
Fluid properties		
Density	ρ	998.2 kg/m^3
Viscosity	μ	1.003 mPa.s
Specific Heat	c_p	4182 J/kg.K
Flow properties		
Average velocity	V	0.05 m/s
Outlet (gauge) pressure	p_{out}	0 Pa
Reynolds number	Re	9952
Mass flow rate	$\dot{m}$	1.57 kg/s
Heat properties		
Heat generation	q	0.1 MW/m^3
Wall conductivity	k	202.4 W/m.K
Inlet Temperature	T_{in}	298 K

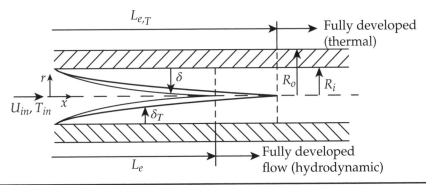

FIGURE 5.4   Developing hydrodynamic and thermal boundary layers in the entry region of the pipe flow.

fluid flow velocity, V, fluid properties (density ρ, dynamic viscosity μ, thermal conductivity k, specific heat c_p), interfacial area between the solid and the fluid, $\mathcal{A} = 2\pi R_i \mathcal{L}$, in addition to the temperature difference between the inner wall face and the fluid bulk. Such a dependence can be determined by solving the governing equations, subject to specified initial and boundary conditions, for only simplified problems involving laminar flows. For more complex problems, especially those involving turbulent flows, a convenient and approximate approach for modeling the heat transfer is to use Newton's law of cooling, which expresses the convective heat flux in terms of the difference between the surface temperature and the mean (bulk) temperature as

$$\vec{q_s}(x) = h(x) \, [T_s(x) - T_m(x)] \tag{5.11}$$

where $h(x)$ is the local convection heat transfer coefficient, $T_s(x) = T(R_i, x)$, and $T_m(x)$, the mean (bulk) temperature, is the average temperature at a given cross-section that yields the same rate of internal energy carried by the flow, so that

$$T_m(x) = \frac{1}{\dot{m}c_p} \int_0^{R_i} 2\pi\rho u(r,x) c_p T(r,x) r dr \tag{5.12}$$

where $\dot{m} = \int_0^{R_i} 2\pi\rho u(r,x) r dr$ is the mass flow rate.

5.3 Governing Equations and Boundary Conditions

For a steady, incompressible fluid, and in the absence of body forces, the continuity and momentum equations can be expressed as

$$\nabla \cdot \vec{u} = 0 \tag{5.13}$$

$$\rho \vec{u} \cdot \nabla \vec{u} = -\nabla p + \mu \nabla^2 \vec{u} \tag{5.14}$$

where the viscosity μ is assumed to be constant over the temperature range encountered in the problem. Since the density ρ is constant, it follows that the velocity vector field is divergence-free, that is, $\nabla \cdot \vec{u} = 0$, and from Eq. (1.75), $du = c_v dT$, where u is the specific internal energy. The energy equation in the fluid region, $r < R_i$, simplifies from Eq. (1.69) to

$$\rho c_v \vec{u} \cdot \nabla T = k \nabla^2 T + \mu \Phi \tag{5.15}$$

where we assumed that the specific heat c_v is constant over the temperature range encountered in the problem.

In the solid region, $R_i < r < R_o$, the energy equation is

$$0 = k_w \nabla^2 T + q \tag{5.16}$$

where q is the heat generation per unit volume per unit time [J/(s.m^3)]. Analyzing the heat conduction in the pipe wall is not of major importance to the convection problem in hand. However, we discuss it in detail here in order to show Fluent's capability in solving heat conduction problem, as described in Section 5.3.1 below.

In what follows, we start by modeling heat conduction in the pipe wall, where we will present approximate expressions for the temperature distribution in the pipe wall. Then, we discuss the forced convection heat transfer in the pipe. We present the governing equations in differential form, simplified by neglecting the terms whose physical contribution to the balance is significantly weaker than other terms. We will then apply the conservation of energy and momentum for an infinitesimally thin cylindrical control volume to express the mean flow temperature in terms of the wall heat flux and the pressure drop. Then, we briefly discuss heat transfer in the entry region and the associated challenges. A solution for the temperature distribution for laminar fully developed pipe flow, subject to constant wall heat flux, is then presented, followed by correlation for Nusselt number in the turbulence regime. Finally we put everything together to solve the axial profiles of wall heat flux, the bulk temperature and the wall temperatures.

5.3.1 Heat Conduction in the Pipe Wall

Steady state heat transfer by conduction in the pipe wall is governed by Eq. (5.16), which in axisymmetric cylindrical coordinates assumes the form

$$k_w \left[\frac{1}{r} \frac{\partial}{\partial r} \left(r \frac{\partial T}{\partial r} \right) + \frac{\partial^2 T}{\partial x^2} \right] = -q, \quad 0 \le x \le \mathcal{L}, R_i \le r \le R_o \tag{5.17}$$

subject to the boundary conditions

$$\frac{\partial T}{\partial r} = 0 \text{ at } 0 \le x \le \mathcal{L}, r = R_o \tag{5.18}$$

$$\frac{\partial T}{\partial x} = 0 \text{ at } x = 0, R_i \le r \le R_o \tag{5.19}$$

$$\frac{\partial T}{\partial x} = 0 \text{ at } x = \mathcal{L}, R_i \le r \le R_o \tag{5.20}$$

$$k_w \frac{\partial T}{\partial r} = q_s^{\rightarrow}(x) \text{ at } 0 \le x \le \mathcal{L}, r = R_i \tag{5.21}$$

The first three conditions correspond to the outer side of the pipe wall being thermally insulated. The fourth condition is a representation of the continuity of the heat flux at the inner side of the pipe wall, which is equal to the flux of heat carried away by the convecting fluid. The mean radial temperature in the pipe wall is

$$T_w(x) = \frac{2\pi}{\mathcal{A}_w} \int_{R_i}^{R_o} T(r, x) r dr \tag{5.22}$$

where $\mathcal{A}_w = \pi(R_o^2 - R_i^2)$ is the cross-sectional area of the pipe wall. Integrating Eq. (5.17) over the cross-sectional area of the pipe wall yields an equation for $T_w(x)$ as follows

$$\int_{R_i}^{R_o} k_w \left(\frac{1}{r} \frac{\partial}{\partial r} \left(r \frac{\partial T}{\partial r} \right) + \frac{\partial^2 T}{\partial x^2} \right) 2\pi r dr = -\int_{R_i}^{R_o} q 2\pi r dr$$

$$k_w \int_{R_i}^{R_o} \frac{\partial}{\partial r} \left(r \frac{\partial T}{\partial r} \right) dr + k_w \mathcal{A}_w \frac{d^2 T_w(x)}{dx^2} = -q\mathcal{A}_w$$

$$k_w \left(\left(r \frac{\partial T}{\partial r} \right)_{R_o} - \left(r \frac{\partial T}{\partial r} \right)_{R_i} + \mathcal{A}_w \frac{d^2 T_w(x)}{dx^2} \right) = -q\mathcal{A}_w$$

Using boundary conditions Eqs. (5.18) and (5.21) yields

$$-R_i q^{\rightarrow}(x) + k_w \mathcal{A}_w \frac{d^2 T_w(x)}{dx^2} = -q\mathcal{A}_w \tag{5.23}$$

Equation (5.23) is an expression of the conservation of energy in the control volume consisting of an annular element of inner and outer radii R_i and R_o and extending from x to $x + dx$, as depicted in Figure 5.5. Solving for $T_w(x)$ requires knowledge of the heat flux $q^{\rightarrow}(x)$ which is dependent on the flow behavior and fluid properties, in addition to the wall temperature, thus capturing the coupling between the energy and momentum equation.

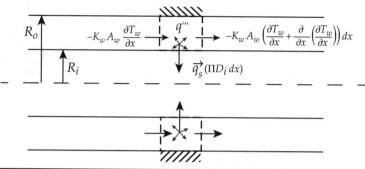

Figure 5.5 The control volume for the energy balance in the pipe wall consists of an annulus of inner and outer radii R_i and R_o, extending from x to x + dx.

An approximate expression for the temperature distribution of the pipe wall can be obtained by approximating $\partial^2 T/\partial x^2 \simeq d^2 T_w/dx^2$,

$$\frac{1}{r}\frac{\partial}{\partial r}\left(r\frac{\partial T}{\partial r}\right) = -\frac{R_i q^{\rightarrow}(x)}{k_w A_w} \tag{5.24}$$

which yields

$$T(r,x) \simeq T_s(x) + \frac{R_i q^{\rightarrow}(x)}{k_w A_w}\left(\frac{R_i^2 - r^2}{4} + \frac{R_o^2}{2}\ln\frac{r}{R_i}\right) \tag{5.25}$$

where $T_s(x) = T(R_i, x)$ is the temperature at the inner side of the pipe wall at axial location x. A relation between $T_s(x)$ and $T_w(x)$ can be obtained by integrating Eq. (5.25) over the cross-sectional area of the pipe wall, which yields

$$T_w(x) \simeq T_s(x) + \frac{R_i q^{\rightarrow}(x)}{k_w A_w}\left(\frac{R_i^2 - R_o^2}{8} - \frac{R_o^2}{4} + \frac{R_o^4 \ln\left(\frac{R_i}{R_o}\right)}{2(R_i^2 - R_o^2)}\right) \tag{5.26}$$

In this derivation, we considered the case when the flux $q^{\rightarrow}(x)$ at the pipe internal wall is known such that, when $T_s(x)$ is computed by Fluent, Eq. (5.26) can be used to compute the average solid wall temperature $T_w(x)$. Another assumption is to assume that $T_s(x)$ is known and compute $q^{\rightarrow}(x)$ accordingly.

5.3.2   Forced Convection of Internal Flow

We now turn attention to studying the heat transfer in the internal pipe flow considered in this problem. Most of the heat generated inside the pipe wall will be transferred to the fluid via forced convection heat transfer mode, and a small part will be conducted within the wall itself.

As the flow enters the pipe, a thermal boundary layer will start developing, in addition to the hydrodynamic boundary layer. The thermal boundary layer is the flow region near the wall where the thermal effects of the wall are felt. As the flow moves downstream along the pipe axis both boundary layers grow due to diffusion and mixing until they reach the centerline, at which point the fully developed flow conditions are reached, that is, $\partial u/\partial x = 0$.

At steady conditions, and in the absence of heat sources in the flow region, the velocity and temperature in the flow region are governed respectively by the continuity, momentum, and energy equations, which, in axisymmetric cylindrical coordinates, are expressed as

$$\frac{\partial u}{\partial x} + \frac{1}{r}\frac{\partial(ru)}{\partial r} = 0 \tag{5.27}$$

$$u\frac{\partial u}{\partial x} + v\frac{\partial u}{\partial r} = -\frac{1}{\rho}\frac{\partial p}{\partial x} + \frac{\nu}{r}\frac{\partial}{\partial r}\left(r\frac{\partial u}{\partial r}\right) \tag{5.28}$$

$$0 \simeq \frac{\partial p}{\partial r} \tag{5.29}$$

$$u\frac{\partial T}{\partial x} + v\frac{\partial T}{\partial r} = \frac{\alpha}{r}\frac{\partial}{\partial r}\left(r\frac{\partial T}{\partial r}\right) \tag{5.30}$$

where we assumed that the radial component of the momentum equation simplifies, for $\frac{R_i}{\mathcal{L}} \ll 1$, to Eq. (5.29), based on the scaling $v \sim u\frac{R_i}{\mathcal{L}}$ which follows from the continuity equation (Eq. (5.27)). We also assumed in the x-component of the momentum equation (Eq. (5.28)) and in the energy equation (Eq. (5.30)), that diffusion of momentum and heat in the radial direction dominates that in the axial direction, that is, $\left|\frac{\partial^2 u}{\partial x^2}\right| \ll \left|\frac{1}{r}\frac{\partial}{\partial r}\left(r\frac{\partial u}{\partial r}\right)\right|$, $\left|\frac{\partial^2 T}{\partial x^2}\right| \ll \left|\frac{1}{r}\frac{\partial}{\partial r}\left(r\frac{\partial T}{\partial r}\right)\right|$, which follows from the condition $\left(\frac{R_i}{\mathcal{L}}\right)^2 \ll 1$, in accordance with the boundary layer approximation.

Another relation between $q^{\rightarrow}(s)$ and $T_m(x)$ may be obtained by carrying out energy balance for the control volume shown in Figure 5.6.

At steady state and for constant density of the fluid, the mass flow rate is constant, so that $\dot{m} = \int_{\mathcal{A}} \rho u \, d\mathcal{A} = \dot{m}^{\leftarrow}(x) = \dot{m}^{\rightarrow}(x+dx)$, where $\mathcal{A}$ is the pipe cross-section $\mathcal{A} = \pi D_i^2/4$. For this axisymmetric steady flow, the axial velocity component is a function of x and r only, and $d\mathcal{A} = 2\pi r \, dr$. Applying the law of conservation of energy (first law of thermodynamics) at steady state yields

$$0 + \dot{m}\left(\mathfrak{h}_m(x+dx) - \mathfrak{h}_m(x)\right) = q_s^{\rightarrow}\pi D_i dx \tag{5.31}$$

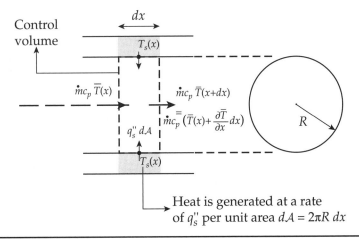

FIGURE 5.6 Control volume extends from x to $x + dx$.

where $\mathfrak{h}_m$ is the mean flow enthalpy defined as the mass flow average over the pipe cross-section,

$$\mathfrak{h}_m = \frac{\int_A \rho u \mathfrak{h} \, dA}{\dot{m}} \tag{5.32}$$

where the rate at which viscous forces $(\vec{f}_v = \mu \nabla^2 \vec{u})$ are doing work on the fluid in the control volume, $\dot{W}_f^{\leftarrow} = \int_V \vec{u} \cdot \vec{f}_v \, dV$, is zero because of the no-slip boundary condition at the wall and the fact that the viscous force and velocity are normal to each other at the pipe inlet and outlet. Note also that changes in kinetic and potential energy have been neglected in Eq. (5.31), as their contribution to the energy balance is negligible. Noting that the density ρ is constant, and that the specific enthalpy is related to the specific internal energy as $\mathfrak{h} = \mathfrak{u} + p/\rho$, then $d\mathfrak{h} = d\mathfrak{u} + \frac{dp}{\rho}$. The first law of thermodynamics simplifies to the statement that the rate of heat loss by the heated wall to the fluid is balanced by a gain in the mean enthalpy of the fluid. Assuming constant volume specific heat, and expressing $d\mathfrak{u}_m = c_v dT_m$, Eq. (5.31) yields

$$\dot{m} c_v \frac{dT_m}{dx} + \frac{\dot{m}}{\rho} \frac{d\bar{p}}{dx} = \vec{q_s} \pi D_i \tag{5.33}$$

Integrating from the pipe inlet to axial position x yields

$$T_m(x) = T_{in} + \frac{p_{in} - p(x)}{\rho c_v} + \frac{\pi D_i}{\dot{m} c_v} \int_0^x \vec{q_s}(\xi) d\xi \tag{5.34}$$

where $p_{in} - p(x)$ is the pressure drop from the pipe inlet to axial location x, which can be determined by solving the Navier-Stokes equations.

Applying the momentum balance for the same control volume, shown in Figure 5.6,

$$(p_{in} - p(x))\mathcal{A} = \tau_w \pi D_i \mathcal{L} \tag{5.35}$$

Finding the heat flux and the shear at the inner side of the wall requires solving the temperature and velocity profiles in the pipe or the use of correlations put together by engineers based on experimental data for specific flow configuration over specific ranges of operating parameters.

5.3.3 Entry Region

In the entry region, Eqs. (5.27)–(5.29) have to be solved for the velocity (u, v), which are then used in the energy equation (Eq. (5.30)) to solve for T. These equations are not possible to solve analytically, even when the flow is laminar. There are some available expressions for the axial variation of the heat transfer coefficient in the entry region for laminar flows under the restrictive assumption of developing thermal boundary layer in a hydrodynamically fully developed region [36]. The distance from the pipe inlet at which the flow reaches the fully developed thermal conditions depends on the pipe diameter D, fluid properties ρ, μ, k, c_p, and flow velocity V. This dependence can be expressed as $L_{e,T} = f(D, \rho, \mu, k, c_p, V)$, or in dimensionless form as $\frac{L_{e,T}}{D} = \mathcal{F}(\text{Re}, \text{Pr})$. The Reynolds number, Re, and the Prandtl number, Pr, are defined in Section 5.1.2 where in the case of internal flow the diameter D of the pipe is used as the characteristic length L. For laminar flow, the thermal entry length is expressed as [36]:

$$\frac{L_{e,T}}{D} = 0.05 \, \text{Re} \, \text{Pr} \tag{5.36}$$

For turbulent flow, the thermal entry length is weakly dependent on the Prandtl number

$$\frac{L_{e,T}}{D} = 10 \tag{5.37}$$

where for our problem, $D_i = 2R_i$. A correlation for the hydrodynamic entry length is given by Eq. (4.37).

5.3.4 Fully Developed Region

Laminar Flow
In the fully developed region $(x > \max(L_e, L_{e,T}))$, the radial velocity profile does not depend on x, that is, $\partial u/\partial x = 0$. Under this condition, Eq. (5.27) yields $v = 0$, and Eqs. (5.28)–(5.30) simplify to

$$0 = -\frac{1}{\rho}\frac{dp}{dx} + \frac{\nu}{r}\frac{\partial}{\partial r}\left(r\frac{\partial u}{\partial r}\right) \tag{5.38}$$

$$u\frac{\partial T}{\partial x} = \frac{\alpha}{r}\frac{\partial}{\partial r}\left(r\frac{\partial T}{\partial r}\right) \tag{5.39}$$

For laminar flows, an analytical solution can be obtained as follows. Equation (5.38) can be integrated subject to the no-slip boundary condition at the wall and the symmetry condition at the center-line to yield the Poiseuille-Hagen solution

$$u_l(r) = 2V\left(1 - \frac{r^2}{R_i^2}\right) \tag{5.40}$$

where the mean velocity is related to the pressure drop, $\Delta p = p_{in} - p_{out}$, along the pipe as $V = \frac{R_i^2 \Delta p}{8\mu\mathcal{L}}$.

For the case of constant heat flux at the inner surface of the wall, it can be shown that at fully developed conditions, $\partial T(r,x)/\partial x = \text{constant}$. If follows that the solution of Eq. (5.39) subject to the boundary conditions is

$$T(r,x) = \frac{\vec{q_s}R_i}{k}\left[\left(\frac{r}{R_i}\right)^2 - \frac{1}{4}\left(\frac{r}{R_i}\right)^4\right] + \frac{2\alpha\vec{q_s}}{kVR_i}x \tag{5.41}$$

The temperature at the wall $(r = R_i)$ increases linearly with x,

$$T_s(x) = \frac{\vec{q_s}}{k}\left(\frac{3R_i}{4} + \frac{2\alpha}{VR_i}x\right) \tag{5.42}$$

Substituting the expressions for T and u from Eqs. (5.41) and (5.40) yields a linear variation for T_m with x

$$T_m(x) = \frac{\vec{q_s}}{k}\left(\frac{7R_i}{24} + \frac{2\alpha}{VR_i}x\right) \tag{5.43}$$

The dependence of the convective heat flux on the difference between the surface temperature and the mean temperature is commonly expressed as Newton's law of cooling, which expresses locally the heat flux at the inner pipe wall into the flow as,

$$\vec{q_s}(x) = h(x)\left(T_s(x) - T_m(x)\right) \tag{5.44}$$

where the local convection heat transfer coefficient $h(x)$ captures all the parameters that affect the convective heat transfer, and as such, is a function of the flow velocity V and fluid properties ρ, μ, k, and c_p. Noting that $q_s^{\rightarrow}(x) = q_s^{\rightarrow}$, and substituting the linear expressions for T_s and T_m from Eqs. (5.42) and (5.43), yields a constant heat transfer coefficient

$$h = \frac{24k}{11R_i} \Rightarrow \text{Nu} = \frac{48}{11} \tag{5.45}$$

where $D = 2R_i$ and the Nusselt number, $\text{Nu} = \frac{hD}{k}$, is a dimensionless number that measures the ratio of convective transfer to conductive heat transfer in the fluid.

Turbulent Flow

Noting that the flow velocity at the wall is zero due to the no-slip condition, the heat transfer between the fluid and the solid in the region very close to the wall is conductive. As such, the rate of heat transfer depends on the area, A, the fluid conductivity, k, and the temperature gradient at the wall, which depends on the fluid velocity and the fluid properties, in addition to the wall and fluid temperatures. The dependence of h on the flow conditions and fluid properties is commonly expressed in dimensionless form as

$$h = f(\rho, V, D, \mu, k, c_p) \Rightarrow \text{Nu} = \mathcal{F}(\text{Re}, \text{Pr}) \tag{5.46}$$

where the Nusselt number, $\text{Nu} = \frac{hD}{k}$, measures the ratio of convective transfer to conductive heat transfer. Dependence on the flow is explicitly captured through the Reynolds number, $\text{Re} = \frac{\rho V D}{\mu}$, which measures the ratio of inertia to viscous effects. Prandtl number, $\text{Nu} = \frac{\nu}{\alpha} = \frac{c_p \mu}{k}$, which is the ratio of the kinematic diffusivity to the thermal diffusivity, is the key parameter for characterizing the ratio of the thermal boundary layer thickness to the hydrodynamic boundary layer thickness. The thermal and hydrodynamic boundary layers that form at the surface of the hot plate, of the example shown in Figure 5.2, play an important role in heat transfer as they affect the rate at which the fluid is moving and subsequently the temperature gradient at the wall. Except for simple flows, such as laminar flow over a flat plate, the dependence of Nu on Re and Pr, and possibly other dimensionless numbers, is empirically established from experimental measurements.

For fully developed turbulent flow in a tube, the average convection coefficient, h, can be estimated from the frequently used correlation for Nusselt number in the Dittus-Boelter equation [37]:

$$\text{Nu} = 0.023\,\text{Re}^{0.8}\,\text{Pr}^n \tag{5.47}$$

where the properties are evaluated at the fluid mean (bulk) temperature and $n = 0.3$ if the hot fluid is passing through the tube, otherwise $n = 0.4$. Equation (5.47) is applicable for both constant temperature and heat flux conditions, for the ranges $0.7 \leq \text{Pr} \leq 160$, $\text{Re} \geq 10,000$, $L/D \geq 10$. The Nusselt number is $\text{Nu} = \frac{hD}{k}$, Reynolds number is $\text{Re} = \frac{VD}{\nu}$, and Prandtl number is $\text{Pr} = \frac{\nu}{\alpha}$, where $\nu = \mu/\rho$ and $\alpha = k/(\rho c_p)$. Unless otherwise specified, all fluid properties are evaluated at the mean (bulk) temperature of the fluid.

Solving for the temperature distributions in the fluid and in the pipe wall requires solving the coupled Eqs. (5.23), (5.26), (5.34), and (5.44) for the distributions along the pipe's axis of $T_m(x), T_w(x), T_s(x), h(x)$, and $q(x)$. Note that the number of unknowns (five) exceeds the number of equations (four) by one. An additional equation that relates the heat transfer coefficient to the fluid properties and flow velocity is needed.

Approximation

In the fully developed region, $h(x)$ can be approximated using Eq. (5.47) as

$$h \simeq 0.023 \frac{k}{D_i} \text{Re}^{0.8} \text{Pr}^{0.4} \qquad (5.48)$$

Another approximation that significantly simplifies the solution is neglecting the axial diffusion terms in Eq. (5.23),[1] which yields a constant heat flux at the inner pipe surfaces

$$\vec{q_s} \simeq \frac{q \mathcal{A}_w}{R_i} \qquad (5.49)$$

This approximation states that all the heat generated inside the pipe wall is advected by the moving fluid in the pipe. This approximation is justifiable on the grounds that heat transfer in the wall by conduction along the flow direction is negligible compared to heat transfer to the fluid, that is,

$$\frac{k_w \mathcal{A}_w}{R_i} \left| \frac{d^2 T_w(x)}{dx^2} \right| << |\vec{q}(x)|$$

$$\Rightarrow \frac{k_w \mathcal{A}_w}{R_i} \frac{T_w}{\mathcal{L}^2} << h\Delta T$$

$$\Rightarrow \frac{h \mathcal{L}^2 R_i}{k_w \mathcal{A}_w} \frac{(T_s - T_m)}{T_w} >> 1$$

For constant heat flux, $\vec{q_s}$, Eq. (5.33) can be solved to yield a linear profile for the axial variation of the mean temperature

$$T_m(x) = T_i + \frac{\vec{q_s} \pi D}{\dot{m} c_p} x = T_i + \frac{2\pi q \mathcal{A}_w}{\dot{m} c_p} x \qquad (5.50)$$

where T_i is the flow temperature at the pipe inlet. The wall surface temperature can then be estimated from Eq. (5.44)

$$T_s(x) = T_m(x) + \frac{\vec{q_s}}{h} \qquad (5.51)$$

where $T_m(x)$ is given by Eq. (5.50) and h by Eq. (5.48). These approximate profiles for $T_m(x)$ and $T_s(x)$, along with the value of heat transfer coefficient will be compared with those predicted by the detailed numerical simulations using Fluent, presented next.

5.4 Modeling Using Fluent

The numerical domain for this problem is shown in Figure 5.7.

We will create the two-dimensional computational domain for the calculation then we will mesh it by dividing it into cells. We will then set up the model by assigning material properties, applying boundary conditions, modeling the heat source and specifying initial conditions. Finally, we will choose the solver and run the calculation.

Note that the problem can be simplified by modeling only the fluid domain. The heat generated in the solid pipe can be modeled as boundary conditions at the wall of the

---

[1]This assumption is valid for a large Peclet number, which is the ratio of the advective and diffusive transport rates.

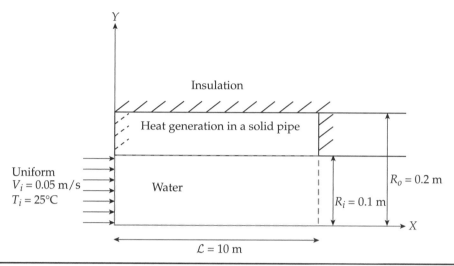

FIGURE 5.7 Fluent model description.

fluid domain. However, we are modeling the solid pipe for the purpose of introducing the user to modeling solid bodies in Fluent.

Launch ANSYS Workbench. Double-click **Fluid Flow (Fluent)** to create a new **Fluid Flow (Fluent)** analysis into the **Project Schematic** and rename it Chapter5.

Save the workbench project in your working directory and name it Heat Transfer.

5.4.1 Geometry

In this section, the geometry of the model will be created using ANSYS **DesignModeler**.

Figure 5.8 is a flowchart for the process of creating the geometry using **DesignModeler**.

Instructions to build the geometry using SpaceClaim are included in Appendix D.

The model is a two-dimensional axisymmetric model resembling the model created in Chapter 3. We will create the fluid domain first and then we will add the domain for the solid pipe surrounding the fluid region.

Right-click **Geometry** and select **New DesignModeler Geometry...** from the drop-down list to launch Design Modeler.

1. **Set the Units**
   Click **Units** and select **Meter** if it is not selected.

2. **Create the Fluid Domain**
   Click **XYPlane** under the **Tree Outline**.
   Click **Look at Face/Plane/Sketch** to display the XY plane on the screen.
   Click **Sketching** to switch to the sketching mode. Click **Draw** → Click **Rectangle** by two points. Click the origin to select it as the first point. Make sure the pointer of the mouse displays the letter **P** before selecting the origin. Drag the cursor and click anywhere in the first quadrant to place the opposite corner of the rectangle.
   Click **Dimensions** → Click **General** and select the upper edge of the rectangle, then drag the cursor in the positive Y direction. Click anywhere to place the dimension above the upper edge. Repeat the process for the right edge of the

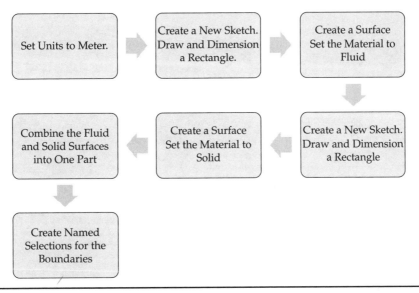

FIGURE 5.8 Flowchart of DesignModeler.

rectangle but drag the cursor in the positive X direction. Adjust the dimensions of the rectangle in **Details View** by changing the value of **H1** to 10 and the value of **V2** to 0.1.

Right-click anywhere inside the **Graphics** window and select **Zoom to Fit**.

Move the dimensions to be closer to the rectangle by clicking **Dimensions** → clicking **Move**. Click the dimensions one at a time and move to the desired location.

Click **Modeling** button, located next to **Sketching**, and expand **XYPlane** in **Tree Outline** view. **Sketch1** is the rectangle created. Right-click **Sketch1** and select **Rename**. Type MyFluidRectangle as the new name of the sketch. Press **Enter**.

We will now create the surface inside **MyFluidRectangle** sketch. Click **Concept** and then select **Surfaces From Sketches**. Select **MyFluidRectangle** in **Tree Outline** view and click **Apply** next to **Base Objects** in **Details View**.

Notice the lightning bolt at the bottom left of the screen. A message is displayed next to it with hints on how to complete the process of creating the surface.

Click **Generate** to create the surface. The rectangle turns gray.

Expand **1 Part, 1 Body** in the **Tree Outline** to view the created surface named by default **Surface Body**. Rename it my-fluid by right-clicking it and selecting **Rename**. In **Details View** of **my-fluid**, the surface created is **Solid** by default. Click **Fluid/Solid** to activate the option for the cell. An arrow next to **Solid** will appear. Click the arrow to access the drop-down list. Change the material from **Solid** to **Fluid**.

The screen should look like Figure 5.9.

3. **Create the Solid Domain**
Click **New Sketch** shown in Figure 5.9 to create a new sketch in the currently active plane.

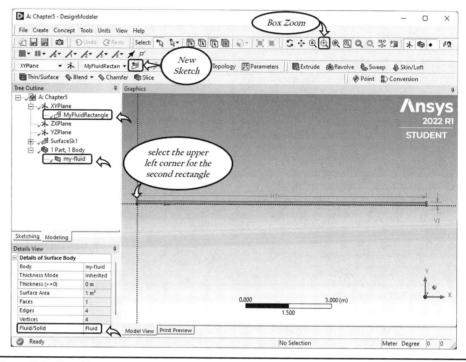

FIGURE 5.9 New Sketch and change material from Solid to Fluid.

If the **New Sketch** icon is grayed out, click **XYPlane** under the **Tree Outline** to make it visible. Expand **XYPlane** in **Tree Outline** to view the new sketch named **Sketch2**. A plane can hold any number of sketches. Make sure **Sketch2** is selected before you start drawing.

Without using **New Sketch**, any drawing made will be automatically added to the currently active sketch, **MyFluidRectangle** in this case. Also note that if a drawing is being generated on a plane that has no sketches, a new sketch will be automatically created to hold the drawing.

Click **Sketching** to switch to the sketching mode. Click **Draw** → Click **Rectangle** by two points.

It is helpful to use the **Box Zoom** icon shown in Figure 5.9 and zoom closer to the top left corner of the rectangle, right above the origin. After zooming in, click the **Box Zoom** icon again to deactivate it.

Click the top left corner of the existing rectangle as the first point as shown in Figure 5.9. Make sure the pointer of the mouse displays the letter **P** before selecting the corner. Drag the cursor in the positive X and Y directions and click anywhere to place the opposite corner of the rectangle above **MyFluidRectangle**.

Click **Dimensions** → Click **General** and select the upper edge of the top rectangle. Drag the cursor in the positive Y direction and click anywhere to place the dimension above the upper edge. Repeat the process for the right edge of the top rectangle but drag the cursor in the positive X direction. Adjust the dimensions of the rectangle in **Details View** by changing the value of **H3** to 10 and the value of **V4** to 0.1.

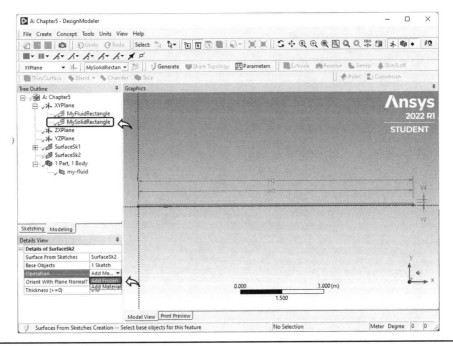

FIGURE 5.10   Add Frozen Operation.

Click **Modeling** button and expand **XYPlane** in **Tree Outline** view. Rename **Sketch2** by right-clicking it and selecting **Rename**. Type `MySolidRectangle` as the new name of **Sketch2**. Press **Enter**.

We will now create the surface inside the **MySolidRectangle** sketch. Click **Concept** then select **Surfaces From Sketches**. Select **MySolid Rectangle** in **Tree Outline** view as shown in Figure 5.10 and click **Apply** next to **Base Objects** in **Details View**. Click **Operation** shown in Figure 5.10 and change it from **Add Material** to **Add Frozen**.

Add Frozen will allow the user to add material as a separate body, which can be very useful when the model contains different materials.

Click **Generate**.

The top rectangle turns gray and the surface is created. Note that the shade of gray color is different for the solid surface than the fluid surface indicating they are made of different materials.

Expand **2 Parts, 2 Bodies** in the **Tree Outline** and rename the created surface my-solid by right-clicking **Surface Body** and selecting **Rename**. Type `my-solid` and press **Enter**.

4. **Form One Part**

Hold the **Ctrl** button and select **my-fluid** and **my-solid** under **2 Parts, 2 Bodies** in the **Tree Outline**. Right-click and select **Form New Part**. Expand **Part** under **1 Part, 2 Bodies** to view the two bodies as shown in Figure 5.11. Note that in the **Details View** of **Part** the cell next to **Fluid/Solid** displays **Mixed**.

Forming one part of the two bodies will allow them to share the boundary and therefore connect their mesh in the following section.

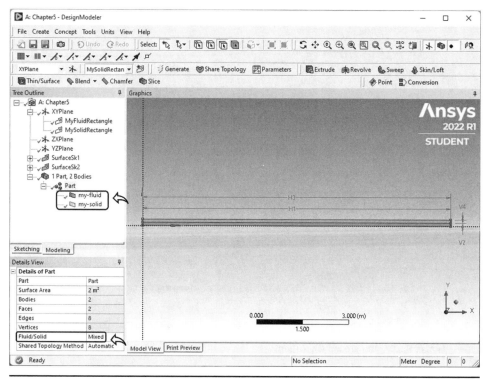

FIGURE 5.11 Form One Part.

5. **Create Named Selection**

We will name all the boundaries of the model where boundary conditions need to be applied.

In the selection toolbar, activate the **Selection Filter: Edges** icon. Zooming in and out may be needed for the selection of the edges.

Select the left edge of the fluid rectangle and right-click anywhere in the **Graphics** window. Select **Named Selection** from the list of options. Click **Apply** next to **Geometry** in **Details View**. Change the default name **NamedSel1**, next to **Named Selection**, by selecting it and typing `inlet`. Click **Generate**.

In the same manner name the outlet of the fluid. Select the right edge of the fluid rectangle, and right-click anywhere in the **Graphics** window. Select **Named Selection** from the list of options. Click **Apply** next to **Geometry** in **Details View**. Change the default name **NamedSel2** to `outlet`. Click **Generate**.

Create a **Named Selection** on each of the three edges of the solid part of the model according to its locations: `top-wall`, `right-wall`, and `left-wall`. Remember to Click **Generate** after naming each boundary.

Create a **Named Selection** at the bottom edge of the fluid rectangle and name it `centerline`.

The created names for the boundaries appear in the **Tree Outline** view.

Note there is no need to name the interface between **my-solid** and **my-fluid** as no boundary conditions are applied at the interface.

6. **Save the Project**

Click **File** and select **Save Project**. Close the DesignModeler application by clicking **File** and select **Close DesignModeler**.

The **Geometry** in Workbench has now a green check symbol next to it indicating the application is up to date. We are ready to start meshing the model.

Geometry Summary

* Click **Units**. Select **Meter**.

* Click **XYPlane**. Click **Look At Face/Plane/Sketch**.

* Click **Sketching**. Click **Draw** → Click **Rectangle**. Select the origin as the first point. Drag the cursor and place the opposite corner in the first quadrant.

* Click **Dimensions** → Click **General**. Select the horizontal edge of the rectangle. Drag the cursor and click anywhere to place the dimension. Repeat for the vertical edge. Adjust the value of **H1** to 10 and the value of **V2** to 0.1. Press **Enter**.

* Click **Modeling**. Expand **XYPlane** and rename **Sketch1** MyFluid-Rectangle.

* Click **Concept** and select **Surfaces From Sketches**. Select **MyFluidRectangle** for **Geometry**. Click **Generate**.

* Expand **1 Part, 1 Body** and rename the created body my-fluid. Select **Fluid** next to **Fluid/Solid**.

* Click **New Sketch**.

* Click **Sketching**. Click **Draw** → Click **Rectangle**. Select the top-left corner of **MyFluidRectangle** as the first point. Drag the cursor in the positive X and Y directions and place the opposite corner anywhere above **MyFluidRectangle**.

* Click **Dimensions** → Click **General**. Select the horizontal edge of the upper rectangle. Drag the cursor and click anywhere in the **Graphics** window to place the dimension. Repeat for the vertical edge. Adjust the value of **H3** to 10 and the value of **V4** to 0.1. Press **Enter**.

* Click **Modeling**. Expand **XYPlane** and rename **Sketch2** MySolid-Rectangle.

* Click **Concept** and select **Surfaces From Sketches**. Select **MySolidRectangle** for **Geometry** and **Add Frozen** for **Operation**. Click **Generate**.

* Expand **2 Parts, 2 Bodies** and rename the created body my-solid.

* Control-select **my-fluid** and **my-solid** under **2 Parts, 2 Bodies**. Right-click and select **Form New Part**.

* Activate **Selection Filter: Edges**

* Select the left edge of the fluid rectangle. Right-click and select **Named Selection**. Click **Apply** next to **Geometry** and type inlet. Click **Generate**.

* Repeat and create a **Named Selection** on the right edge of the fluid rectangle. Name it outlet.

* Repeat and create a **Named Selection** on the upper edge of the solid rectangle. Name it `top-wall`.

* Repeat and create a **Named Selection** on the right edge of the solid rectangle. Name it `right-wall`.

* Repeat and create a **Named Selection** on the left edge of the solid rectangle. Name it `left-wall`.

* Repeat and create a **Named Selection** on the lower edge of the fluid rectangle. Name it `centerline`.

* Click **File** and select **Save Project**.

* Click **File** and select **Close DesignModeler**.

5.4.2 Mesh

Open the Mesh application by double-clicking **Mesh** in ANSYS Workbench **Project Schematic**.

A flowchart of the steps needed to create the mesh is displayed in Figure 5.12.

1. **Default Mesh**
   Right-click **Mesh** in **Outline** view. Select **Generate Mesh**. Once Mesh is loaded, click **Mesh** in **Outline** view.

   Zoom in and rotate the view to examine the mesh. Click the **Z** axis in the triad at the bottom of the screen to redirect the view to the XY plane along the Z axis. Right-click in the geometry window and select **Zoom To Fit**. Note that the fluid and solid regions share the same nodes at the interface. The screen should look like Figure 5.13.

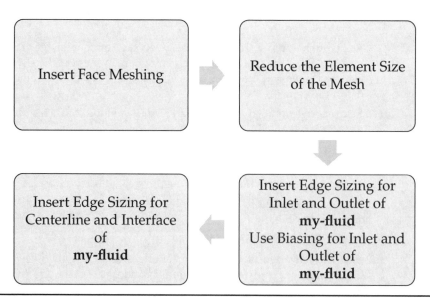

FIGURE 5.12 Flowchart of Mesh.

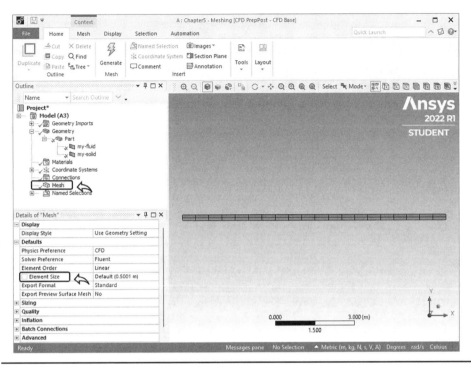

FIGURE 5.13 Element Size in Details of "Mesh."

Expand **Statistics** in **Details of "Mesh"** to check the number of **Nodes** and **Elements**. This is a very coarse mesh so we will refine it.

2. **Insert Face Meshing**

Right-click **Mesh** in **Outline** view and select **Insert** ⇒ **Face Meshing**.

Click **No Selection** next to **Geometry** in **Details of "Face Meshing"** - **Mapped Face Meshing**. Activate **Face** selection filter. Hold the **Ctrl** button and select the two rectangular surfaces (solid and fluid) in the geometry window. Click **Apply** next to **Geometry** in **Details of "Face Meshing"** - **Mapped Face Meshing**. A green check mark will appear next to **Face Meshing** in **Outline** view. **2 Faces** will appear next to **Geometry** in **Details of "Face Meshing"** - **Mapped Face Meshing** confirming the selection of the geometry.

Face Meshing creates a structured more uniform mesh with less distorted elements using less nodes.

3. **Reduce Element Size**

Click **Mesh** in **Outline** view. In **Details of "Mesh"**, change the default value of **Element Size** under **Defaults** from **0.5001**, shown in Figure 5.13, to 0.008. Right-click **Mesh** in **Outline** view and select either **Update** or **Generate Mesh**.

In our case, **Update** and **Generate Mesh** produce the same result, however, there is a difference between them. **Generate Mesh** produces the mesh and is useful when we are exploring different settings on the mesh. **Update** will update the geometry if it needs to be updated, refreshes the geometry and generates the mesh.

Updating the mesh in Mesh application will eliminate the need to update **Mesh** in Workbench **Project Schematic**.

Once meshing is completed, click **Mesh** in **Outline** view to see the generated refined mesh. Expand **Statistics** in **Details of "Mesh"** to see the number of **Nodes** and **Elements** in the mesh.

4. **Insert Edge Sizing at the Inlet and the Outlet**

Right-click **Mesh** in **Outline** view and select **Insert** ⇒ **Sizing**.

Activate **Edge** selection filter. Hold the **Ctrl** button and select the inlet and outlet edges of the fluid.

If the user is having difficulty selecting the inlet and outlet edges, click **Box Zoom** and zoom in near the inlet of the fluid. Activate **Edge** selection filter. Select the inlet edge of the fluid. Click **Zoom To Fit** then click **Box Zoom** and zoom in near the outlet of the fluid. Activate **Edge** selection filter, hold the **Ctrl** button and select the outlet edge of the fluid.

Click **Apply** next to **Geometry** in **Details of "Sizing"-Sizing**. The inserted **Sizing** in the **Outline** view changes to **Edge Sizing** as shown in Figure 5.14.

Select the cell next to **Element Size** and type 0.0013. Press **Enter**.

Click **Behavior** and change it to **Hard**.

Click **Bias Type**. The bias feature is very useful as the user can cluster the nodes along a specific location on a specific edge. From the drop-down list select the third type which is fine to coarse to fine. Select **Bias Factor** for the **Bias Option**. Click **Please Define**, type 25 and press **Enter**. Click **Zoom To Fit** to adjust the view in the geometry window. See Figure 5.14.

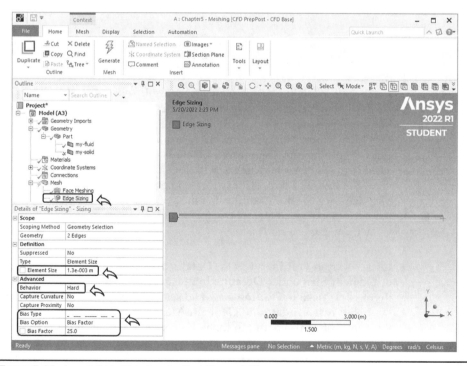

FIGURE 5.14 Insert Edge Sizing and adjust Element Size.

The bias feature will help reduce the total number of nodes in the model as refinement can be performed on specific locations. The **Bias Factor** is defined as the ratio of the largest element size to the smallest element size between the nodes. The ratio of the size of successive nodes is controlled by two parameters: the number of divisions and the bias factor. If the **Type** of sizing is set to **Element Size**, the number of divisions is the ratio of the edge size to the element size.

5. **Insert Edge Sizing at the Interface and at the Centerline**
   Right-click **Mesh** in **Outline** view and select **Insert** ⇒ **Sizing**.
   Activate **Edge** selection filter. Hold the **Ctrl** button and select the interface and the centerline as shown in Figure 5.15. Click **Apply** next to **Geometry** in **Details of "Sizing"-Sizing**. Select the cell next to **Element Size** and type 0.0019. Press **Enter**. Click **Behavior** and change it to **Hard**.

   Right-click **Mesh** in **Outline** view and select **Update**. Click **Mesh** to view the refined mesh, shown in Figure 5.16. Zoom in to check the details of the mesh.

   Note there is **1 Message** in the status bar since we used the **Update** option for the mesh. The message can be viewed by clicking it. See Figure 5.16. It displays **"The mesh translation to Fluent was successful."** indicating a successful mesh translation to Fluent. Reading the messages can be helpful when debugging a failed mesh. Expand **Quality** in **Details of "Mesh"** and select **Aspect Ratio** next to **Mesh Metric**. Note that the maximum value of the aspect ratio, **Max**, is **10.79**. The cells with the highest aspect ratios are in the fluid region near the interface and the centerline where biasing was used to refine the mesh.

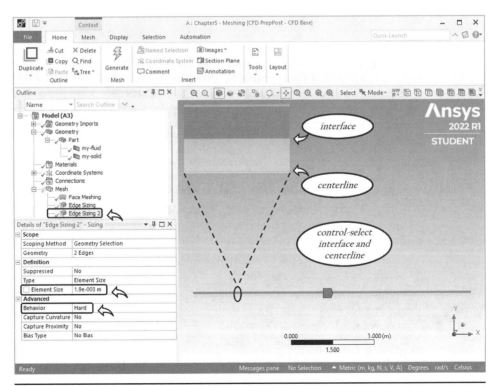

FIGURE 5.15   Control-select Interface and Centerline.

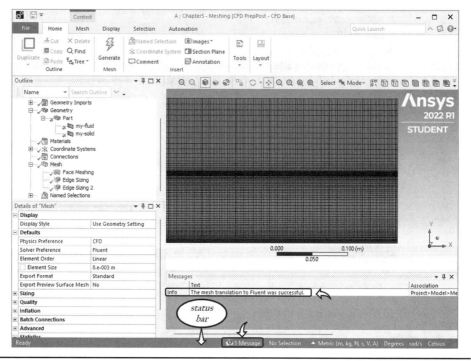

FIGURE 5.16 Insert Edge Sizing and status bar.

6. **Close Mesh Application**

Click **File** and select **Save Project** from the drop-down list.

Click **File** and select **Close Meshing** to close the ANSYS Mesh application and return to **Project Schematic**.

A green check symbol next to **Mesh** indicates the model is ready for the ANSYS **Setup** Application. Right-click **Mesh** and select **Update** if needed.

Mesh Summary

* Right-click **Mesh** and select **Generate Mesh**.

* Right-click **Mesh** and select **Insert ⇒ Face Meshing**. Activate **Face** selection filter and control-select the solid and fluid rectangles in the geometry window for **Geometry**.

* Click **Mesh**. Change the **Element Size** under **Defaults** to 0.008. Right-click **Mesh** and select **Generate Mesh**.

* Right-click **Mesh** and select **Insert ⇒ Sizing**. Activate **Edge** selection filter and control-select the inlet and outlet edges of the fluid rectangle for **Geometry**. Set the **Element Size** to 0.0013 and the **Behavior** to **Hard**. Select the third option next to **Bias Type** and set the **Bias Factor** to 25.

* Right-click **Mesh** and select **Insert ⇒ Sizing**. Activate **Edge** selection filter and control-select the interface and the centerline edges of the fluid

rectangle for **Geometry**. Set the **Element Size** to `0.0019` and the **Behavior** to **Hard**. Right-click **Mesh** and select **Update**.

* Click **File** and select **Save Project**.
* Click **File** and select **Close Meshing**.

5.4.3 Setup

Figure 5.17 is a flowchart of the steps needed to set up the model in Fluent. In ANSYS workbench **Project Schematic**, double-click **Setup** to launch Fluent. Enable **Double Precision** in the dialog box. Increase the number of **Solver Processes** depending on the available processors for usage. Click **Start**.

This will launch Fluent. We will go in order of what needs to be modified in the **Outline View** to setup the problem.

1. **General**

    Expand **Setup** → Double-click **General**. In the **Task Page**, click **Display...** under **Mesh** to display the **Surfaces** that make our mesh, as shown in the **Mesh Display** dialog box in Figure 5.18.

    Fluent recognizes **my-fluid** and **my-solid** as interior regions and are therefore named **interior-part-my-fluid** and **interior-part-my-solid**. **part-my-fluid** is the interface between **my-fluid** and **my-solid** and was generated automatically by Fluent. **part-my-fluid-shadow** was also generated automatically by Fluent since the interface belongs to two bodies.

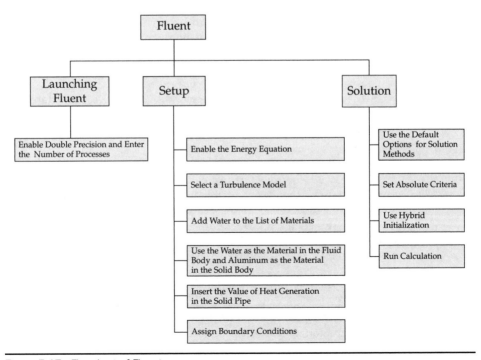

FIGURE 5.17 Flowchart of Fluent.

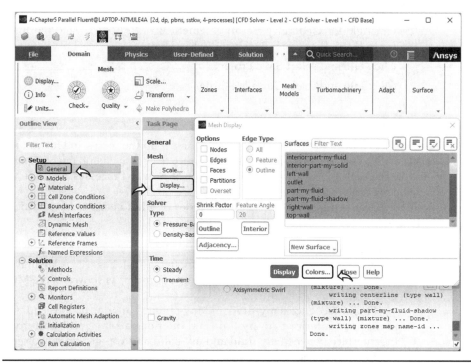

FIGURE 5.18   Mesh Display dialog box.

All surfaces are selected and displayed. Deselect some of the surfaces and click **Display**. Displaying selected surfaces to view them in the graphics window can be useful when there is a need to distinguish the different surfaces. Changing the default color of the mesh can also be done by clicking **Colors...** in the **Mesh Display** dialog box.

Close the **Mesh Display** dialog box.

Select **Axisymmetric** for **2D Space** under **Solver**. Keep the default options of **Pressure-Based** for the **Type**, **Absolute** for the **Velocity Formulation**, and **Steady** for **Time**.

2. **Models**

Expand **Setup** → Double-click **Models**.

Double-click **Energy - Off** in the **Task Page** view. The energy equation is disabled by default in Fluent. Enable **Energy Equation** as shown in Figure 5.19 to activate it. Click **OK**.

The energy equation covers both the conduction and the convection heat transfer in the model. The radiation modeling is separate and will not be covered in this model. The user is advised to check the Fluent user's manual for radiation modeling.

The k-epsilon turbulence model will be used.

In the **Task Page** view, double-click **Viscous - SST k-omega** as shown in Figure 5.19. Choose **k-epsilon (2 eqn)** from the **Model** list. The default values for **Cmu**, **C1-Epsilon**, and **C2-Epsilon** are used. Select **Enhanced Wall Treatment** under **Near-Wall Treatment** to capture the viscous sublayer inside the turbulent boundary layer. Click **OK**.

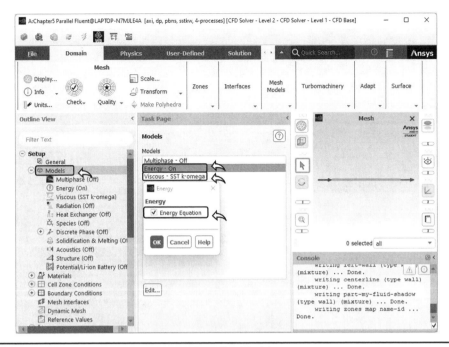

3. **Materials**

Expand **Setup** → Double-click **Materials**.

In the **Task Page** view, click **Create/Edit...** button to open the **Create/Edit Materials** dialog box. Click **Fluent Database....** Scroll to the bottom of the **Fluent Database Materials** list and select **water-liquid (h2o<l>)**. Click the **Copy** button. Close the **Fluent Database Materials** dialog box. Close the **Create/Edit Materials** dialog box.

The added material of **water-liquid** from the **Fluent Database...** appears under **Materials** in the **Task Page**.

4. **Cell Zone Conditions**

Expand **Setup** → Double-click **Cell Zone Conditions**.

Expand **Fluid** under **Cell Zone Conditions**. The domain for the model is named **my-fluid** in ANSYS DesignModeler and is listed as **part-my-fluid.1**.

Double-click **part-my-fluid.1** in the **Task Page** view. Change the **Material Name** assigned to it from **air** to **water-liquid** in the **Fluid** dialog box. Click **Apply** then click **Close** to close the **Fluid** dialog box.

Double-click **part-my-solid** in the **Task Page** view. The material assigned for the pipe is **aluminum** since it is the default solid material in Fluent. To model the heat generation in the solid pipe, enable **Source Terms** in the **Solid** dialog box and click **Source Terms** tab as shown in Figure 5.20. Click **Edit...** under **Source Terms**. Increase the **Number of Energy sources** from **0** to **1**. In the **Energy sources** dialog box, click the arrow shown in Figure 5.21 and select **constant** from the list of options for the heat source. Enter 100000 next to [W/m^3] and click **OK**. The **Energy sources** dialog box closes. The **Solid** dialog box displays **1 source** of **Energy**. The source of energy is generating 100,000 W/m^3 inside

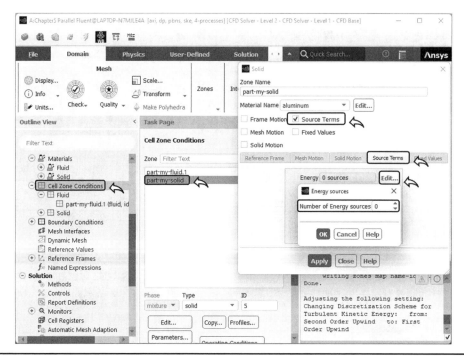

FIGURE 5.20 Enable Source Terms and Edit Energy sources.

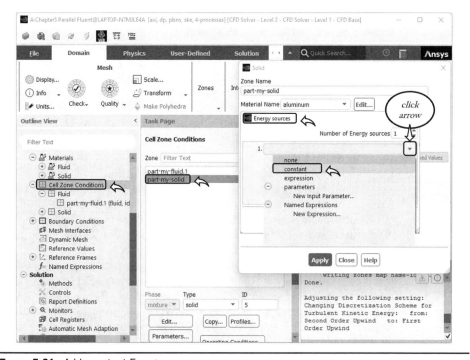

FIGURE 5.21 Add constant Energy sources.

the solid pipe. Note that two-dimensional axisymmetric models are treated as three-dimensional models in Fluent, therefore the heat rate is per unit volume. Click **Apply** and then click **Close**.

5. **Boundary Conditions**

Expand **Setup** → Double-click **Boundary Conditions**.

We will step through all the boundaries listed in alphabetical order in the **Task Page** view and make changes if needed.

Select **centerline** from the list of **Boundary Conditions** in the **Task Page**. Change its **Type** to **axis**. Click **Apply** then click **Close** to close the **Axis** dialog box.

Select **inlet**. Make sure its **Type** is **velocity-inlet**. Click **Edit...**, Choose **Magnitude, Normal to Boundary** as the **Velocity Specification Method**. Set the **Velocity Magnitude [m/s]** value to 0.05 and click **Apply**. Click **Thermal** tab next to the **Momentum** tab. The default value of **Temperature [K]** is **300**. Change it to 298 which is the fluid inlet temperature. Note that the **Thermal** tab is active because the **Energy** equation is enabled. Click **Apply** and close the **Velocity Inlet** dialog box.

The **interior-part-my-fluid** and **interior-part-my-solid** are the interiors of the domain, and we will not make changes to them.

Select **left-wall**. Make sure its **Type** is **wall**. Click **Edit...**. Confirm the **Thermal** conditions for the wall are set to 0 for **Heat Flux [W/m²]** meaning it is an insulated wall. Click **Apply**, then click **Close**.

Select **outlet**. Make sure its **Type** is **pressure-outlet**. Click **Edit...**. Confirm the value of **Gauge Pressure [Pa]** at the outlet is 0. Click **Apply** then click **Close**.

part-my-fluid and **part-my-fluid-shadow** are the internal interfaces between the fluid and solid regions and are stationary walls coupled thermally by default to allow the transfer of heat from one region to another. In the **Task Page** view, select **part-my-fluid** and click **Edit...**. Make sure the **Thermal** boundary conditions are coupled as shown in Figure 5.22. Click **Apply** and then click **Close**.

The **right-wall** and **top-wall** boundary conditions are similar to the **left-wall** boundary conditions. After selecting each and clicking **Edit...** confirm that the **Thermal** conditions are set to 0 for **Heat Flux [W/m²]** to reflect the insulation boundary conditions.

6. **Methods**

Expand **Solution** → Double-click **Methods**. Keep the default options for the **Solution Methods**:

Pressure-Velocity Coupling → **Scheme** → **Coupled**.

Pressure-Velocity Coupling → **Flux Type** → **Rhie-Chow: distance based**.

Spatial Discretization → **Gradient** → **Least Squares Cell Based**.

Spatial Discretization → **Pressure** → **Second Order**.

Spatial Discretization → **Momentum** → **Second Order Upwind**.

Spatial Discretization → **Turbulent Kinetic Energy** → **First Order Upwind**.

Spatial Discretization → **Turbulent Dissipation Rate** → **First Order Upwind**.

Spatial Discretization → **Energy** → **Second Order Upwind**.

Scroll down if needed and make sure **Pseudo Time Method** is enabled.

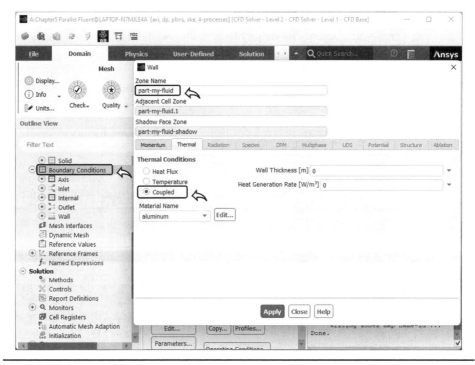

FIGURE 5.22 Thermal Boundary Conditions.

7. **Residuals**
   Expand **Solution** → Expand **Monitors** → Double-click **Residual**.
   Note that in addition to the residuals for the continuity, x-velocity, and y-velocity, we have three additional residuals corresponding to the three additional equations for energy, k, and epsilon.
   Set the **Absolute Criteria** to be 1e-14 for all six equations to be solved. In addition, disable **Check Convergence** for all six equations. By disabling **Check Convergence**, the calculation will not stop until the number of iterations is reached, regardless of the **Absolute Criteria**. Click **OK** to close the **Residual Monitors** dialog box.

8. **Initialization**
   Expand **Solution** → Double-click **Initialization**.
   Click **Initialize** in the **Task Page**. The **Console** will display a message when **Hybrid Initialization** is done.

9. **Run Calculation**
   Expand **Solution** → Double-click **Run Calculation**.
   In the **Task Page** view, set the **Number of Iterations** under **Parameters** to 1000. Click the **Calculate** button under **Solution Advancement**.
   The calculation can be stopped once the residuals for all equations reach a minimum value that does not change. In this case, we can stop the calculations after 300 iterations. To stop the calculation, click the **Stop** button shown in Figure 5.23. A **Calculation complete** message appears. Click **OK**.

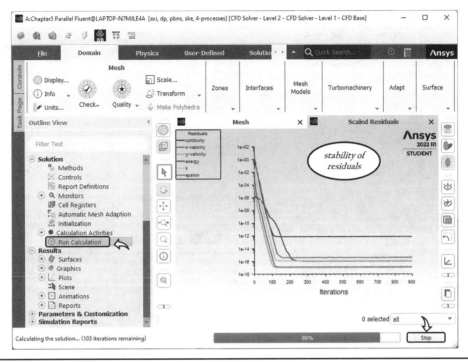

FIGURE 5.23 Stop calculation button.

10. **Flux Reports**

Expand **Results** → Double-click **Reports**.

In the **Task Page** view, double-click **Fluxes** or select **Fluxes** and click **Set Up...** to check the conservation of mass.

In the **Flux Reports** dialog box, select **Mass Flow Rate** for **Options**, select **inlet** and **outlet** from the list of **Boundaries** to check the conservation of mass. Click **Compute**.

The **Net Results** is displayed in the dialog box and also it is displayed in the **Console** window as shown in Figure 5.24. This net mass imbalance is an indication the model converged to a solution. However, the converged solution needs to be verified.

To check the conservation of energy, select **Total Heat Transfer Rate** under **Options**, while the **Flux Reports** dialog box is still available. Click **Select All Shown** to select all the boundaries, as shown in Figure 5.25. Click **Compute**.

The **Net Results** is displayed in the **Flux Reports** dialog box and shown in Figure 5.25. This **Net Results** for the **Total Heat Transfer Rate** is an indication of the conservation of energy.

The **User Source**, which is the total amount of heat generated by the solid body, is equal to **94,247.78** W and is displayed in the dialog box.

Note that the heat generation in the solid pipe is $q = 100,000$ W/m^3. The volume of the solid pipe is

$$\mathcal{V} = \pi \left(R_o^2 - R_i^2 \right) \mathcal{L},$$

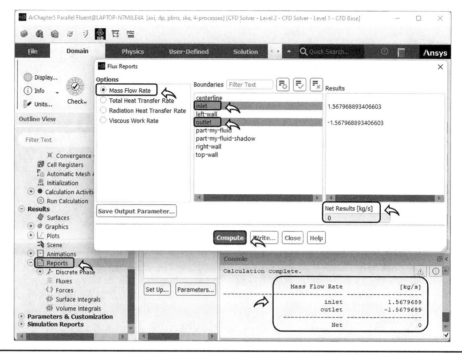

FIGURE 5.24   Mass Flow Rate Flux Reports.

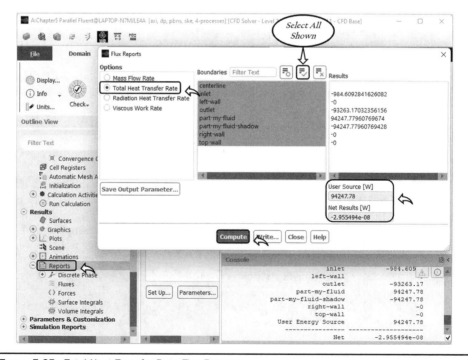

FIGURE 5.25   Total Heat Transfer Rate Flux Reports.

where R_o = the outer radius of the solid pipe, R_i = the inner radius of the solid pipe, and $\mathcal{L}$ = length of the solid pipe.

$$\mathcal{V} = \pi \times \left((0.2 \text{ m})^2 - (0.1 \text{ m})^2 \right) \times 10 \text{ m}$$

$$\mathcal{V} = 0.942478 \text{ m}^3$$

The rate of heat generated by the solid pipe is

$$Q = q \times \mathcal{V},$$

Q = 94,247.8 W which is in agreement with the **User Source** value.

Click **Close** to close the **Flux Reports**.

11. **Mass-Weighted Average Temperature**
    The mass-weighted average of the temperature at a cross-section inside the pipe is computed in Fluent by dividing the summation of the value of the temperature multiplied by the absolute value of the dot product of the area at the cross-section and momentum vectors by the summation of the absolute value of the dot product of the area and momentum vectors.

    The mass-weighted average of the temperature at an axial distance x inside the pipe is equivalent to the mean temperature value $T_m(x)$ as defined by Eq. (5.50).

    We will report the Fluent calculated **Mass-Weighted Average** temperature of the fluid inside the pipe at the cross-sections at x=8m, x=9m, and x=10m (outlet).

    To report the calculated **Mass-Weighted Average** temperature at x=8m, we need to create a vertical line at x=8m inside the two-dimensional model.

    Click **Domain** in the ribbon. Click **Surface** and select **Create** ⇒ **Line/Rake....** Type x=8m under **New Surface Name**. Enter 8 and 0 next to **x0** and **y0** respectively. Enter 8 and 0.1 next to **x1** and **y1** respectively. See Figure 5.26. Click **Create**. The name under **New Surface Name** changes indicating that the line **x=8m** has been created.

    While the **Line/Rake Surface** dialog box is still open, create the second line at x=9m. Select the name under **New Surface Name** and type x=9m. Enter 9 and 0 next to **x0** and **y0** respectively. Enter 9 and 0.1 next to **x1** and **y1** respectively. Click **Create** and click **Close** to close the **Line/Rake Surface** dialog box.

    Expand **Results** → Double-click **Reports**. In the **Task Page**, double-click **Surface Integrals** as shown in Figure 5.27. Select **Mass-Weighted Average** from the drop-down list under **Report Type**. Select **Temperature...** from the upper drop-down list and **Static Temperature** from the lower drop-down list under **Field Variable**. Select **x=8m, x=9m**, and the **outlet** from the list of **Surfaces**. Click **Compute**.

    The results appear in the **Console Window** where the **Mass-Weighted Average** temperature at x = 8 m is **309.5 K**, the **Mass-Weighted Average** temperature at x = 9 m is **310.94 K** and the **Mass-Weighted Average** temperature at the outlet is **312.37 K**. Click **Close** to close the **Surface Integrals** dialog box.

    The **Mass-Weighted Average** values of the temperature presented in this section will be used for the verification of the problem.

12. **Save the Project and Close Fluent**
    Save the project by clicking **File** and selecting **Save Project** from the drop-down list. Close the Fluent application by clicking **File** and selecting **Close Fluent** and return to Workbench.

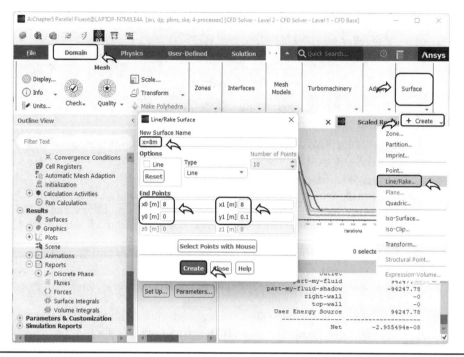

FIGURE 5.26 Create a Line.

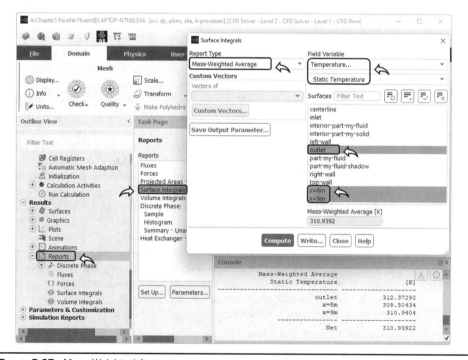

FIGURE 5.27 Mass-Weighted Average.

Setup Summary

* Expand **Setup** → Double-click **General**. Select **Axisymmetric** for **2D Space** and make sure **Pressure-Based** is selected for **Type**, **Absolute** is selected for **Velocity Formulation** and **Steady** is selected for **Time**.

* Expand **Setup** → Double-click **Models**. Double-click **Energy-Off**. Enable **Energy Equation** and click **OK**. Double-click **Viscous SST k-omega**. Select **k-epsilon (2eqn)** for **Model** and **Enhanced Wall Treatment** for **Near-Wall Treatment**. Click **OK**.

* Expand **Setup** → Double-click **Materials**. Click **Create/Edit....** Click **Fluent Database...**, select **water-liquid (h2o<l>)** and click **Copy**. Close the dialog boxes.

* Expand **Setup** → Double-click **Cell Zone Conditions**. Double-click **part-my-fluid.1** and select **water-liquid** next to **Material Name**. Click **Apply** then click **Close**. Double-click **part-my-solid** and enable **Source Terms**. Click **Edit...** under **Source Terms** and set the **Number of Energy sources** to 1. Select **constant** from the drop-down list of the **Energy Sources** dialog box and type 100000 next to [W/m^3]. Click **OK**. Click **Apply** then click **Close**.

* Expand **Setup** → Double-click **Boundary Conditions**. Select **Centerline** and select **axis** as the **Type**. Click **Apply** then click **Close**. Double-click **inlet** and set the **Velocity Magnitude [m/s]** to 0.05. Click the **Thermal** tab and set the **Temperature [K]** to 298. Click **Apply** then click **Close**.

* Expand **Solution** → Double-click **Methods**. Use the default options **Solution Methods**.

* Expand **Solution** → Expand **Monitors** → Double-click **Residual**. Set the **Absolute Criteria** for all equations to 1e-14. Disable **Check Convergence** for all equations. Click **OK**.

* Expand **Solution** → Double-click **Initialization**. Click **Initialize**.

* Expand **Solution** → Double-click **Run Calculation**. Set the **Number of Iterations** to 1000. Click **Calculate**.

* Expand **Results** → Double-click **Reports**. Double-click **Fluxes**. Select **Mass Flow Rate** for **Options**. Select **inlet** and **outlet** for **Boundaries**. Click **Compute**. Select **Total Heat Transfer Rate** for **Options** and click **Select All Shown** for **Boundaries**. Click **Compute**. Click **Close**.

* Click **Domain** in the ribbon and click **Surface**. Select **Create** ⇒ **Line/Rake....** Type x=8m under **New Surface Name**, 8 next to **x0**, 0 next to **y0**, 8 next to **x1** and 0.1 next to **y1**. Click **Create**. Type x=9m under **New Surface Name**, 9 next to **x0**, 0 next to **y0**, 9 next to **x1** and 0.1 next to **y1**. Click **Create**. Click **Close**.

* Expand **Results** → Double-click **Reports**. Double-click **Surface Integrals**. Select **Mass-Weighted Average** under **Report Type**. Select **Temperature...** and **Static Temperature** for **Field Variable**. Select **outlet x=8m, x=9m** from **Surfaces**. Click **Compute**. Click **Close**.

5.4.4 Results

In ANSYS workbench **Project Schematic**, double-click **Results** to launch CFD-Post where the ANSYS Fluent results are automatically loaded. The screen should look like Figure 5.28.

Since an axisymmetric model was specified in Fluent, a wedge reflecting a three-dimensional model is displayed when CFD-Post is launched. In the list of **part my fluid.1**, **periodic 1** and **periodic 1 shadow** are added to the surfaces created by the user. Also **periodic 2** and **periodic 2 shadow** are listed under **part my solid**. A periodic surface in Fluent always comes in pair and appears with its shadow. Check the surfaces listed by enabling and disabling them in the **Outline** view as they become highlighted in the viewer area.

Double-click **Default Transform** under **User Locations and Plots** in the **Outline** view, shown in Figure 5.28. In **Details of Default Transform**, disable **Instancing Info From Domain**. Enable the **Apply Reflection** as shown in Figure 5.29. Scrolling down the **Details of Default Transform** may be needed to enable **Apply Reflection**. Select **ZX Plane** next to **Method**. Click **Apply**.

The reflection of the model is applied and the full size two-dimensional pipe is being displayed in the **3D Viewer**.

By applying the reflection in **Default Transform**, all created plots by the user will be reflected about the ZX plane by default, unless the user modifies the view for a particular plot. Click the **Z** axis in the triad to display the model in the XY plane. The screen should look like Figure 5.29.

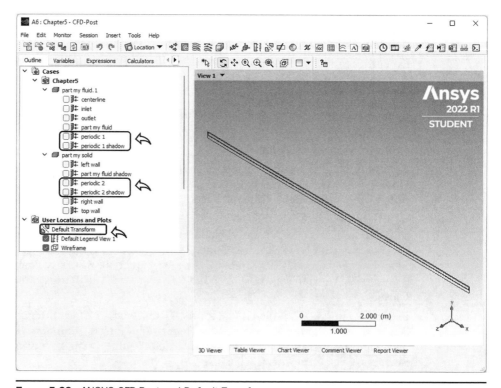

FIGURE 5.28 ANSYS CFD-Post and Default Transform.

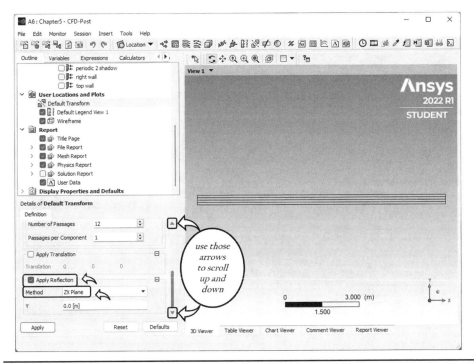

FIGURE 5.29 Apply Reflection in Details of Default Transform.

1. **Temperature Contours**
   Click **Contour**, shown in Figure 5.30, to create the temperature contours. Type My Temp as the name of the **Contour**. Click **OK**.

   In **Details of My Temp**, click **Location editor**, shown in Figure 5.30, to open the **Location Selector** dialog box displaying the complete list of available locations. Hold the **Ctrl** button and select **periodic 1** and **periodic 2**. Click **OK**. Select **Temperature** next to **Variable**. Change the # **of Contours** to 50.

   Click the **View** tab in **Details of My Temp** shown in Figure 5.31. Scroll down if needed and enable **Apply Scale**. Change the Y direction scale in the middle cell to 5. Click **Apply**.

   The **Apply Instancing Transform** at the bottom of **Details of My Temp** will allow the user to enable/disable the reflection of the model about the ZX plane.

   Double-click **Default Legend View 1** under **User Locations and Plots** in the **Outline** view. In **Details of Default Legend View 1**, click **Appearance** and change **Precision** under **Text Parameters** to **Fixed**. Click **Apply**.

   Disable **Wireframe**, shown in Figure 5.31. The **Wireframe** is not scaled and therefore disabling it is recommended for a better view of **My Temp** contour.

   The screen should look like Figure 5.31.

2. **Temperature distribution of the Inner Side of the Pipe Wall as a function of x**
   To display the results of the temperature on the inner side of the wall of the pipe in CFD-Post, we need to create the line that represents the inner side of the wall, the interface between the fluid and the solid region, where the values of the temperature are sampled.

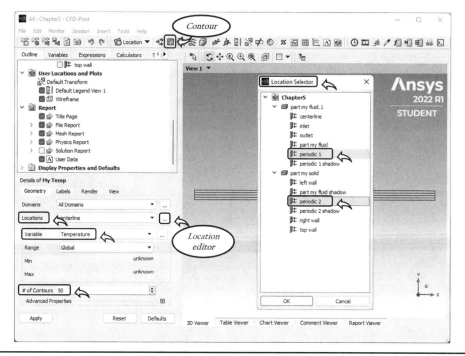

FIGURE 5.30   Location editor and Location Selector.

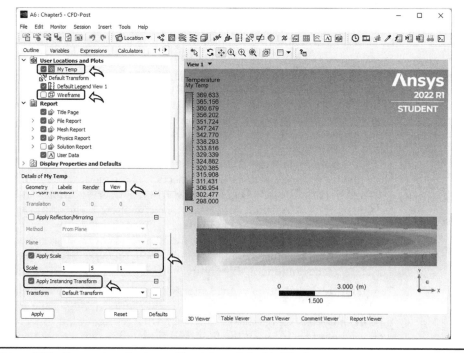

FIGURE 5.31   Temperature Contours, View tab, and disable Wireframe.

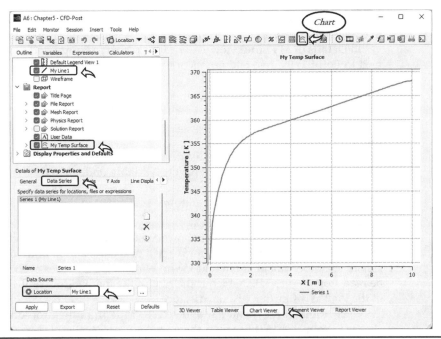

FIGURE 5.32 **Chart Viewer** tab and chart of the temperature at the inner side of the wall as a function of axial distance.

Click **Location** and select **Line** from the list of options. Type My Line1 next to **Name** and click **OK**. In **Details of My Line1** view, enter (0,0.1,0) and (9.99,0.1,0), the two end points of the line. Note that the solution created by Fluent is at the cell centers, justifying the x-coordinate for the second point.

Increase the number of samples to 100 to create a smooth line and click **Apply**.

The line is created and appears under **User Location and Plots** in **Outline** view.

Click **Chart** in the toolbar as shown in Figure 5.32 or click **Insert** → select **Chart**. Type My Temp Surface next to **Name** and click **OK**.

In **Details of My Temp Surface** view, disable **Display Title** under **General**.

Under **Data Series**, click the drop-down list next to **Location** and select **My Line1**. This is the location along which we are plotting the variable.

Under **X Axis**, select **X** from the drop-down list next to **Variable** to specify we are plotting the temperature as a function of axial distance. **X** is located toward the bottom of the drop-down list.

Under **Y Axis**, select **Temperature** from the drop-down list of **Variable** to specify we are plotting the temperature variable.

Click **Apply**. The screen should look like Figure 5.32.

The plot created is listed under **Report** and can be accessed at any time by clicking **Chart Viewer** tab at the bottom of the **3D Viewer** area.

3. **Temperature of the Fluid at the Outlet of the Pipe as a function of the Radial Distance**

To display the results of the temperature at the outlet, we need to create a vertical line at the outlet of the pipe to plot the temperature along this vertical line.

Click **Location** and select **Line** from the list of options. Type `My Line2` next to **Name** and click **OK**.

In **Details of My Line2** view, enter $(10,0,0)$ next to **Point 1** and $(10,0.1,0)$ next to **Point 2**, the two end points of the line. Change the number of samples to `100` and click **Apply**.

Click **Chart** in the toolbar and enter `My Temp Outlet`. Click **OK**.

In **Details of My Temp Outlet** view, disable **Display Title** under **General**. Under **Data Series**, click the drop-down list next to **Location** and select **My Line2**.

Select **Y** from the drop-down list next to **Variable** under **X Axis**. Select **Temperature** from the drop-down list next to **Variable** under **Y Axis**.

Click **Apply**.

Figure 5.33 displays the temperature at the outlet of the pipe.

4. **Wall Temperature at Specific Locations**

We will demonstrate in this section how to use the **Probe** tool to check variables at a specific location in the domain. We will probe the temperature of the inner side of the pipe wall at x=8m and x=9m. Note that the values probed can also be read from Figure 5.32, which is the plot of the temperature of the inner side of the Wall as a function of axial distance.

Click **Tools** in the menu bar and select **Probe** from the drop-down list, or click the **Probe** icon in the toolbar shown in Figure 5.34. Disable **My Temp** under **User Locations and Plots**

At the bottom of the **3D viewer**, set the coordinates next to **Probe At** to 8, 0.1 and 0 respectively, and choose **Temperature** as the variable to be probed.

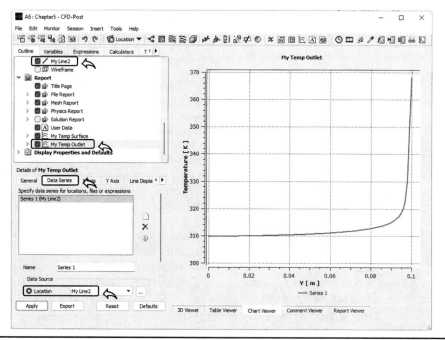

FIGURE 5.33 Chart of temperature of the fluid at the outlet as a function of radial distance.

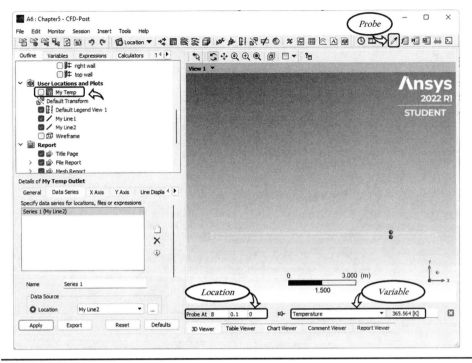

Figure 5.34 Probe tool.

The cell next to **Temperature** displays **365.56 [K]** as shown in Figure 5.34. This is the wall temperature at x=8m.

To determine the temperature of the solid surface at x=9m, set the coordinates next to **Probe At** to 9, 0.1 and 0 respectively and choose **Temperature** as the variable to be probed. The cell next to **Temperature** displays **366.98 [K]**.

5.5 Verification

5.5.1 Grid Independent Study

A grid independent study, using Richardson Extrapolation scheme described in Section 2.9, on the convection heat transfer coefficient was performed. The Mesh statistics for this study are summarized in Table 5.2. The convergence metrics are reported at nine points at the interface between the fluid and the solid. The local order of accuracy ranges between 0.33 and 7.31 with a global average p_{av} = 1.36. Oscillatory convergence, considered to be one of the limitations of the Richardson Extrapolation scheme as discussed in

TABLE 5.2 Mesh Statistics

Mesh # j	Quality	N_j	h_j(mm)	$r_{j,j-1}$	max(y^+)
1	Fine	1,514,880	1.149		0.74
2	Medium	860,337	1.525	1.33	1
3	Coarse	500,080	2	1.31	1.3

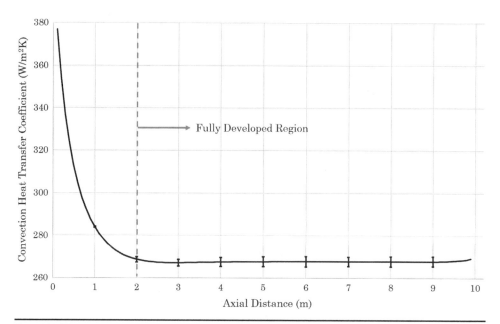

FIGURE 5.35 Amplified error bars for the heat transfer coefficient.

Section 2.9, occurred at only one point belonging to the thermal entry region. Figure 5.35 displays the convection heat transfer coefficient profile along with its error bars. We note here that the error bars are amplified by a factor of 10 for display purposes. The maximum magnitude of the error is estimated to be ± 0.244 W/(m^{2}K), which corresponds to a maximum discretization uncertainty of 0.18%.

5.5.2   Comparison with Empirical Correlations

Mean Temperature

The Fluent-calculated mass-weighted average values of the fluid temperature inside the pipe at the cross-sections at x=8m, x=9m, and x=10m were reported in the **Mass-Weighted Average Temperature** in Section 5.4.3 and were equal to 309.5K, 310.94K, and 312.37 K respectively.

The mass-weighted average of the temperature at an axial distance x inside the pipe is equivalent to the mean temperature value, $T_m(x)$ as defined by Eq. (5.50):

$$T_m(x) = T_{in} + \frac{\vec{q}_s \pi D}{\dot{m} c_p}, x$$

where $T_{in} = 298K$, $\vec{q}_s = \frac{Q}{\pi D \mathcal{L}} = \frac{94248}{\pi \times 0.2 \times 10} = 15,000 \frac{W}{m^2}$, $D = 0.2m$, $c_p = 4182 \frac{J}{kg.K}$, and $\dot{m} = 1.57 \frac{kg}{s}$.

Using Eq. (5.50), the mean temperature of the fluid is equal to 309.48 K at x=8m, 310.92 K at x=9m, and 312.35 K at x=10m.

The analytical values for the mean temperatures at the three different locations are in excellent agreement with the solution obtained by Fluent with a maximum relative percent error less than 0.1%.

Heat Transfer Coefficient

The heat transfer coefficient can be evaluated in Fluent using Eq. (5.44), $\vec{q}_s(x) = h(x)$ $(T_s(x) - T_m(x))$, where $\vec{q}_s(x)$ is the heat flux into the flow at the inner pipe wall and is equal to 15,000 W/m² (as shown above), $T_s(x) = T(R_i, x)$ is the surface temperature at the wall, and $T_m(x)$ is the mean temperature at a given cross-section perpendicular to the axis of the pipe.

We calculate the convection heat transfer coefficient at two locations x=8m and x=9m in the fully developed region inside the pipe.

The surface temperatures T_s at x=8m and x=9m were determined using the **Probe** tool in **Wall Temperature at Specific Locations** in Section 5.4.4. They are equal to 365.56 K and 366.98 K, respectively.

The mean temperatures T_m at x=8m and x=9m were calculated using the **Mass-Weighted Average Temperature** in Fluent in Section 5.4.3 and were equal to 309.50 K and 310.94 K, respectively.

Therefore, the convection heat transfer coefficients are

$$h_{x=8m} = \frac{q''}{T_s - T_m} = 267.57 \frac{W}{m^2 K}$$

$$h_{x=9m} = \frac{q''}{T_s - T_m} = 267.66 \frac{W}{m^2 K}$$

The values for the heat transfer coefficient at the two locations are within 0.03% which is an indication the model converged since the heat transfer coefficient remains constant for a thermally fully developed flow in a pipe with constant heat flux.

Nusselt Number

The Nusselt number $\mathrm{Nu} = \frac{hD}{k}$ at x=8m can be calculated using Fluent where h is the heat transfer coefficient at x=8 evaluated in this section and equal to 267.57 $\frac{W}{m^2 K}$. Therefore, $\mathrm{Nu}_{x=8m} = \frac{h_{x=8m} \times D}{k_{x=8m}} = \frac{267.57 \times 0.2}{0.6} = 89.19$.

For fully developed turbulent flow in a tube, the Nusselt number can be estimated from the Dittus-Boelter correlation, Eq. (5.47), where the properties are evaluated at the fluid mean (bulk) temperature and $n = 0.4$ since cold fluid is passing through the pipe.

Using water properties at 298 K, $\mathrm{Re}_{x=8m} = 9952$ and $\mathrm{Pr}_{x=8m} = 6.99$ which lead to a local Nusselt number $\mathrm{Nu}_{x=8m} = 79.04$.

The relative percent error in the Nusselt number between the solution obtained by Fluent and the empirical solution is around 13%.

CHAPTER **6**

Three-Dimensional Fluid Flow and Heat Transfer Modeling in a Heat Exchanger

List of Symbols	
Temperature	T
Overall heat transfer coefficient of the heat exchanger	U
Mass flow rate	$\dot{m}$
Constant pressure specific heat of hot fluid	c_p
Volume	$\mathcal{V}$
Pressure	p
Log mean temperature difference	ΔT_m
Rate of heat transfer	q
Heat capacity rate	C
Thermal resistance	R_t
Fouling resistance	R_f
Wall resistance	R_w
Diameter	D
Convection heat transfer coefficient	h
Thermal conductivity of the tube wall	k_w
Pipe length	$\mathcal{L}$
Nusselt number	Nu
Reynolds number	Re
Prandtl number	Pr
Wall roughess	ϵ
Density	ρ
Viscosity	μ
Velocity	V

This chapter covers the modeling of a three-dimensional heat exchanger. The heat exchanger is made of concentric pipes where the hot fluid flows through an inner pipe and the cold fluid flows in the same direction through an outer pipe, hence the name parallel flow concentric heat exchanger. The heat is transferred from the hot fluid to the cold fluid through the wall of the inner pipe.

Unlike the problem presented in Chapter 5 where an analytical solution exists, the heat exchanger problem does not have an analytical solution. Computational Fluid Dynamics (CFD) plays an important role in industrial heat transfer applications as we rely on modeling to predict the amount of heat being transferred in a heat exchanger.

The instructions to build a three-dimensional Fluent model of the concentric pipes heat exchanger will be presented. The results of the simulations will also be presented along a brief verification of the Fluent model.

The Fluent tutorials presented in the previous chapters must be completed before starting the tutorial in this chapter.

6.1   Introduction to Heat Exchangers

A heat exchanger is a device that allows the heat transfer between two fluids that are separated by a solid wall. Heat exchangers are found in many engineering applications including heating, ventilation, air conditioning, waste heat recovery, and chemical processing. They are classified based on the fluid flow arrangement and the way they are constructed.

The shell-and-tube is a common type of heat exchanger where a bundle of tubes are enclosed in a shell. The fluid flowing inside the tubes and the fluid flowing inside the spaces between the tubes can be parallel or perpendicular to each other.

The concentric tube heat exchanger, also called double pipe heat exchanger, is another type in which one pipe is placed inside another. They are very common, have the advantage of easy maintenance, and are suitable for high pressure fluids and high fouling conditions. A parallel flow concentric tube heat exchanger is one for which the hot and cold fluids enter at the same end and therefore move in the same direction. In a counterflow heat exchanger, the fluids enter at the opposite ends.

Consider the parallel flow concentric heat exchanger shown in Figure 6.1 where $T_{h,i}$ = inlet temperature for hot fluid, $T_{h,o}$ = outlet temperature for hot fluid, $T_{c,i}$ = inlet temperature for cold fluid, $T_{c,o}$ = outlet temperature for cold fluid, where all temperatures are in (K), U = overall heat transfer coefficient of the heat exchanger (W/m^{2}K), and $\mathcal{A}$ = heat transfer surface area (m^2).

To design heat exchangers, it is essential to relate the total heat transfer to inlet and outlet temperatures, the overall heat transfer coefficient and the surface area through which the heat transfer is happening.

In the analysis that follows in this section, we are assuming negligible heat transfer between the heat exchanger and the surroundings (perfectly insulated), negligible changes in potential and kinetic energy between inlet and outlet flows, negligible conduction heat transfer in the walls in the direction parallel to the flow, and constant heat transfer coefficient. The rate of heat exchanged between the hot and cold fluids at steady operation is equal to the rate of enthalpy gained by the cold stream and to the rate of enthalpy lost by the hot stream

$$q = \dot{m}_h c_{p,h} \left(T_{h,i} - T_{h,o} \right) \tag{6.1}$$

$$= \dot{m}_c c_{p,c} \left(T_{c,o} - T_{c,i} \right) \tag{6.2}$$

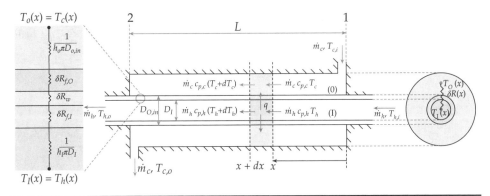

Figure 6.1 Schematic of a concentric heat exchanger.

where $\dot{m}$ is the mass flow rate, c_p is the constant pressure specific heat, the indices h and c refer respectively to hot and cold streams, and the indices i and o refer to the inlets and outlets. Depending on the working fluid and operating conditions, the nonnegligible dependence of the specific heat on temperature and pressure is accounted for by using the averaged value over the volume of the fluid. For example, for the hot stream,

$$\bar{c}_{p,h} = \int_{\mathcal{V}_h} c_p(T, p)\, d\mathcal{V} \tag{6.3}$$

The rate of heat exchanged between the hot and the cold streams may also be expressed in terms of the overall heat transfer coefficient U

$$q = U\mathcal{A}\Delta T_m \tag{6.4}$$

The product $U\mathcal{A}$ is such that

$$U\mathcal{A} = U_h \mathcal{A}_h = U_c \mathcal{A}_c \tag{6.5}$$

where $\mathcal{A}_h$ and $\mathcal{A}_c$ are respectively the areas on the hot and cold sides through which heat transfer is taking place, and U_h and U_c are the corresponding heat transfer coefficients. The log-mean temperature difference (LMTD) between the two streams, ΔT_m, is obtained by balancing the heat transfer over the infinitesimal section, $x - x + dx$, along the flow direction, as depicted in Figure 6.1, for the parallel-flow heat exchanger.

The infinitesimal rate of heat transferred from the hot fluid to the cold fluid is equal to the rate of heat lost by the hot fluid and to the rate of heat gained by the cold fluid, such that

$$dq = -\dot{m}_h c_{p,h} dT_h \tag{6.6}$$

$$= \dot{m}_c c_{p,c} dT_c \tag{6.7}$$

Expressed in terms of the heat transfer coefficient,

$$dq = U\, d\mathcal{A}\, (T_h - T_c) \tag{6.8}$$

Equations (6.6) and (6.7) yield

$$dT_h - dT_c = -\frac{dq}{\dot{m}_h c_{p,h}} - \frac{dq}{\dot{m}_c c_{p,c}} = -dq \left(\frac{1}{C_h} + \frac{1}{C_c} \right) \tag{6.9}$$

where $C_h = \dot{m}_h c_{p,h}$ and $C_c = \dot{m}_c c_{p,c}$ are the heat capacity rates of the hot and cold fluids respectively. Substituting dq for its value in Eq. (6.8), and denoting $\Delta T = T_h - T_c$, Eq. (6.9) becomes

$$d(\Delta T) = -U \, d\mathcal{A} \, \Delta T \left(\frac{1}{C_h} + \frac{1}{C_c} \right) \tag{6.10}$$

Dividing both sides of the equation by ΔT, and integrating over the area of the heat exchanger

$$\ln \left(\frac{\Delta T_2}{\Delta T_1} \right) = - \left(\frac{1}{C_h} + \frac{1}{C_c} \right) \int_{\mathcal{A}} U \, d\mathcal{A} \tag{6.11}$$

where the indices 1 and 2 refer to the ends of the heat exchanger at $x = 0$ and $x = \mathcal{L}$, respectively. Denoting $\overline{U}$ as the mean overall heat transfer coefficient, satisfying $\overline{U}\mathcal{A} = \int_{\mathcal{A}} U \, d\mathcal{A}$, and using $C_h = q/(T_{h,i} - T_{h,o})$ and $C_c = q/(T_{c,o} - T_{c,i})$ from Eqs. (6.1) and (6.2), and rearranging, we get

$$q = \overline{U} A \, \Delta T_m, \text{ where } \Delta T_m = \frac{\Delta T_2 - \Delta T_1}{\ln (\Delta T_2 / \Delta T_1)} \tag{6.12}$$

For a parallel flow heat exchanger, the cold and hot fluid inlets are on the same side, so that $\Delta T_1 = T_{h,i} - T_{c,i}$ and $\Delta T_2 = T_{h,o} - T_{c,o}$. For a counter flow heat exchanger, the cold and hot fluid inlets are on opposite sides, so that $\Delta T_1 = T_{h,i} - T_{c,o}$ and $\Delta T_2 = T_{h,o} - T_{c,i}$.

The most challenging part of the problem is determining the heat transfer coefficient $\overline{U}$. Referring to Figure 6.1, the product of the local heat transfer coefficient, $U(x)$, and the area $d\mathcal{A}(x)$, at location x is the inverse of $\delta R(x)$, the resistance to heat transfer across the area $d\mathcal{A}(x)$ between the inner and outer fluids at temperatures $T_h(x)$ and $T_c(x)$ respectively. This resistance consists of the sum of the following resistances in series: convection heat transfer resistance on the cold (outer) side, $1/[h(x)d\mathcal{A}(x)]_O$, where h is the convection heat transfer coefficient, conduction across the tube wall δR_w, and convection heat transfer resistance on the hot (inner) side, $1/[h(x)d\mathcal{A}(x)]_I$ [37]. In practice, two additional thermal resistances, called the fouling factors $R_{f,I}$ and $R_{f,O}$, are introduced on the hot (inner) and cold (outer) sides to account for the increase in resistance due to the deposition of scale on the inner and outer surfaces of the wall separating the hot and cold fluids. Then,

$$U(x) \, d\mathcal{A} = \frac{1}{\delta R(x)} \tag{6.13}$$

where

$$\delta R(x) = \frac{1}{(h(x)d\mathcal{A}(x))_O} + \delta R_{f,O} + \delta R_w + \delta R_{f,I} + \frac{1}{(h(x)d\mathcal{A}(x))_I} \tag{6.14}$$

For the parallel flow heat exchanger depicted in Figure 6.1, $d\mathcal{A}(x)_O = \pi D_{O,in} dx$, $d\mathcal{A}(x)_I = \pi D_I dx$ and $\delta R_w = \ln(D_{O,in}/D_I)/(k_w 2\pi dx)$, where k_w is the thermal conductivity of the tube wall separating the two fluids and D_I and $D_{O,in}$ are respectively the inner and outer diameters of the tube. Then,

$$U(x) \, d\mathcal{A} = \frac{1}{\dfrac{1}{\overline{h}_O(x)\pi D_{O,in}dx} + \delta R_{f,O} + \dfrac{\ln(D_{O,in}/D_I)}{k_w \, 2\pi dx} + \delta R_{f,I} + \dfrac{1}{\overline{h}_I(x)\pi D_I dx}} \tag{6.15}$$

So that

$$
\begin{aligned}
\overline{U}\,\mathcal{A} &= \overline{U}_c\,\mathcal{A}_c = \overline{U}_h\,\mathcal{A}_h \\
&= \int_{\mathcal{A}} \frac{1}{\delta R(x)} d\mathcal{A} \\
&= \int_0^{\mathcal{L}} \frac{1}{\frac{1}{h_O(x)\pi D_{O,in}} + \delta R_{f,O} + \frac{\ln(D_{O,in}/D_I)}{k_w\,2\pi} + \delta R_{f,I} + \frac{1}{h_I(x)\pi D_I}}\,dx
\end{aligned} \tag{6.16}
$$

and if all the properties do not change along the length of the heat exchanger,

$$
\begin{aligned}
\overline{U}\,\mathcal{A} &= \overline{U}_O\,\mathcal{A}_O = \overline{U}_I\,\mathcal{A}_I \\
&= \frac{1}{\frac{1}{h_O\pi D_{O,in}\mathcal{L}} + R_{f,O} + \frac{\ln(D_{O,in}/D_I)}{k_w\,2\pi\mathcal{L}} + R_{f,I} + \frac{1}{h_I\pi D_I\mathcal{L}}}
\end{aligned} \tag{6.17}
$$

where $\mathcal{A}_O = \pi D_{O,in}\mathcal{L}$ and $\mathcal{A}_I = \pi D_I \mathcal{L}$, where $\mathcal{L}$ is the length of the interface of the inner and outer pipes. In practice, the total fouling factors, $R_{f,O}$ and $R_{f,I}$, denoting the additional resistance due to the scale deposited on the inner and outer wall of the central pipe, are measured experimentally. If we model fouling as scale layers at the cold and hot sides of the tube wall of respective thicknesses and thermal conductivities of $t_{O,s}$, $k_{O,s}$ and $t_{I,s}$, $k_{I,s}$, then

$$
R_{f,O} = \frac{\ln[(D_{O,in} + t_{O,s})/D_{O,in}]}{k_{O,s}\,2\pi\mathcal{L}} \tag{6.18}
$$

$$
R_{f,I} = \frac{\ln[(D_I)/(D_I - t_{I,s})]}{k_{I,s}\,2\pi\mathcal{L}} \tag{6.19}
$$

The challenging aspect of the accurate estimation of the overall heat transfer coefficient is that the convection heat transfer coefficients on the inner and outer sides of the central pipe wall are functions of the flow conditions in addition to the thermal properties of the fluid [37]. The flow conditions are established by the balance of forces governed by the momentum equations (the Navier-Stokes equations). Note that, in general, the momentum and energy equations are coupled. For an ideal gas, they are coupled through the ideal gas law that relates the pressure, density, and temperature. In some cases, such as liquid flows with viscosity weakly temperature dependent, coupling is one way; i.e., the temperature distribution depends on the velocity field, but not the other way around. While the convection heat transfer coefficient, h, can be determined analytically for some simple cases, such as laminar flow over a flat plate with constant wall temperature, it must be determined experimentally or using CFD simulations for complex flows.

For fully developed turbulent flow in a pipe, a frequently used correlation for Nusselt number is the Dittus-Boelter equation (Eq. (5.47)), $\mathrm{Nu} = 0.023\,\mathrm{Re}^{0.8}\mathrm{Pr}^n$. The properties are evaluated at the fluid bulk temperature and $n = 0.4$ for heating, $n = 0.3$ for cooling. The Dittus-Boelter equation is applicable for the ranges $0.7 \leq \mathrm{Pr} \leq 160$, $\mathrm{Re} \geq 10{,}000$, $\mathcal{L}/D \geq 10$ and can be used to estimate h, the convection heat transfer coefficient where in evaluating Nu and Re, the characteristic length is the hydraulic diameter $D_H = 4\mathcal{A}/P_w$. $\mathcal{A}$ is the cross-sectional area and P_w is the wetted perimeter.

Equations (6.1), (6.2), and (6.12) are usually used to perform heat exchanger analysis when the inlet and outlet temperature are easily determined or known. Once the LMTD is determined using Eq. (6.12), the heat exchanged between the two fluids can be determined using Eq. (6.2) (or Eq. (6.1)) for a given value of C_c (or C_h). This in turn allows determining C_h (or C_c) and the product $U\mathcal{A}$. If the inlet and outlet temperatures are not easily determined, an iterative procedure involving LMTD can be used. Alternatively, it is easier to use the effectiveness-NTU (Number of Transfer Units) method.

In the example modeled in this chapter, the outlet temperatures are not specified. Inlet temperature and inlet mass flow rates for hot and cold fluids are given. In this case, it is easier to use the effectiveness-NTU or NTU method. The effectiveness of the heat exchanger is the ratio of q, the actual rate of heat transferred between the hot and cold streams, to q_{max}, the maximum possible heat transfer rate

$$\varepsilon = \frac{q}{q_{max}} \tag{6.20}$$

where the maximum possible heat transfer rate is given by

$$q_{max} = C_{min}\left(T_{h,i} - T_{c,i}\right) \tag{6.21}$$

where C_{min} is the minimum of C_c and C_h and $T_{h,i} - T_{c,i}$ is the maximum possible temperature difference (realized through an infinitely long heat exchanger) experienced by the stream with the smaller heat capacity rate. Equations (6.20) and (6.21) allow us to determine the actual heat transferred in terms of the inlet conditions as

$$q = \varepsilon C_{min}\left(T_{h,i} - T_{c,i}\right) \tag{6.22}$$

where the effectiveness depends on the ratio of the heat capacity rates and the NTU

$$\varepsilon = f(C_r, \text{NTU}) \tag{6.23}$$

where $C_r = \frac{C_{min}}{C_{max}}$, C_{max} is the maximum of C_c and C_h, and the NTU is

$$\text{NTU} = \frac{\overline{U}\mathcal{A}}{C_{min}} \tag{6.24}$$

For double-pipe heat exchangers, the effectiveness is

$$\varepsilon = \frac{1 - e^{-\text{NTU}(1+C_r)}}{1 + C_r} \qquad \text{for parallel flow} \tag{6.25}$$

$$\varepsilon = \frac{1 - e^{-\text{NTU}(1-C_r)}}{1 - C_r e^{-\text{NTU}(1-C_r)}} \qquad \text{for counter flow} \tag{6.26}$$

In some cases, one would want to determine NTU, given a target effectiveness. Equations (6.25) and (6.26) can be inverted to yield

$$\text{NTU} = -\frac{\ln[1 - \varepsilon(1 + C_r)]}{1 + C_r} \qquad \text{for parallel flow} \tag{6.27}$$

$$\text{NTU} = -\frac{1}{C_r - 1}\ln\left(\frac{\varepsilon - 1}{\varepsilon C_r - 1}\right) \qquad \text{for counter flow} \tag{6.28}$$

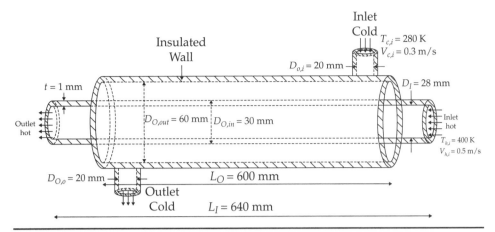

FIGURE 6.2   The domain and boundary conditions.

6.2   Problem Statement

The heat exchanger shown in Figure 6.2 is a schematic of the heat exchanger problem solved in this chapter.

It is a double pipe heat exchanger where the inner pipe is held concentrically inside the larger diameter outer pipe. Hot fluid with a velocity of 0.5 m/s at the entrance is flowing inside the inner pipe, while the cold fluid enters the outer pipe with a velocity of 0.3 m/s. The hot and cold fluids enter the heat exchanger at temperature of 400 K and 280 K, respectively. Additional fluid parameters and the dimensions of the heat exchanger are provided in Table 6.1.

A fluid based on a modified water liquid is the fluid used in our model. The thermal conductivity of water liquid was increased to enhance and better demonstrate the heat transfer between the hot and cold fluids since the size of the heat exchanger model is limited by the number of allowable elements in the student version of ANSYS.

The parameters of the heat exchanger we modeled in this section are displayed in Table 6.1.

6.3   Modeling Using Fluent

This section will guide the reader step by step on how to build a model of the heat exchanger problem using Fluent, with the assumption that the reader has completed the tutorials in the previous chapters. Instructions covered in the previous chapters will not be repeated.

Launch ANSYS Workbench. Double-click **Fluid Flow (Fluent)** to create a new **Fluid Flow (Fluent)** analysis into the **Project Schematic** and rename it Chapter6.

Save the workbench project in a working directory and name it Heat Exchanger.

6.3.1   Geometry

In this section, the geometry of the model will be created using ANSYS **DesignModeler**. Figure 6.3 is a flowchart for the process of creating the geometry using **DesignModeler**. Instructions to build the geometry using SpaceClaim are included in Appendix D.

TABLE 6.1 Heat Exchanger Model Dimensions, Properties, and Flow Parameters

Variable	Symbol	Value
Outer pipe dimensions		
Inlet diameter	$D_{O,i}$	20 mm
Outlet diameter	$D_{O,o}$	20 mm
Pipe length	$\mathcal{L}_O$	600 mm
Annulus inner diameter	$D_{O,in} = 2R_{O,in}$	30 mm
Annulus outer diameter	$D_{O,out} = 2R_{O,out}$	60 mm
Wall roughness	ϵ	0
Inner pipe dimensions		
Pipe length	$\mathcal{L}_I$	640 mm
Pipe diameter	$D_I = 2R_I$	28 mm
Wall roughness	ϵ	0
Cold and hot fluid properties		
Density	ρ	998.2 kg/m^3
Viscosity	μ	1.003 mPa.s
Specific heat	c_p	4182 J/kg.K
Thermal conductivity	k	5.97 W/m.k
Flow properties: cold fluid flow		
Inlet velocity	$V_{c,i}$	0.3 m/s
Reynolds number at inlet	$\mathrm{Re}_{c,i}$	5,971
Inlet temperature	$T_{c,i}$	280 K
Mass flow rate	$\dot{m}_c$	0.094 kg/s
Velocity in annular section	V_c	0.044 m/s
Reynolds number in annular section	Re_c	1,327
Outlet (gauge) pressure	p_o	0 Pa
Heat capacity rate	C_c	393 W/K
Flow properties: hot fluid flow		
Inlet velocity	$V_{h,i}$	0.5 m/s
Reynolds number	Re_h	13,933
Inlet temperature	$T_{h,i}$	400 K
Mass flow rate	$\dot{m}_h$	0.31 kg/s
Outlet (gauge) pressure	p_o	0 Pa
Heat capacity rate	C_h	1,296 W/K

Figure 6.4 shows the final geometry of the model that we will generate in this section. The heat exchanger exhibits symmetry around the YZ plane, and therefore we will model half the heat exchanger to minimize the computational efforts. We will start by creating the full model of the outer pipe along with its inlet and outlet, then we will create the concentric inner pipe. We will fill the inner pipe with the hot fluid body and the space

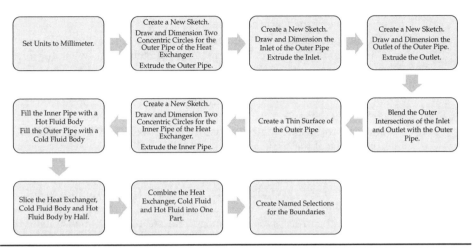

FIGURE 6.3 Flowchart of DesignModeler.

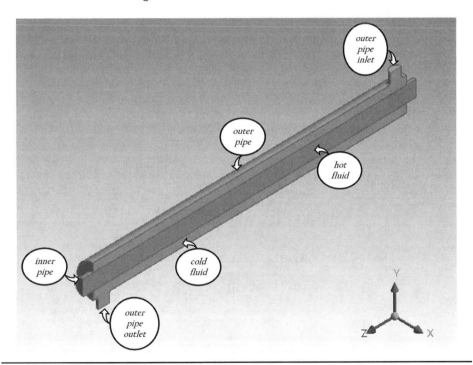

FIGURE 6.4 Heat exchanger geometry.

between the inner pipe and outer pipe with the cold fluid body. The detailed steps are listed in this section along with their summaries.

Right-click **Geometry** and select **New DesignModeler Geometry...** from the drop-down list to launch DesignModeler.

1. **Set the Units**
   Click **Units** and select **Millimeter** if it is not already selected.

2. **Create the Outer Pipe Cylinder**

Click **XYPlane** under the **Tree Outline**. Click **Look at Face/ Plane/Sketch** to display the XY plane in the **Graphics** window. Click **Sketching** to switch to the sketching mode. Click **Draw** → Click **Circle** to draw a circle by its center.

The full list of shapes under **Draw** may not be visible without scrolling the list.

Note that a new sketch is automatically being created since there are no existing sketches in the **XYPlane**.

Make sure the pointer of the mouse displays the letter **P** before selecting the origin as the center of the circle. Drag the cursor and click to place the circle anywhere on the screen. Once the first circle is drawn, click the origin again. Drag the cursor and click at any point inside or outside the first circle.

Click **Dimensions** → Click **General** from **Sketching Toolboxes** and select the outer circle. Drag the cursor and click to place the dimension anywhere on the screen. Repeat the process for the inner circle. Adjust the dimensions of the circles in **Details View** by changing the values of **D1** and **D2** to 60 and 30, respectively. See Figure 6.5.

Click **Modeling** button and expand **XYPlane** in **Tree Outline** view. **Sketch1** is the annulus created. Right-click **Sketch1** and select **Rename**. Type MyOuterPipe as the new name of the sketch and press **Enter**. The screen should look like Figure 6.5.

Extrude MyOuterPipe by clicking **Extrude** in the 3D Feature toolbar shown in Figure 6.5. Note that the features in the 3D Feature toolbar that require bodies to operate are greyed out because the current model has only surfaces.

Extrude can also be accessible through the **Create** menu by clicking **Create** and selecting **Extrude**.

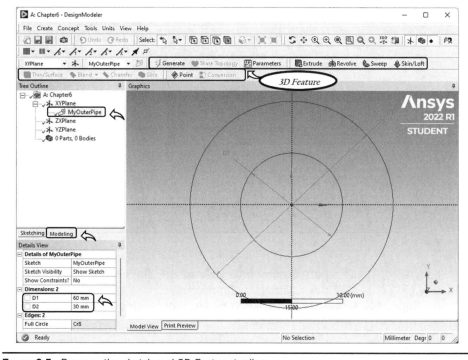

FIGURE 6.5 Rename the sketch and 3D Feature toolbar.

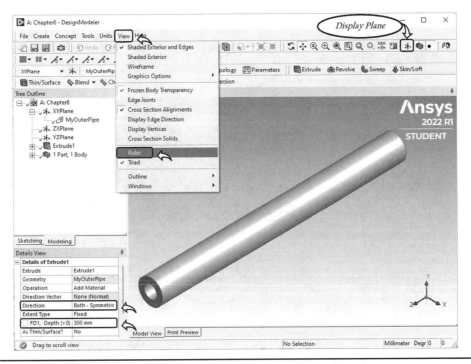

FIGURE 6.6   Extrude direction and depth, uncheck Ruler, and Display Plane icon.

Extrude1 appears in the **Tree Outline** view with its details in **Details View**.

Select **MyOuterPipe** in **Tree Outline** view. Click **Apply** next to **Geometry** in **Details View** to select MyOuterPipe sketch as the geometry to be extruded. Click **Direction** and change it to **Both - Symmetric**. Enter 300 next to **FD1, Depth (>0)** as the depth of the extrusion as shown in Figure 6.6. Click **Generate**. Click the cyan **ISO** ball. Zoom in and out and rotate the pipe to examine it.

In Figure 6.6, the ruler is hidden in the **Graphics** window. Click **View** and uncheck the **Ruler** to hide it. Note that we can hide **MyOuterPipe** sketch in the **Graphics** window by deactivating the icon **Display Plane** shown in Figure 6.6.

3. **Create Cold Fluid Inlet**

We will create the inlet of the outer pipe shown in Figure 6.4. We will draw a circle in the ZX plane and extrude it to create the inlet. Figure 6.7 will guide us through the process.

Click **ZXPlane** under the **Tree Outline**. Click **Look at Face/Plane/Sketch** to display the ZX plane on the screen. Click **Sketching** to start a new sketch in this plane. Click **Draw** → Click **Circle** by its center. To select the center of the circle, click any point along the Z axis located to the left of the origin. Make sure the pointer of the mouse displays the letter **C** before selecting the center of the circle. Note that the Z axis may not be visible but the letter **C** will appear on the screen, detecting a curve, when the cursor is on Z axis. Drag the cursor to place the circle anywhere in the **Graphics** window.

Click **Dimensions** → Click **General** from **Sketching Toolboxes** and select the circle, then drag the cursor and click anywhere to place the dimension.

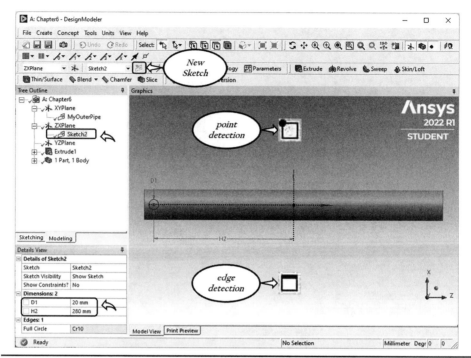

FIGURE 6.7   Adjust D1 and H2 values, point versus edge detection, and New Sketch Icon.

Click **Dimensions** → Click **Horizontal** to dimension the horizontal distance between the center of the circle created and the X axis (vertical axis), named **H2** and shown in Figure 6.7. Select the center of the circle as the first point. Be careful when selecting the center of the circle by monitoring the pointer of the mouse. The pointer will be different whether it detects a point or an edge, as shown in Figure 6.7. The bottom left side of the screen displays a message asking to **Select second point or 2D Edge for Horizontal dimension**. Select the vertical axis (X axis) then drag the mouse and click anywhere away from the cylinder.

Adjust the dimensions created in **Details View** by changing the value of **D1** to 20 and **H2** to 280.

Note that the user should not be confused with repeated names for dimensions because the number for dimension names begins with 1 for each plane. Also note the letter **D** is used for diameters and the letter **H** is used for horizontal dimensions.

Click **Modeling** button and expand **ZXPlane** in **Tree Outline** view. **Sketch2** is the circle created.

Click **Extrude**. Select **Sketch2** in **Tree Outline** view and click **Apply** next to **Geometry** in **Details View** to extrude the inlet of the pipe. Set the **Direction** to **Normal**. Set the depth of the extrusion **FD1, Depth(>0)** to 50 and click **Generate**.

Rotate the pipe to view the extruded inlet.

4. **Create Cold Fluid Outlet**
   We will create the outlet of the outer pipe in the same manner as the inlet but will position it on the right side of the origin and will extrude it in the reverse direction.
   Click **ZXPlane** under **Tree Outline** view. Click **Look at Face/Plane/Sketch**. Click **New Sketch** shown in Figure 6.7. **Sketch3** is created in the **ZXPlane**. Note that it is

important to create a new sketch for the outlet otherwise the outlet will be part of **Sketch2** and cannot be extruded in the opposite direction.

Click **Sketching** to change to the sketching mode. Click **Draw** → Click **Circle** by its center. Place the center of the circle on the **Z** axis at any point located to the right of the origin. Make sure the pointer of the mouse displays the letter **C** when selecting the center of the circle. Drag the cursor and click anywhere in the **Graphics** window.

Click **Dimensions** → Click **General** from **Sketching Toolboxes** and select the circle, then drag the cursor and click anywhere to place the dimension.

Click **Dimensions** → Click **Horizontal** to dimension the horizontal distance between the center of the circle created and the X axis. Select the center of the circle as the first point, then select the X axis. Drag the mouse and click anywhere away from the cylinder.

In **Details View**, set **D3**, the diameter of the circle, to 20. Set **H4**, the distance between the center of the circle and the X axis to 280.

Click **Modeling** button and expand **ZXPlane** in **Tree Outline** view. **Sketch3** is the circle created.

Click **Extrude**. Select **Sketch3** in **Tree Outline** view and click **Apply** next to **Geometry** in **Details View** to extrude the outlet of the pipe. Set the **Direction** to **Reversed**. Set the depth of the extrusion **FD1, Depth (>0)** to 50. Click **Generate**.

Rotate the model to view the outlet added to the outer pipe.

5. **Blend the Intersections**

We will demonstrate the **Fixed Radius Blend** feature by using it to smooth the sharp intersections of the inlet and the outlet with the pipe. The intersections to be smoothed are shown in Figure 6.8. Note that two edges were created at each intersection since this is a three dimensional model. Be careful when selecting the edges to be smoothed as we will only smooth the outer intersections.

Click **Blend** in the 3D Feature toolbar which can also be accessed from the **Create** menu. Select **Fixed Radius** from the drop-down list of **Blend**.

FBlend1 appears in the **Tree Outline** view with its details in **Details View**.

Make sure the **Selection Filter: Edges** is activated before selecting the edges.

Hold the **Ctrl** button and select the outer intersections of the inlet and outlet with the pipe as shown in Figure 6.8. Click **Apply** next to **Geometry** in **Details View**.

If the user is unable to control-select the two edges, the following instructions will help select them one at a time. Click **Apply** after selecting the first edge. Rotate the model and select the second edge. Click **1 Edge** next to **Geometry**. It will change into **Apply** and **Cancel**. Click **Apply**. **2 Edges** appears next to **Geometry** reflecting the selection of the two edges.

Once the two edges are selected, they will be highlighted. Select the default value next to **FD1, Radius(>0)** and type 5 as shown in Figure 6.8. Click **Generate**.

Rotate, zoom in and out to examine the model. Click the cyan **ISO** ball. Click **Zoom to Fit** to adjust the view.

6. **Pull the Outer Pipe**

We will use the feature **Thin/Surface** to pull the outer surfaces of the pipe and its inlet and outlet in the outward direction. Pulling the surfaces 1 mm outward will create a 1-mm thick solid body to be used as the outer pipe of the heat exchanger.

Click **Thin/Surface** in the 3D Feature toolbar which can also be accessed from the **Create** menu.

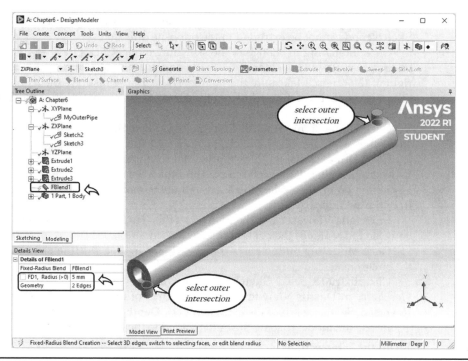

FIGURE 6.8 The outer intersections of the inlet and the outlet with the pipe.

For the **Geometry**, in **Details View**, we need to select the outer faces on the solid body that will be pulled out to create the 1-mm thick outer pipe. The total number of outer faces is seven. In addition to the curved surface of the outer pipe, there are three faces near the inlet and three faces near the outlet. The seven faces are highlighted in Figure 6.9.

Click the **Selection Filter: Faces** if it is not activated. Hold the control button and select the seven faces. Use the **Pan**, **Zoom**, and **Rotate** features in order to change the views while control-selecting all faces. The following paragraph will guide the user with detailed instructions on how to select the seven faces, in case it is needed.

Click the **X** axis in the triad to display the YZ plane in the **Graphics** window. Hold the **Ctrl** button and select all visible faces: the curved face of the inlet, the curved face of the outlet, the two blended faces, and the curved face of the heat exchanger. Click **Apply** next to **Geometry**. The cell next to **Geometry** displays **5 Faces**. Notice that when we click **Apply**, the color of the selected faces changes from green to cyan as shown in Figure 6.10. We need to add to the selection the two annulus side faces of the cylinder of the outer pipe. Click the **Z** axis in the triad. Select the annulus face highlighted in Figure 6.11. Notice that two planes are displayed at the bottom left corner of the **Graphics** window as shown in Figure 6.11. The two planes indicate two features, one visible and one invisible, detected when selecting the annulus. Hold the **Ctrl** button and select both planes corresponding to the two annulus faces of the outer pipe. The color of the planes turns red when selected. Click **5 Faces** next to **Geometry**. **5 Faces** will change into **Apply** and **Cancel**. Click **Apply**. The number of faces in the cell next to **Geometry** will update and will display **7 Faces**.

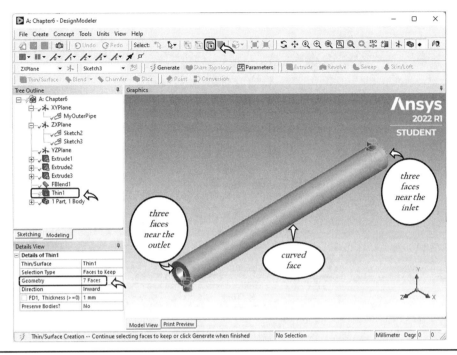

FIGURE 6.9 The seven faces used for Thin/Surface Feature.

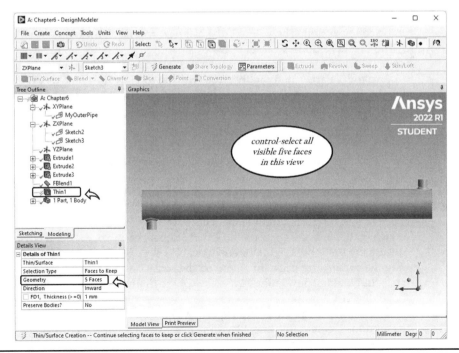

FIGURE 6.10 Control-select the five visible faces in the current view.

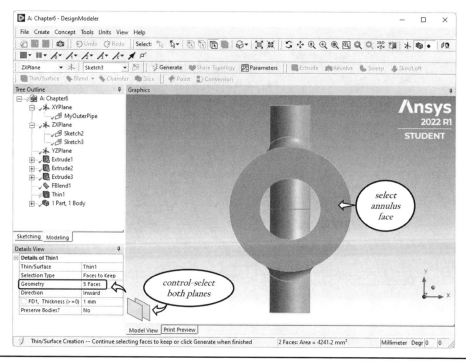

Figure 6.11 Select the two annulus faces and add to Geometry in Details of Thin1.

In **Details View**, click **Direction** and select **Outward** from the drop-down list of the cell next to it. The value of the cell next to **FD1, Thickness (>=0)** should be 1. The **Preserve Bodies?** is set to **No** by default and will keep it this way as we no longer need the original solid body. Click **Generate**.

Expand **1 Part, 1 Body** in the **Tree Outline** to view the created body named by default **Solid**. Rename it my-hx. In **Details View** of **my-hx**, the **Fluid/Solid** option is set to **Solid** by default.

7. **Create the Inner Pipe**

Click **XYPlane** under the **Tree Outline**. Click **Look at Face/Plane/Sketch** to display the XY plane in the **Graphics** window. Click **New Sketch**. **Sketch4** is created in the **XYPlane**.

Switch to the sketching mode by clicking **Sketching**. Click **Draw** → Click **Circle** to draw circle by its center. Make sure the pointer of the mouse displays the letter **P** before selecting the origin as the center of the circle. Drag the cursor and click anywhere to place the circle on the screen. Repeat to create a second circle inside or outside the first one. Click **Dimensions** and select **General**. Select the outer circle, drag the cursor, and click anywhere on the screen. Repeat the process for the inner circle. In **Details View**, adjust the value of the diameter **D3** to 30 and the value of the diameter **D4** to 28.

Click **Modeling** button and expand **XYPlane** in **Tree Outline** view. **Sketch4** is the annulus created. Rename it MyInnerPipe as shown in Figure 6.12.

Click **Extrude** in the 3D Feature toolbar. Select **MyInnerPipe** in **Tree Outline** view and click **Apply** next to **Geometry** in **Details View** to select the annulus sketch

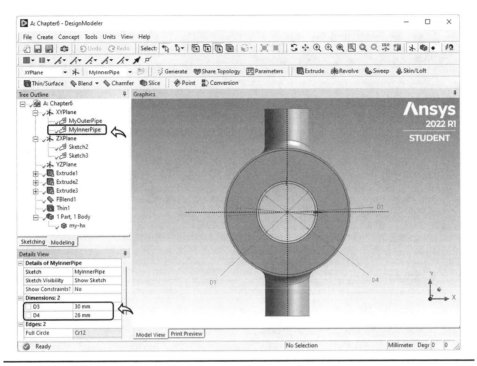

FIGURE 6.12 Build annulus and extrude to create the inner pipe.

of the inner pipe as the geometry to be extruded. Change the **Direction** to **Both - Symmetric**. Enter 320 next to **FD1, Depth (>0)** as the depth of the extrusion and click **Generate**. This will create a 1-mm thick inner pipe for the heat exchanger.

The heat exchanger geometry is now completed. It is a 1-mm thick solid body with a void inside its inner pipe and a void between the inner pipe and the outer pipe. The next step is to create the geometry for the hot fluid inside the inner pipe and the cold fluid between the inner pipe and the outer pipe.

8. **Create Hot Fluid Body**
   We will use the **Fill** by cavity feature to fill the void region inside the inner pipe. The **Fill** feature is located in **Tools** in the menu toolbar. Click the cyan **ISO** ball in the triad. Click **Zoom to Fit** to have a full isometric view of the heat exchanger. Click **Tools** and select **Fill** from its drop-down list. Fill1 appears in the **Tree Outline** view with its details in **Details View**. The **Extraction Type** is **By Cavity**. Make sure **Selection Filter: Faces** is activated. Select the inner face of the inner pipe as shown in Figure 6.13 and click **Apply** next to **Faces** in **Details View**. The color of the selected face changes into cyan after we click **Apply**. The hot fluid region is created by enclosing the cavity for the faces selected. Click **Generate**.

   Expand **2 Parts, 2 Bodies** in the **Tree Outline** to view the created body named **Solid** by default. Rename it my-hot-fluid. In **Details View** of **my-hot-fluid**, make sure the **Fluid/Solid** option is set to **Fluid**.

9. **Create the Cold Fluid Body**
   Adjust the view, if needed, by clicking the cyan **ISO** ball and clicking **Zoom to Fit**. Click **Tools** and select **Fill**. Fill2 appears in the **Tree Outline**. Make sure **Extraction**

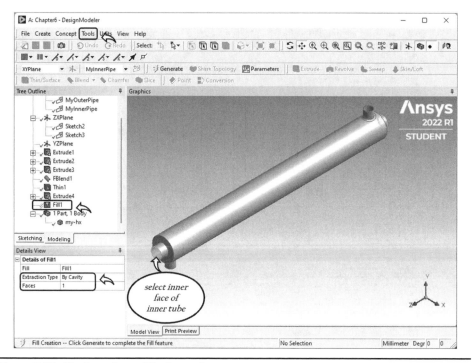

Figure 6.13 The inner face of the inner pipe.

Type is **By Cavity**. Use **Selection Filter: Faces** and select the inner face of the inlet of the outer pipe shown in Figure 6.14. Click **Apply** next to **Faces** in **Details View**. The color of the inner face of the inlet turns cyan and the cell next to **Faces** in **Details View** displays **1**. Use the **Rotate** icon to rotate the model until the inner face of the outlet is visible. Activate the **Selection Filter: Faces** and select the inner face of the outlet of the outer pipe as shown in Figure 6.15. Click **1** next to **Faces** in **Details View** to update the number of faces selected. **1** will turn into **Apply** and **Cancel**. Click **Apply**. The color of the inner face of the outlet turns cyan indicating it is added, and the cell next to **Faces** displays **2** reflecting the addition of the inner face of the outlet.

Click **Generate** to create the cold fluid region.

Expand **3 Parts, 3 Bodies** in the **Tree Outline** to view the created third body named **Solid**. Rename it `my-cold-fluid` and make sure its **Fluid/Solid** option is set to **Fluid**.

10. **Slice the Geometry**

The model exhibits symmetry around the YZ plane, therefore, we will model half of the created geometry.

Click **Slice** in the 3D Feature toolbar as shown in Figure 6.16.

Slice1 appears under the **Tree Outline** view. Make sure **Slice Type** is set to **Slice by Plane**. Select **YZPlane** in **Tree Outline** view and click **Apply** next to **Base Plane** in **Details View** of **Slice1**. Make sure **Slice Targets** is set to **All Bodies**. Click **Generate**.

The 3 bodies are sliced into 6 bodies. Expand **6 Parts, 6 Bodies** to view the list of the six bodies.

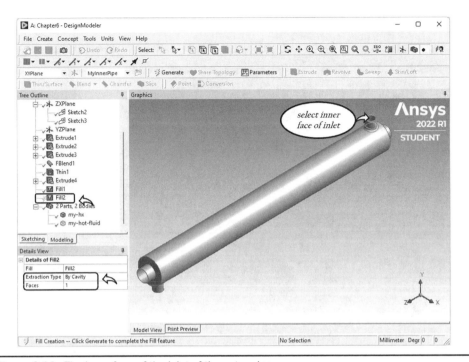

FIGURE 6.14   The inner face of the inlet of the outer pipe.

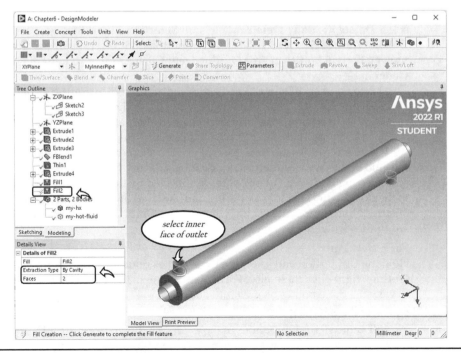

FIGURE 6.15   The inner face of the outlet of the outer pipe.

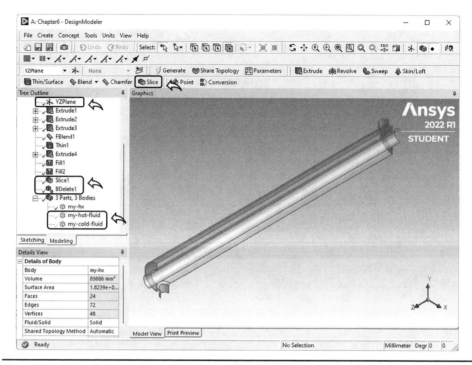

FIGURE 6.16 Slice tool, invisible fluid bodies, and final geometry.

We will delete the top three bodies under **6 Parts, 6 Bodies**. Click **Create** and select **Delete** ⇒ **Body Delete**. Make sure **Selection Filter: Bodies** is activated. In **Details View** of BDelete1, hold the **Ctrl** button and select the upper three bodies under **6 Parts, 6 Bodies** in **Tree Outline** view. Click **Apply** next to **Bodies** then click **Generate**.

Deleted bodies, **BDelete1**, can be restored, if desired, by right-clicking **BDelete1** and selecting **Delete**. This is an advantage of using the **Delete** feature.

The suppress feature can also be used as an alternative way to delete bodies and can be used by right-clicking a body and selecting **Suppress Body**.

Figure 6.16 shows the final geometry of the sliced heat exchanger. In this figure, the fluid bodies were hidden by right-clicking the **my-hx** body in **Tree Outline** view and selecting **Hide All Other Bodies** as reflected by the invisible check marks next to the fluid bodies. Explore by hiding and showing different bodies from the list of **3 Parts, 3 bodies**. Right-click anywhere in the **Graphics** window and select **Show All Bodies** when done exploring.

11. **Form New Part**
    Hold the **Ctrl** button and select the three bodies under **3 Parts, 3 Bodies** in the **Tree Outline**, right-click then select **Form New Part**. Expand **Part** under **1 Part, 3 Bodies** to view the three bodies formed into one part.

12. **Create Named Selections**
    We will name all the boundaries of the model, as shown in Figure 6.17. To make the bodies opaque, click **View** and uncheck **Frozen Body Transparency**. This will help the user with the selection of the faces.

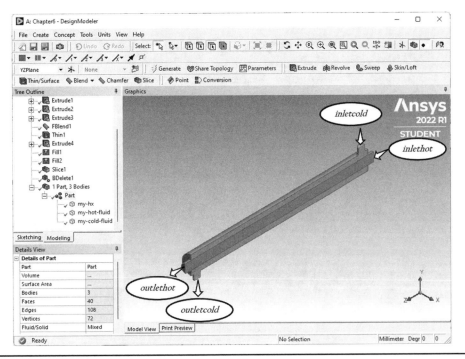

FIGURE 6.17 Inlets and outlets of hot and cold fluids.

Adjust the view by clicking the cyan **ISO** ball and **Zoom to Fit**.

In the selection toolbar, click the **Selection Filter: Faces** button. Rotating and zooming in and out may be needed for the selection of the faces.

We will start with the inlet of the cold fluid face and the outlet of the hot fluid face since they are easy to select in the current view of Figure 6.17.

Select the inlet of the cold fluid. Right-click anywhere in the **Graphics** window. Select **Named Selection** from the list. Click **Apply** next to **Geometry**. Change the default name **NamedSel1**, next to **Named Selection**, by selecting it and typing `intlet-cold`. Click **Generate**.

Select the outlet of the hot fluid. Right-click anywhere in the **Graphics** window. Select **Named Selection** from the list. Click **Apply** next to **Geometry**. Change the default name **NamedSel2**, next to **Named Selection**, by clicking and typing `outlet-hot`. Click **Generate**.

Rotate the model to access the outlet of the cold fluid face. Create a **Named Selection** on the outlet of the cold fluid face and name it `outlet-cold`.

Adjust the view to access the inlet of the hot fluid face. Create a **Named Selection** on the inlet of the cold fluid face and name it `inlet-hot`.

Recall to click **Generate** after naming each boundary.

The last named selection is the symmetry plane shown in Figure 6.18. There are five faces in this plane that we will select and name `symmetry`.

In Figure 6.18, it is clear that the cold fluid in the outer pipe is made of two faces, one above and one below the Z axis. In the same manner, the solid heat exchanger is also made of two faces but they are thin and therefore hard to select. The fifth face belongs to the hot fluid at the center of the symmetry plane.

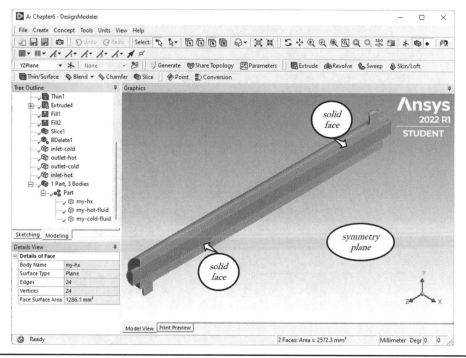

FIGURE 6.18 Symmetry plane made of five faces.

Hold the **Ctrl** button, select the five faces and name them `symmetry`. Zooming in will make it easier to select the solid faces. Click **Generate**.

Detailed instructions on how to select the five faces are provided below in case the user encounters difficulty selecting them.

Right-click **my-hx** body under **1 Part, 3 Bodies** in **Tree Outline**. Select **Hide All Other Bodies** from the list of options.

Select the curved face of the inner pipe shown in Figure 6.19. The four planes shown at the bottom left of the **Graphics** window indicate four faces detected at this location. Make sure when selecting the curved face to click away from the edges of the heat exchanger. If the face is not selected correctly, the number of planes displayed will not be equal to four.

Hold the **Ctrl** button and select the four planes. The color of the planes turns red when selected. Right-click anywhere in the **Graphics** window and select **Hide Face(s)**.

Hold the **Ctrl** button and select the two thin faces made from the solid part of the heat exchanger. Hiding some faces in the **Graphics** window may make this selection easier. Right-click anywhere in the **Graphics** window. Select **Named Selection** from the list. Click **Apply** next to **Geometry**. Change the default name **NamedSel5**, next to **Named Selection**, by selecting it and typing `symmetry`. The screen should look like Figure 6.20.

We will add the remaining three faces that correspond to the cold and hot fluids to `symmetry`. Right-click anywhere inside the **Graphics** window and select **Show All Faces**. Right-click again inside the **Graphics** window and select **Show All Bodies**. Hold the **Ctrl** button and select the three fluid faces: one inner hot fluid face

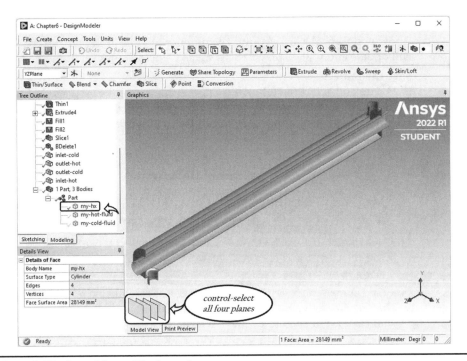

FIGURE 6.19   Curved face of the innerpipe.

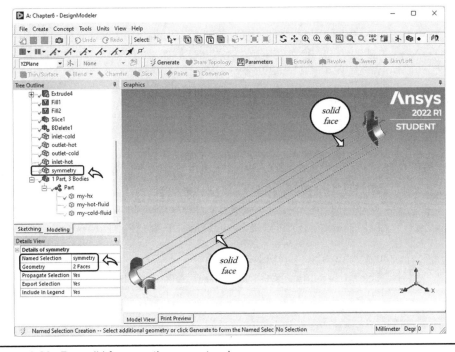

FIGURE 6.20   Two solid faces on the symmetry plane.

and two outer cold fluid faces. Click **2 Faces** next to **Geometry** which changes to **Apply** and **Cancel**. Click **Apply**. The number of faces in the cell next to **Geometry** will update and will display **5 Faces**. Click **Generate** after selecting the five faces.

13. **Save the Project**
Click **File** and select **Save Project**. Close or minimize the ANSYS DesignModeler application. The **Geometry** in **Project Schematic** Workbench now has a green check symbol next to it indicating the application is up to date. We are ready to start meshing the model.

Geometry Summary

* Click **Units**. Select **Millimeter**.

* Click **XYPlane**. Click **Look At Face/Plane/Sketch**.

* Click **Sketching**. Click **Draw** → Click **Circle**. Select the origin as the center. Drag the cursor and click anywhere in the **Graphics** window. Repeat and draw a second circle.

* Click **Dimensions** → Click **General**. Select the outer circle and place the dimension anywhere on the screen. Repeat for the inner circle. Adjust the values of **D1** and **D2** to 60 and 30 respectively.

* Click **Modeling**. Expand **XYPlane** and rename **Sketch1** MyOuterPipe.

* Click **Extrude**. Select **MyOuterPipe** and click **Apply** next to **Geometry**. Set the **Direction** to **Both-Symmetric** and **FD1, Depth (>0)** to 300. Click **Generate**.

* Click **ZXPlane**. Click **Look at Face/Plane/Sketch**.

* Click **Sketching**. Click **Draw** → Click **Circle**. Click any point on the Z axis to the left of the origin as the center of the circle. Drag the cursor and click anywhere in the **Graphics** window.

* Click **Dimensions** → Click **General**. Select the circle, and place the dimension anywhere on the screen. Click **Dimensions** → Click **Horizontal**. Select the center of the circle, then select the X axis. Place the dimension anywhere on the screen. Set the values of **D1** and **H2** to 20 and 280 respectively.

* Click **Modeling** and expand **ZXPlane**.

* Click **Extrude**. Select **Sketch2** for **Geometry**. Set the **Direction** to **Normal** and **FD1, Depth(>0)** to 50. Click **Generate**.

* Click **ZXPlane**. Click **Look at Face/Plane/Sketch**. Click **New Sketch**.

* Click **Sketching**. Click **Draw** → Click **Circle**. Click any point on the Z axis to the right of the origin as the center of the circle. Drag the cursor and click anywhere in the **Graphics** window.

* Click **Dimensions** → Click **General**. Select the circle, and place the dimension anywhere on the screen. Click **Dimensions** → Click **Horizontal**. Select the center of the circle, then select the X axis. Place the dimension anywhere on the screen. Set the values of **D3** and **H4** to 20 and 280 respectively.

* Click **Extrude**. Select **Sketch3** for **Geometry**. Set the **Direction** to **Reversed** and **FD1, Depth(>0)** to 50. Click **Generate** and click the cyan **ISO** ball in the triad.

* Click **Blend** and select **Fixed Radius**. Activate **Selection Filter: Edges**. Control-select the outer intersections of the inlet and the outlet with the pipe. Click **Apply** next to **Geometry**. Set **FD1, Radius(>0)** to 5 and click **Generate**.

* Click **Thin/Surface**. Control-select the outer seven faces on the solid body: the curved surface of the outer pipe, three faces near the inlet and three faces near the outlet. Click **Apply** next to **Geometry**. Set **Direction** to **Outward** and **FD1, Thickness (>=0)** to 1. Click **Generate**.

* Expand **1 Part, 1 Body** and rename **Solid** my-hx.

* Click **XYPlane**. Click **Look At Face/Plane/Sketch**. Click **New Sketch**.

* Click **Sketching**. Click **Draw** → Click **Circle**. Select the origin as the center. Drag the cursor and click anywhere in the **Graphics** window. Repeat and draw a second circle.

* Click **Dimensions** → Click **General**. Select the outer circle and place the dimension anywhere on the screen. Repeat for the inner circle. Set the values of **D3** and **D4** to 30 and 28 respectively.

* Click **Modeling** and rename **Sketch4** MyInnerPipe.

* Click **Extrude**. Select **MyInnerPipe** for **Geometry**. Set **Direction** to **Both-Symmetric** and **FD1, Depth (>0)** to 320. Click **Generate** and click the cyan **ISO** ball in the triad.

* Click **Tools** and select **Fill**. Make sure **Selection Filter: Faces** is activated and select the inner face of the inner pipe. Click **Apply** next to **Faces** and click **Generate**. Under **2 Parts, 2 Bodies**, rename **Solid** my-hot-fluid.

* Click **Tools** and select **Fill**. Control-select the inner faces of the inlet and outlet of the outer pipe. Click **Apply** next to **Faces** and click **Generate**. Under **3 Parts, 3 Bodies**, rename **Solid** my-cold-fluid.

* Click **Slice**. Ensure **Slice Type** is set to **Slice by Plane**. Select **YZPlane** and click **Apply** next to **Base Plane**. Make sure **Slice Targets** is set to **All Bodies** and click **Generate**.

* Click **Create** and select **Delete** ⇒ **Body Delete**. Control-select the upper three bodies under **6 Parts, 6 Bodies**. Click **Apply** next to **Bodies** and click **Generate**.

* Control-select the three bodies under **3 Parts, 3 Bodies**. Right-click and select **Form New Part**.

* Activate **Selection Filter: Faces**.

* Select the inlet of the cold fluid. Right-click and select **Named Selection**. Click **Apply** next to **Geometry**. Type inlet-cold next to **Named Selection** and click **Generate**.

* Repeat and create a **Named Selection** on the outlet of the hot fluid. Name it `outlet-hot`.
* Repeat and create a **Named Selection** on the outlet of the cold fluid. Name it `outlet-cold`.
* Repeat and create a **Named Selection** on the inlet of the hot fluid. Name it `inlet-hot`.
* Create a **Named Selection** on the five faces in the symmetry plane. Name it `symmetry`.
* Click **File** and select **Save Project**.
* Click **File** and select **Close DesignModeler**.

6.3.2 Mesh

Open the Mesh application by double-clicking **Mesh** in ANSYS Workbench **Project Schematic**.

Figure 6.21 is a flowchart of the steps needed to mesh the model.

1. **Set the Units to Millimeter**
   Click **Metric** in the status bar shown in Figure 6.22. Select the third option which is **Metric (mm, kg, N, s, mV, mA)**.

2. **Generate Default Mesh**
   Right-click **Mesh** in **Outline** view. Select **Generate Mesh**. Once the meshing is complete, click **Mesh** in **Outline** view to see the generated mesh, if it is not visible. Expand **Statistics** in **Details of "Mesh"** to see the number of **Nodes** and **Elements** in the mesh.

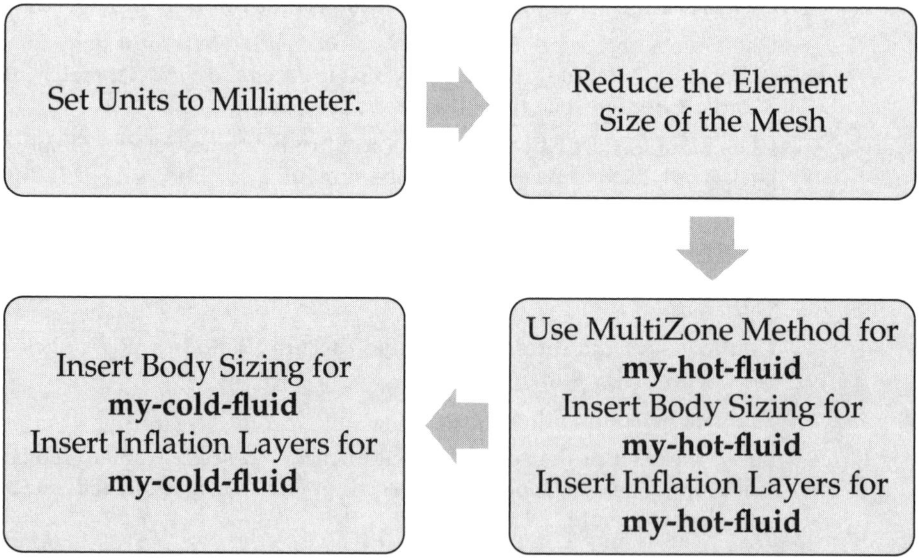

Set Units to Millimeter.

Reduce the Element Size of the Mesh

Use MultiZone Method for **my-hot-fluid** Insert Body Sizing for **my-hot-fluid** Insert Inflation Layers for **my-hot-fluid**

Insert Body Sizing for **my-cold-fluid** Insert Inflation Layers for **my-cold-fluid**

FIGURE 6.21 Flowchart of Mesh.

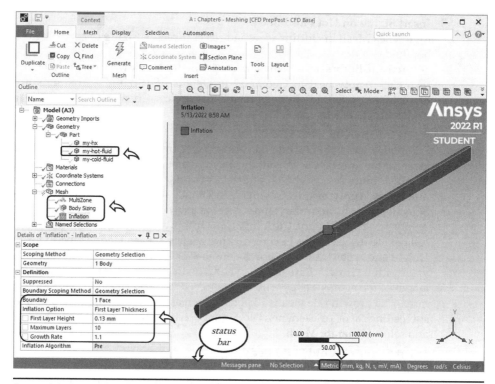

FIGURE 6.22 Details of inflation on hot fluid body.

We will refine the mesh by adding body sizing and inflation layers to the fluid bodies. We will use hexahedral elements for the hot fluid body and tetrahedral elements for the cold fluid body.

3. **Reduce Element Size**
   Click **Mesh** in **Outline** view. In **Details of "Mesh"**, change **Element Size** under **Defaults** to 3 . 2. Scrolling up may be needed to access **Defaults**. Right-click **Mesh** in **Outline** view and select **Generate Mesh**.
   Once completed, the generated refined mesh will be displayed.

4. **Insert Method**
   Expand **Geometry** in **Outline** view. Expand **Part** to view all bodies that make our **Part**. Right click **my-hot-fluid** and select **Hide All Other Bodies**.
   Hiding all other bodies in the model makes it easier to select the hot fluid in the geometry window.
   Right-click **Mesh** in **Outline** view, and select **Insert** ⇒ **Method**.
   In **Details of "Automatic Method" - Method**, click **No Selection** next to **Geometry**. Activate **Body** selection filter. Select the hot fluid body in the geometry window and click **Apply** next to **Geometry**. Click **Method** and select **MultiZone** from the drop-down list next to it.
   The **MultiZone** mesh method provides an automatic distribution of mapped and unmapped regions where hexahedral mesh is used whenever possible inside the region.

5. **Insert Sizing on Hot Fluid Body**
   Right-click **Mesh** in **Outline** view and select **Insert** ⇒ **Sizing** to insert sizing for the hot fluid.

   Activate **Body** selection filter. In the geometry window, select the hot fluid body, then click **Apply** next to **Geometry** in **Details of "Sizing" - Sizing**. Notice that **Sizing** changes to **Body Sizing** in the **Outline** view. Change the **Element Size** from its default value which is **3.2** to 3.1.

6. **Insert Inflation on Hot Fluid Body**
   Right-click **Mesh** in **Outline** view and select **Insert** ⇒ **Inflation** to add inflation layers at the interface of the hot fluid with the heat exchanger.

   Click **Body** selection filter and select the hot fluid body in the geometry window. Click **Apply** next to **Geometry** in **Details of "Inflation"-Inflation**. Click **No Selection** next to **Boundary** and select the curved face of the hot fluid body. Click **Apply** next to **Boundary**. Click **Inflation Option** and from the drop-down list in the cell next to it select **First Layer Thickness**. Click **Please Define** next to **First Layer Height** and type 0.13. Increase the number of **Maximum Layers** to 10 and set the value of the **Growth Rate** to 1.1. The **Growth Rate** represents the increase in the length of the element edge with each succeeding layer of elements from the interface. A 1.1 growth rate means 10 percent increase in element edge length with each succeeding layer of elements. See Figure 6.22. Right-click **Mesh** in **Outline** view and select **Generate Mesh**.

7. **Insert Sizing on Cold Fluid Body**
   Right-click anywhere in the geometry window and select **Show All Bodies**. Right-click **my-cold-fluid** under **Geometry** →**Part** in **Outline** view and select **Hide All Other Bodies**.

   Right-click **Mesh** in **Outline** view and select **Insert** ⇒ **Sizing** to insert sizing for the cold fluid.
   Activate the **Body** selection filter. Select the cold fluid body in the geometry window. Click **Apply** next to **Geometry** in **Details of "Sizing" - Sizing**. Change the **Element Size** from its default value which is **3.2** to 3.1.

8. **Insert Inflation on Cold Fluid Body**
   Right-click **Mesh** in **Outline** view and select **Insert** ⇒ **Inflation** to add inflation layers at the interface of the cold fluid with the heat exchanger.

   Activate the **Body** selection filter. Select the cold fluid body in the geometry window. Click **Apply** next to **Geometry** in **Details of "Inflation 2"-Inflation**. Click **No Selection** next to **Boundary**. Hold the **Ctrl** button and select the curved face at the inlet, the curved face at the outlet, the blended faces near the inlet and outlet, the inner curved face, the outer curved face, and the side faces (annulus) of the cold fluid, as shown in Figure 6.23. Detailed instructions for selecting the eight faces are provided at the end of this section in case the reader encounters difficulty.

   The cell next to **Boundary** displays **8 Faces**.

   Click **Inflation Option** and from the drop-down list in the cell next to it select **First Layer Thickness**. Click **Please Define** and type 0.12. Increase the number of **Maximum Layers** to 10 and set the value of the **Growth Rate** to 1.1. The screen should look like Figure 6.23.

   Right-click **Mesh** in **Outline** view and select **Update**. Right-click anywhere in the geometry window and select **Show All Bodies**. Click **X** axis in the triad and

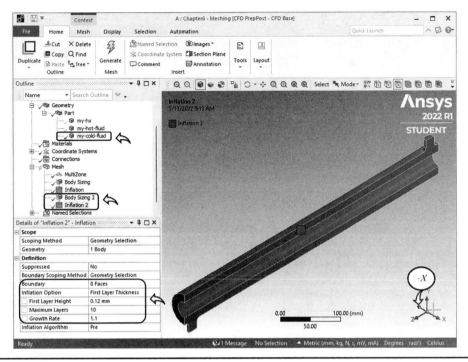

FIGURE 6.23 Details of inflation on cold fluid body.

zoom in to view the final mesh. In **Details of "Mesh"**, expand **Quality** as shown in Figure 6.24. Click **Mesh Metric** and from the drop-down list next to it select **Aspect Ratio**. Note the maximum value of the aspect ratio, **Max**, is 44.48. Expand **Statistics** in **Details of "Mesh"** to make sure the number of **Nodes** and **Elements** in the mesh is within the allowable number for the student version of ANSYS.

Figure 6.24 displays the final Mesh of the heat exchanger model.

Detailed instructions for selecting the eight faces are provided below.

Click **-X** axis in the triad, shown in Figure 6.23, to display the back view. Click **Zoom To Fit** in the graphics toolbar. Hold the **Ctrl** button and select all faces displayed in the back view, as shown in Figure 6.25. Click **Apply** next to **Boundary**. The color of the selected faces will change into dark red. The cell next to **Boundary** displays **5 Faces**.

Click the cyan **ISO** ball in the triad and click **5 Faces** next to **Boundary**. It will change into **Apply** and **Cancel**. Hold the **Ctrl** button and select the inner curved face of the cold fluid body. Click **Apply** next to **Boundary**. The cell next to **Boundary** displays **6 Faces**. Note that the faces selected previously will be deselected if we don't hold the **Ctrl** button.

Click **Z** axis in the triad and adjust the view by clicking **Zoom To Fit**. Click **6 Faces** next to **Boundary**. Hold the **Ctrl** button and select the annulus face shown in Figure 6.26. At the bottom left of the geometry window, two planes are displayed. Hold the **Ctrl** button and select the second plane. Make sure that both planes are red. Click **Apply** next to **Boundary**. A total of **8 Faces** are now selected as reflected in the cell next to **Boundary**.

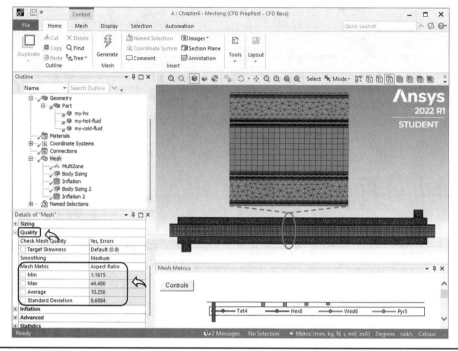

FIGURE 6.24   Mesh of the heat exchanger.

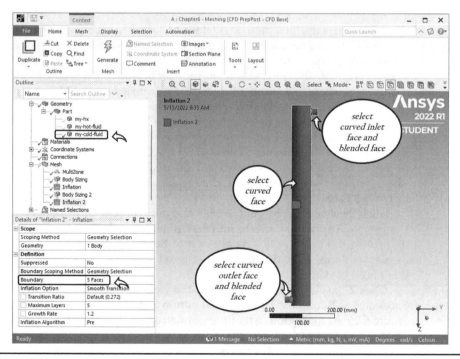

FIGURE 6.25   Eight faces selected as boundaries for the inflation on cold fluid body.

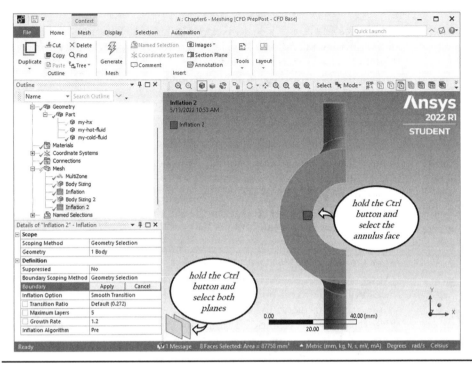

FIGURE 6.26   Selection of the boundaries for the inflation on cold fluid body.

9. **Close Mesh Application**

Click **File** and select **Save Project** from the drop-down list.

Close or minimize the mesh application and return to Workbench. Right click **Mesh** and select **Update**, if needed. A green check symbol next to **Mesh** indicates the model is ready for the ANSYS **Setup** Application.

Mesh Summary

* Click **Metric** in the status bar and select **Metric (mm, kg, N, s, V, A)**.

* Right-click **Mesh** and select **Generate Mesh**.

* Click **Mesh** and change the **Element Size** under **Defaults** to 3.2. Right-click **Mesh** and select **Generate Mesh**.

* Expand **Geometry** → Expand **Part**. Right-click **my-hot-fluid** and select **Hide All Other Bodies**.

* Right-click **Mesh**, and select **Insert** ⇒ **Method**. Activate **Body** selection filter and select the hot fluid body in the geometry window for **Geometry**. Select **MultiZone** next to **Method**.

* Right-click **Mesh** and select **Insert** ⇒ **Sizing**. Make sure **Body** selection filter is activated and select the hot fluid body in the geometry window for **Geometry**. Set the **Element Size** to 3.1.

∗ Right-click **Mesh** and select **Insert** ⇒ **Inflation**. Make sure **Body** selection filter is activated and select the hot fluid body in the geometry window for **Geometry**. Make sure **Face** selection filter is activated and select the curved face of the hot fluid body for **Boundary**. Select **First Layer Thickness** next to **Inflation Option**. Set the **First Layer Height** to 0.13, the **Maximum Layers** to 10 and the **Growth Rate** to 1.1. Right-click **Mesh** and select **Generate Mesh**.

∗ Right-click anywhere in the geometry window and select **Show All Bodies**.

∗ Under **Geometry** → **Part**, right-click **my-cold-fluid** and select **Hide All Other Bodies**.

∗ Right-click **Mesh** and select **Insert** ⇒ **Sizing**. Make sure **Body** selection filter is activated and select the cold fluid body in the geometry window for **Geometry**. Set the **Element Size** to 3.1.

∗ Right-click **Mesh** and select **Insert** ⇒ **Inflation**. Make sure **Body** selection filter is activated and select the cold fluid body in the geometry window for **Geometry**. Activate **Face** selection filter and control-select the curved faces of the inlet and outlet, the blended faces, the inner curved face, the outer curved face and the side faces of the cold fluid for **Boundary**. Select **First Layer Thickness** next to **Inflation Option**. Set the **First Layer Height** to 0.12, the **Maximum Layers** to 10 and the **Growth Rate** to 1.1. Right-click **Mesh** and select **Update**.

∗ Click **Mesh**, expand **Quality** and select **Aspect Ratio** next to **Mesh Metric**.

∗ Click **File** and select **Save Project**.

∗ Click **File** and select **Close Meshing**.

6.3.3 Setup

Figure 6.27 is a flowchart of the steps needed to set up the model in Fluent. Note that we will use ANSYS **Solution** to plot the results of the problem in this section.

Launch Fluent in ANSYS Workbench by double-clicking **Setup** in **Project Schematic**. Remember to enable **Double Precision** and to increase the number of **Solver Processes** depending on the available processors for usage. Click **Start**.

1. **General**
   Expand **Setup** → Double-click **General**. In the **Task Page** view, keep the default options of **Pressure-Based** for **Type**, **Absolute** for **Velocity Formulation**, and **Steady** for **Time** under **Solver** options.

2. **Models**
   Expand **Setup** → Double-click **Models**. In the **Task Page** view, double-click **Energy - Off**. Enable **Energy Equation** and click **OK**.
   Double-click **Viscous - SST k-omega**. Select **k-epsilon (2 eqn)** from the **Model** list. Select **Enhanced Wall Treatment** from the **Near-Wall Treatment** list. The **Near-Wall Treatment** is very important in solving convective heat transfer problems. The use of **Enhanced Wall Treatment** option in Fluent will capture the turbulent boundary layer while using large mesh near the wall. Click **OK**.

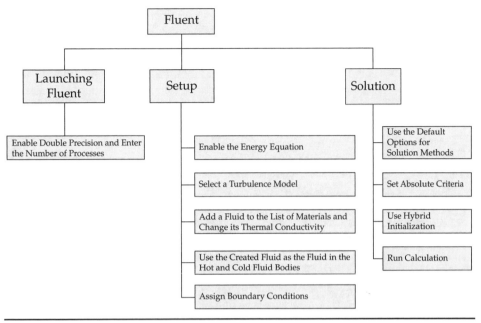

FIGURE 6.27 Flowchart of Fluent.

3. **Materials**

   Expand **Setup** → Double-click **Materials**. In the **Task Page** view, click **Create/ Edit...** button to open the **Create/Edit Materials** dialog box. Make sure **fluid** is selected under **Material Type**. Click **Fluent Database....** Scroll to the bottom of the **Fluent Database Materials** list and select **water-liquid (h2o<l>)**. Click the **Copy** button. Close the **Fluent Database Materials** dialog box. We will create a fluid based on water-liquid called my-fluid, which will have the same exact properties of liquid-water except for the thermal conductivity. In **Create/Edit Materials** dialog box shown in Figure 6.28, change the name from **water-liquid** to my-fluid. Delete **h2o<l>** under **Chemical Formula** since it is unique formula for **water-liquid**. Change the **Thermal Conductivity** from **0.6** to 5.97 to enhance the effect of heat transfer by convection. Click **Change/Create**. Click **Yes** to confirm overwriting **water-liquid** in the dialog box. Close the **Create/Edit Materials** dialog box.

   In the **Task Page** the created material **my-fluid** is listed under **Materials** and is now available to use for our model.

4. **Cell Zone Conditions**

   Expand **Setup** → Double-click **Cell Zone Conditions**. Expand **Fluid** to view the cold fluid and hot fluid regions created in Geometry listed as **part-my-cold-fluid** and **part-my-hot-fluid**.

   Double-click **part-my-cold-fluid** in the **Task Page** view. In the **Fluid** dialog box, change the **Material Name** assigned to it from **air** to **my-fluid**. Click **Apply** then click **Close** to close the **Fluid** dialog box. Repeat the process for **part-my-hot-fluid**.

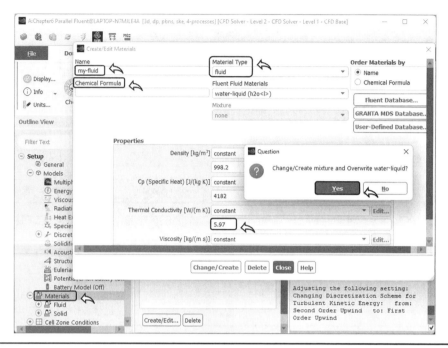

FIGURE 6.28   Create/Edit Materials dialog box.

Double-click **part-my-hx** in the **Task Page** view. The material assigned for the pipe is **aluminum** since it is the default solid material in Fluent. Click **Close** to close the **Solid** dialog box.

5. **Boundary Conditions**

   Expand **Setup** → Double-click **Boundary Conditions** to assign boundary conditions for the listed zones in the **Task Page**.

   Fluent can detect the type of boundaries from their names and group similar boundaries together in the **Outline View**. If the boundaries are not grouped together in the **Task Page** view, click **Toggle Tree View** shown in Figure 6.29. Select **Zone Type** under **Group By** from the list of options which will make it easier to assign boundary conditions.

   Select **inlet-cold** and make sure its **Type** is **velocity-inlet**.

   Click **Edit...** and choose **Magnitude, Normal to Boundary** as the **Velocity Specification Method**. Set the **Velocity Magnitude [m/s]** value to 0.3 and click **Apply**.

   The **Thermal** tab shown in Figure 6.30 is active since the **Energy Equation** is enabled. Set the inlet **Temperature [K]** value of the cold fluid to 280. Click **Apply** then click **Close** to close the **Velocity Inlet** dialog box.

   Repeat for the **inlet-hot** by setting the **Velocity Magnitude [m/s]** value to 0.5 and the **Temperature [K]** to 400.

   The interior parts of the domain do not have the option to make changes to them.

   Click **outlet-cold**. Make sure its **Type** is **pressure-outlet**. Click **Edit...** and confirm the value of **Gauge Pressure [Pa]** at the outlet is 0. Click **Close**. Repeat for the **outlet-hot**.

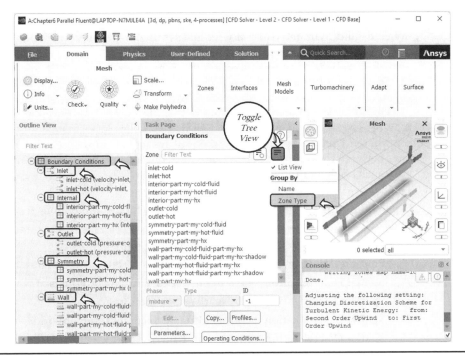

FIGURE 6.29 Toggle Tree View.

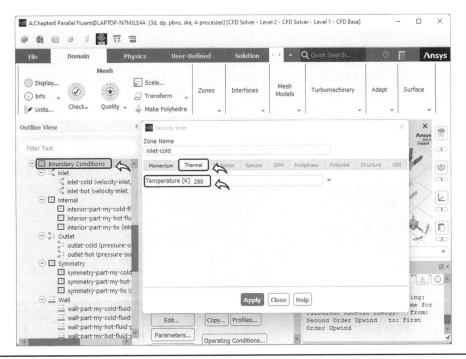

FIGURE 6.30 Thermal tab.

Note the symmetry plane is divided into three zones named **symmetry-part-my-cold-fluid**, **symmetry-part-my-hx**, and **symmetry-part-my-hot-fluid**. Make sure the **Type** for all three zones is **symmetry**.

Select **wall-part-my-hx**. Make sure its **Type** is **wall**. Click **Edit...** and confirm the **Thermal** conditions for the wall are set to 0 for **Heat Flux [W/m²]** reflecting an insulated wall. Click **Apply**, then click **Close**.

The remaining four walls listed under **Wall Boundary Conditions** are the interfaces between the solid and fluid regions in the model. Select them one by one and confirm the **Thermal** conditions for the wall are set to **Coupled**. The internal interfaces between the fluid and solid regions are coupled thermally by default to allow the transfer of heat from one region to another.

6. **Methods**

Expand **Solution** → Double-click **Methods**. Keep the default options for the **Solution Methods**:

Pressure-Velocity Coupling → **Scheme** → **Coupled**.

Pressure-Velocity Coupling → **Flux Type** → **Rhie-Chow: distance based**.

Spatial Discretization → **Gradient** → **Least Squares Cell Based**.

Spatial Discretization → **Pressure** → **Second Order**.

Spatial Discretization → **Momentum** → **Second Order Upwind**.

Spatial Discretization → **Turbulent Kinetic Energy** → **First Order Upwind**.

Spatial Discretization → **Turbulent Dissipation Rate** → **First Order Upwind**.

Spatial Discretization → **Energy** → **Second Order Upwind**.

Scroll down if needed and make sure **Pseudo Time Method** is enabled.

7. **Residuals**

Expand **Solution** → Expand **Monitors** → Double-click **Residual**.

Note that in addition to the residuals for the continuity, x-velocity, y-velocity, and energy equations, we have three additional residuals corresponding to the three additional equations for z-velocity, k, and epsilon.

Change the **Absolute Criteria** to 1e-12 for all equations to be solved. Click **OK** to close the **Residual Monitors** dialog box.

8. **Initialization**

Expand **Solution** → Double-click **Initialization**.

We will use **Hybrid Initialization** for **Initialization Methods**. Click **Initialize** in the **Task Page**. The **Console** will display a message when **Hybrid Initialization** is done.

9. **Run Calculation**

Expand **Solution** → Double-click **Run Calculation**.

In the **Task Page** view, set the **Number of Iterations** to 2000. Click the **Calculate** button under **Solution Advancement**. Click **OK** when the **Calculation complete** message appears.

Figure 6.31 shows the residuals for the mass, momentum, energy, and turbulence equations where convergence was reached after 1,300 iterations.

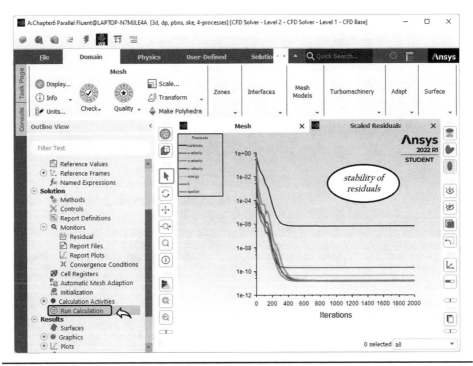

FIGURE 6.31 Scaled Residuals.

10. **Flux Reports**

Although the residuals plot shows the model has converged, we need to make sure the converged solution is the correct solution for the problem. We will check the net mass and heat fluxes through the heat exchanger to ensure that the mass and the energy are balanced.

Expand **Results** → Double-click **Reports**. Double-click **Fluxes** or select **Fluxes** and click **Set Up...** under the group box of **Reports** in the **Task Page** view.

In the **Flux Reports** dialog box, select **Mass Flow Rate** for **Options**, select **inlet-hot** and **outlet-hot** from the list of **Boundaries** to check the mass balance for the inner pipe. Click **Compute**. The mass flow rate value of the hot fluid is **0.1524 kg/s**. Recall that we are modeling half the heat exchanger. Thus, this value is half the actual mass flow rate in the inner pipe (hot fluid) of the heat exchanger.

Click **Deselect All Shown**, shown in Figure 6.32, in the **Flux Reports** dialog box. Select **inlet-cold** and **outlet-cold**. Click **Compute**. The mass flow rate of the cold fluid is **0.0463 kg/s**.

The **Net Results** is displayed in the dialog box. The mass is conserved for both cold and hot fluids and is an indication that the solution has converged.

While the **Flux Reports** dialog box is still open, select **Total Heat Transfer Rate** under **Options**. Click **Select All Shown** to select all the boundaries, as shown in Figure 6.33. Click **Compute**. The net heat imbalance is a very small fraction of the total heat flux through the interfaces, and this is another indication of convergence. Click **Close** to close the **Flux Reports** dialog box.

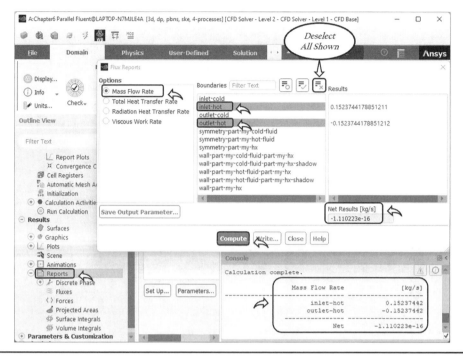

FIGURE 6.32 Flux Reports for Mass Flow Rate.

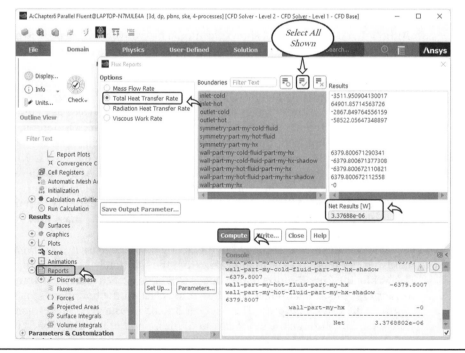

FIGURE 6.33 Flux Reports for Heat Transfer Rate.

Setup Summary

* Expand **Setup** → Double-click **General**. Make sure **Pressure-Based** is selected for **Type**, **Absolute** is selected for **Velocity Formulation** and **Steady** is selected for **Time**.

* Expand **Setup** → Double-click **Models**. Double-click **Energy-Off**. Enable **Energy Equation**. Click **OK**. Double-click **Viscous SST k-omega**. Select **k-epsilon (2 eqn)** for **Model** and **Enhanced Wall Treatment** for **Near-Wall Treatment**. Click **OK**.

* Expand **Setup** → Double-click **Materials**. Click **Create/Edit....** Click **Fluent Database...**, select **water-liquid (h2o<l>)** and click **Copy**. Close the **Fluent Database Materials** dialog box. Type my-fluid under **Name**. Delete **h2o<l>** under **Chemical Formula**. Set the **Thermal Conductivity [W/(m K)]** to 5.97. Click **Change/Create**, then click **Yes**. Click **Close**.

* Expand **Setup** → Double-click **Cell Zone Conditions**. Double-click **part-my-cold-fluid** and select **my-fluid** next to **Material Name**. Click **Apply**, then click **Close**. Double-click **part-my-hot-fluid** and select **my-fluid** next to **Material Name**. Click **Apply**, then click **Close**.

* Expand **Setup** → Double-click **Boundary Conditions**. Double-click **inlet-cold**. Set the **Velocity Magnitude [m/s]** to 0.3, click the **Thermal** tab and set the **Temperature [K]** to 280. Click **Apply**, then click **Close**. Double-click **inlet-hot**. Set the **Velocity Magnitude [m/s]** to 0.5, click the **Thermal** tab and set the **Temperature [K]** to 400. Click **Apply**, then click **Close**.

* Expand **Solution** → Double-click **Methods**. Use the default options for **Solution Methods**.

* Expand **Solution** → Expand **Monitors** → Double-click **Residual**. Set the **Absolute Criteria** for all equations to 1e-12. Click **OK**.

* Expand **Solution** → Double-click **Initialization**. Click **Initialize**.

* Expand **Solution** → Double-click **Run Calculation**. Set the **Number of Iterations** to 2000. Click **Calculate**.

* Expand **Results** → Expand **Reports**.

* Double-click **Fluxes**. Select **Mass Flow Rate**. Select **inlet-hot** and **outlet-hot**. Click **Compute**. Click **Deselect All Shown**. Select **inlet-cold** and **outlet-cold**. Click **Compute**.

* Select **Total Heat Transfer Rate**. Click **Select All Shown**. Click **Compute** and click **Close**.

11. **Temperature Contours on Symmetry Plane**
    Expand **Results** → Double-click **Graphics** in **Outline View**.
    Double-click **Contours** under the group box of **Graphics** in the **Task Page** view. Select the default name under **Contour Name** and type my-temp-symmetry.
    Select **Temperature** from the upper drop-down list and **Static Temperature** from the lower drop-down list under **Contours of**.

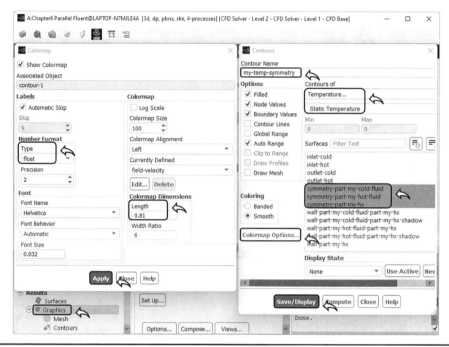

FIGURE 6.34 Contours of Temperature. Colormap Options.

Select **symmetry-part-my-cold-fluid**, **symmetry-part-my-hot-fluid**, and **symmetry-part-my-hx** under **Surfaces**. Disable **Global Range** and click **Colormap Options**.

By disabling the **Global Range**, the minimum and maximum values will be based on the range of temperature values on the selected surfaces, rather than on the entire domain.

In the **Colormap** dialog box, select **float** from the drop-down list of **Type** under **Number Format**. Select the default value of **Length** under **Colormap Dimensions** and type 0.81 and click **Apply** then close the **Colormap** dialog box. See Figure 6.34.

Click **Save/Display** in the **Contours** dialog box. Close the **Contours** dialog box.

Click **Arrange the workspace** shown in Figure 6.35 and deselect **Task Page** and **Console**. Click **X** axis in the triad and then click **Fit to Window** in the view tools toolbar. The screen should look like Figure 6.35.

12. **Temperature Contours at the Hot Fluid Outlet.**
Expand **Results** → Expand **Graphics** → Expand **Contours**.

Right-click **my-temp-symmetry** and select **Copy....** Select the default name under **Contour Name** and type my-temp-outlet-hot. Click **Deselect All Shown**, as shown in Figure 6.36, and select **outlet-hot** from the list of **Surfaces**. Click **Save/Display** and click **Close** to close the **Contours** dialog box. Click **Z** axis in the triad and click **Fit to Window** in the view tools toolbar to adjust the view.

The screen should look like Figure 6.37.

Note that creating a reflection to display the temperature of the full cross section of the pipe can be accomplished only in CFD-Post application as demonstrated in Section 5.4.4.

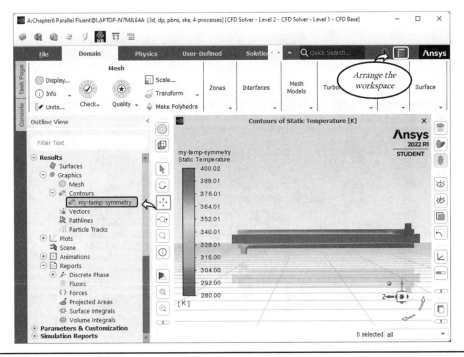

FIGURE 6.35   Arrange the workspace and contours of temperature.

FIGURE 6.36   Contours dialog box and Deselect All Shown surfaces.

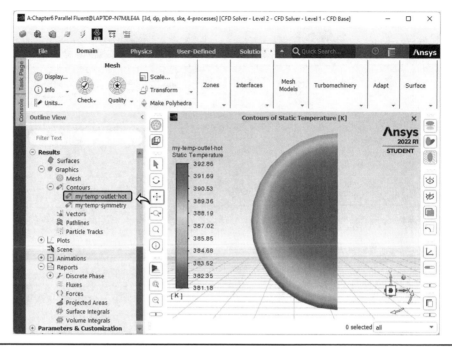

FIGURE 6.37 Contours of temperature at the hot fluid outlet.

13. **Temperature Contours at Cold Fluid Outlet.**
Expand **Results** → Expand **Graphics** → Expand **Contours**. Right-click **my-temp-outlet-hot** and select **Copy....** Select the default name under **Contour Name** and type `my-temp-outlet-cold`. Click **Deselect All Shown**, shown in Figure 6.36, and select **outlet-cold** from the list of **Surfaces**. Click **Save/Display** and click **Close** to close the **Contours** dialog box. Click **Y** axis in the triad and click **Fit to Window** to adjust the view. The screen should look like Figure 6.38.

14. **Velocity Pathlines**
Expand **Results** → Expand **Graphics** → Double-click **Pathlines**.
    Select the default name under **Pathline Name** and type `my-vel-cold`. Select **Velocity** from the upper drop-down list and **Velocity Magnitude** from the lower drop-down list of **Color by**. Select **inlet-cold** from the list of **Release from Surfaces**. Increase the number of **Steps** to `2000` and of **Path Skip** to 7. See Figure 6.39. Click **Colormap Options...** and select **float** for **Type** under **Number Format** and set the **Length** under **Colormap Dimensions** to `0.81`. Click **Apply** and close the **Colormap** dialog box. Click **Save/Display** and close the **Pathlines** dialog box.
    Click **X** axis in the triad and click **Fit to Window** to adjust the view. The screen should look like Figure 6.40.
    Pathlines are the lines traveled by the fluid particles in the fluid domain and help visualize the flow of particles in the domain.
    To animate the pathlines created for the cold fluid, right-click **my-vel-cold** under **Pathlines** in the **Outline View** and select **Edit....** Select **Continuous** under **Pulse Mode** and click **Pulse**. Click **Stop!** to stop the animation.
    Close the **Pathlines** dialog box.

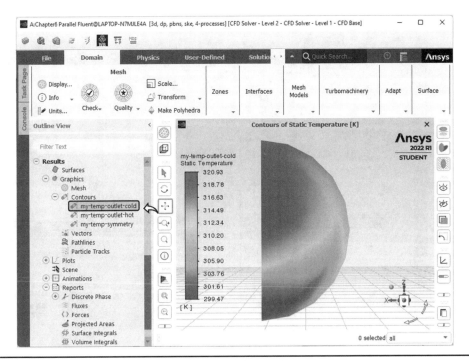

FIGURE 6.38 Contours of temperature at the cold fluid outlet.

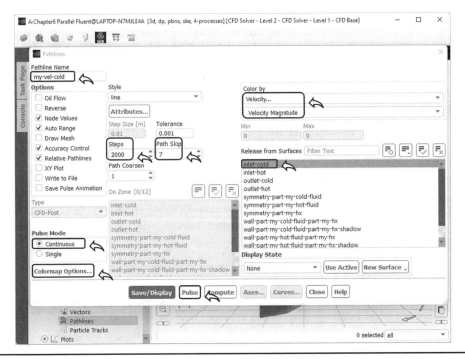

FIGURE 6.39 Pathlines Dialog Box.

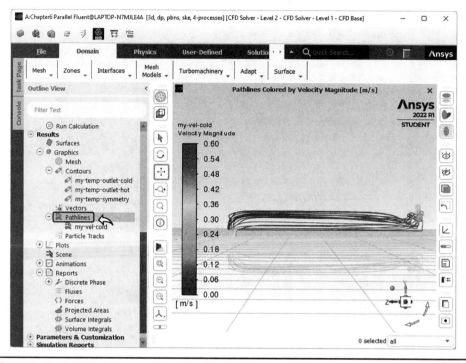

FIGURE 6.40 Pathlines Released from the Cold Fluid Inlet.

15. **Saving the Project**
    Save the project by clicking **File** and selecting **Save Project** from the drop-down list. Close the Fluent application by clicking **File** and selecting **Close Fluent** and return to Workbench.

6.4 Verification

In this section, we will perform a grid independent study using Richardson Extrapolation scheme described in Section 2.9.

6.4.1 Grid Independent Study

Local Values

A grid independent study using Richardson Extrapolation scheme on the temperature of the centerline along the Z axis of the hot fluid was performed. The mesh statistics for the study are summarized in Table 6.2.

TABLE 6.2 Mesh Statistics

Mesh # j	Quality	N_j	h_j (mm)	$r_{j,j-1}$	$\max(y^+)$
1	Fine	6,599,650	0.5203		2.21
2	Medium	1,537,017	0.8457	1.6245	3.29
3	Coarse	493,797	1.2348	1.4601	4.51

TABLE 6.3 Convergence Metrics Using Richardson Extrapolation for the Centerline Temperature

i	z_i (m)	ϕ_{l_1} (K)	ϕ_{l_2} (K)	ϕ_{l_3} (K)	s_i	$(e_a^{21})_i$ %	$(e_a^{32})_i$ %	p_i	$(GCI^{21})_i$ ×100	EB_i ×10^3
1	−0.2	400.00	399.99	399.98	1	0.00125	0.00275	2.45	0.00033	1.33
2	−0.16	399.95	399.92	399.87	1	0.00800	0.01175	1.49	0.00214	8.54
3	−0.12	399.72	399.65	399.56	1	0.01801	0.02252	1.10	0.00481	19.22
4	−0.08	399.22	399.13	399.05	1	0.02179	0.01929	0.30	0.00582	23.22
5	−0.04	398.54	398.47	398.45	1	0.01782	0.00452	2.05	0.00476	18.95
6	0	397.79	397.75	397.80	−1	0.01081	0.01157	0.16	0.00289	11.48
7	0.04	397.03	397.02	397.12	−1	0.00378	0.02418	4.75	0.00101	4.00
8	0.08	396.29	396.30	396.43	1	0.00177	0.03356	7.86	0.00047	1.87
9	0.12	395.58	395.60	395.76	1	0.00556	0.03893	5.31	0.00148	5.87
10	0.16	394.91	394.94	395.11	1	0.00734	0.04127	4.76	0.00196	7.74
11	0.2	394.29	394.32	394.48	1	0.00786	0.04134	4.60	0.00210	8.27
12	0.24	393.71	393.74	393.89	1	0.00737	0.04038	5.60	0.00197	7.74
13	0.28	393.16	393.19	393.34	1	0.00585	0.03917	6.60	0.00156	6.14
14	0.32	392.66	392.68	392.83	1	0.00535	0.03794	7.60	0.00143	5.61

The convergence metrics are reported at 14 points on the centerline of the hot fluid. The z-location of the points ranges between −0.2 m and 0.32 m. The convergence metrics calculated are summarized in Table 6.3. The local order of accuracy ranges between 0.16 and 7.86 with a global average of 3.56. Around 14% of the points exhibited oscillatory convergence behavior.

Figure 6.41 displays the variation of the temperature as a function of axial distance. The error bars are amplified by a factor of 10 for visibility purposes. The maximum value of the error bar is ±0.023 K, and the maximum discretization error is around 0.011%.

Global Values

In this section, we will apply the Richardson Extrapolation scheme on the following global parameters: mass flow rate of hot fluid, mass flow rate of cold fluid, mass weighted average temperature of hot fluid, mass weighted average temperature of cold fluid and the total heat transfer rate between the hot and cold fluids.

The convergence metrics calculated using Richardson Extrapolation scheme on the selected global parameters are summarized in Table 6.4.

We observe that the order of accuracy the global parameters ranges between 1.1 and 3, and that the outlet temperature of the cold fluid showed oscillatory convergence.

Mass Flow Rate Balance

The mass flow rates calculated in Fluent are presented in the flux reports in Section 6.3.3 and displayed in Table 6.4. They correspond to the Fluent model which is half of the actual heat exchanger. Hence, those values should be multiplied by two in order to calculate the mass flow rates for the full heat exchanger. For instance, the coarse grid yields a cold fluid mass flow rate equal to $0.0463 \times 2 = 0.0926$ kg/s, and a hot fluid mass flow rate equal to $0.1524 \times 2 = 0.3048$ kg/s.

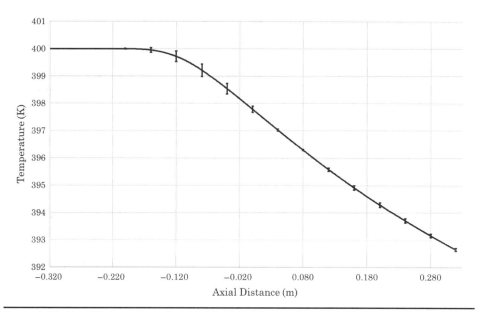

FIGURE 6.41 Amplified error bars for the temperature along the centerline.

TABLE 6.4 Convergence Metrics Using Richardson Extrapolation for the Global Parameters

	ϕ_{h_1}	ϕ_{h_2}	ϕ_{h_3}	s	(e_a^{21})	(e_a^{32})	p	GCI^{21}	EB
Mass Flow Rate Cold (kg/s)	0.0469	0.0467	0.0463	1	0.51	0.91	1.97	0.39	0.00018
Mass flow Rate Hot (kg/s)	0.1535	0.1532	0.1524	1	0.20	0.53	2.83	0.09	0.00013
Mean Temp Outlet - Cold $T_{m,o,c}$ (K)	312.98	313.01	312.96	−1	0.01	0.02	1.11	0.02	0.05829
Mean Temp Outlet - Hot $T_{m,o,c}$ (K)	389.92	389.94	389.99	1	0.01	0.01	2.98	0.00	0.00782
Wall Heat Rate (W)	6473	6447	6378	1	0.41	1.07	2.90	0.16	10.63960

Using $\dot{m} = \rho \times V \times A$, the cold fluid mass flow rate is $\dot{m}_c = 0.0941$ kg/s while the hot fluid mass flow rate is $\dot{m}_h = 0.3073$ kg/s.

The relative percent error between the theoretical value and the coarse grid Fluent solution for the cold fluid is around 1.6%. However, as we refine the grid, the relative percent error decreases to 0.3%.

As for the hot fluid mass flow rate, the relative percent error between the theoretical value and the coarse grid Fluent solution is around 0.81%. Refining the grid decreases this error to 0.1%.

Heat Transfer Balance

The calculated heat absorbed by the cold fluid and the heat lost by the hot fluid are

$$q_h = \dot{m}_h c_{p,h} \left(T_{h_i} - T_{h_o} \right) = 0.3048 \times 4182 \times (400 - 389.99) = 12,759 \text{ W}$$
$$q_c = \dot{m}_c c_{p,c} \left(T_{c_i} - T_{c_o} \right) = 0.0926 \times 4182 \times (312.96 - 280) = 12,764 \text{ W}$$

Note the mass flow rates and the average values of the hot and cold outlet temperatures of the fluids are calculated by Fluent (Table 6.4).

The relative percent difference between the heat absorbed by the cold fluid and the heat lost by the hot fluid is around 0.04%. As we refine the mesh, this error decreases to 0.03%.

CHAPTER 7

Three-Dimensional Fluid Flow and Heat Transfer Modeling in a Heat Sink

List of Symbols	
Heat	q
Thermal conductivity	k
Convection heat transfer coefficient	h
Distance	x
Perimeter	P
Area	A
Excess temperature	θ
Length	L
Width	W
Height	H
Efficiency	η
Thickness of the fin	t
Density	ρ
Specific Heat	c_p
Viscosity	μ

This chapter covers the modeling of a three-dimensional aluminum heat sink surrounded by air. The heat sink is used to cool a Central Processing Unit (CPU) to prevent overheating and keep it below its maximum operating temperature. The heat generated by the CPU is modeled as a heat source at the base of the heat sink, therefore eliminating the need to include the CPU in the model. The heat is transferred by conduction from the base of the heat sink to the fins of the heat sink, and by convection from the fins to the surrounding air.

The heat sink problem does not have an analytical solution. We rely on Computational Fluid Dynamics to predict the amount of heat being transferred in heat sink applications.

The instructions to build the geometry of the heat sink in Fluent will be presented. In addition, the user is provided instructions on how to import a computer aided design (CAD) model (provided to the reader) of the heat sink into Fluent. Instructions to mesh the model, set up the fluent simulations will be presented. The results of the simulations will also be presented along a brief verification of the Fluent model.

7.1   Introduction to Heat Sinks

Heat sinks are devices that can help dissipate heat from a hot source. Heat travels from the hot source through the heat sink by conduction and is transferred by convection to the surrounding fluid. Using one-dimensional, steady-state, conduction analysis for extended surfaces, we calculate the temperature distribution along the fin and the rate of heat transfer from the fin to the surrounding air.

Consider the extended surface (fin) shown in Figure 7.1. Assuming the heat transfer by conduction is dominant in the x-direction (the temperature is only a function of x), constant thermal conductivity of the surface, negligible radiation heat transfer, and uniform convection heat transfer coefficient h over the surface, and applying an energy balance on the differential element dx along the fin, we obtain

$$q_x = q_{x+dx} + dq_{conv} \tag{7.1}$$

Using Fourier's law for conduction

$$q_x = -k \frac{dT}{dx} A_c \tag{7.2}$$

where k is the thermal conductivity of the fin and A_c is the cross-sectional area of the fin, and recalling Newton's law for cooling

$$q_{conv} = h \left(T - T_\infty \right) P dx \tag{7.3}$$

where P is the perimeter of the cross sectional area A_c as shown in Figure 7.1, h is the heat transfer coefficient, and T_∞ is the fluid temperature, along with Taylor's expansion

$$q_{x+dx} = q_x + \frac{dq_x}{dx} dx, \tag{7.4}$$

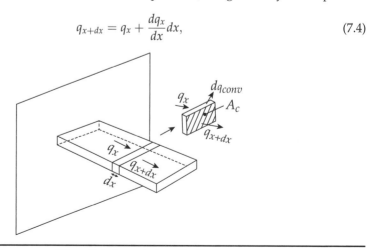

Figure 7.1 Energy balance on a differential element along the fin.

Substituting Eqs. (7.2)-(7.4) into Eq. (7.1) and for the case of a rectangular uniform cross-sectional area, as shown in Figure 7.1, we obtain

$$\frac{d^2T}{dx^2} = \frac{hP}{kA_c} \cdot (T - T_\infty) \tag{7.5}$$

The solution to Eq. (7.5) provides the temperature distribution T as a function of x. To simplify, let $\theta = T - T_\infty$, and $m^2 = \frac{hP}{kA_c}$. Since T_∞ is a constant, then $\frac{d\theta}{dx} = \frac{dT}{dx}$ and $\frac{d^2\theta}{dx^2} = \frac{d^2T}{dx^2}$. Substituting $\theta = T - T_\infty$ into Eq. (7.5), we get

$$\frac{d^2\theta}{dx^2} - m^2\theta = 0 \tag{7.6}$$

The solution to Eq. (7.6) is

$$\theta = C_1 e^{mx} + C_2 e^{-mx} \tag{7.7}$$

The constants C_1 and C_2 are calculated using the boundary conditions for the differential equation. Different boundary conditions on the heat sink will have different solutions for the temperature along the fins [37].

For the case of rectangular fins transferring heat to the fluid by convection through the tips of the fins, we obtain

$$h(T_L - T_\infty) = -k\frac{dT}{dx}\Big|_{x=L} \tag{7.8}$$

Providing one equation for solving for the unknowns. The second equation is derived from the boundary condition at the base where

$$T_{x=0} = T_b \tag{7.9}$$

Using the boundary conditions expressed in Eqs. (7.8) and (7.9) we can solve for C_1 and C_2 and provide the temperature distribution along the fin

$$\frac{\theta}{\theta_b} = \frac{\cosh[m(L-x)] + \left(\frac{h}{mk}\right)\sinh[m(L-x)]}{\cosh(mL) + \left(\frac{h}{mk}\right)\sinh(mL)} \tag{7.10}$$

The rate of heat transfer from the fin to the surrounding must equal the heat conducted at the base of the fin

$$q_f = q_b = -kA_c\frac{dT}{dx}\Big|_{x=0} \tag{7.11}$$

For the case of convection heat transfer at the tip, and using the temperature found in Eq. (7.10), we obtain

$$q_f = \sqrt{hPkA_c}\,\theta_b\,\frac{\sinh(mL) + \left(\frac{h}{mk}\right)\cosh(mL)}{\cosh(mL) + \left(\frac{h}{mk}\right)\sinh(mL)} \tag{7.12}$$

Another way to calculate q_f is through the efficiency η_f

$$\eta_f = \frac{q_f}{q_{max}} = \frac{q_f}{hA_f\theta_b} \tag{7.13}$$

where for a rectangular fin the efficiency is expressed as

$$\eta_f = \frac{\tanh\left(mL_c\right)}{mL_c} \tag{7.14}$$

In Eq. (7.14), L_{cor} is the corrected length of the fin, $L_{cor} = L + \frac{t}{2}$ where t is the thickness of the fin.

7.2   Problem Statement

The heat sink shown in Figure 7.2 is a schematic of the problem solved in this chapter. This example heat sink is used to cool a CPU and keep its maximum temperature under control. The CPU is not part of the model, however its heat generation is being modeled through the boundary conditions applied at the base of the heat sink.

A simplified three-fin heat sink is chosen in order to reduce the computational time needed to predict the temperature of the heat sink. Keep in mind that the larger the number of fins on a heat sink, the more heat transfer through the surface area is obtained.

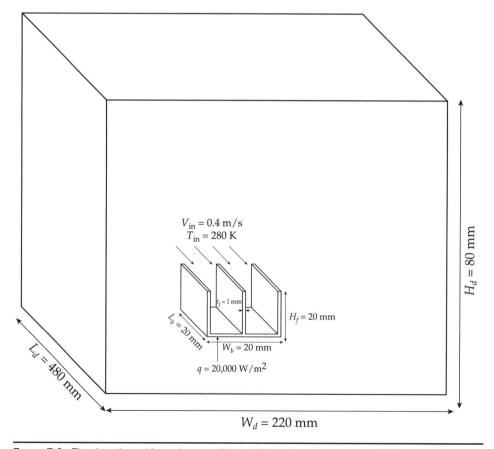

$V_{in} = 0.4\ \text{m/s}$
$T_{in} = 280\ \text{K}$

$t_f = 1\ \text{mm}$

$H_f = 20\ \text{mm}$

$L_b = 20\ \text{mm}$

$W_b = 20\ \text{mm}$

$q = 20{,}000\ \text{W/m}^2$

$H_d = 80\ \text{mm}$

$L_d = 480\ \text{mm}$

$W_d = 220\ \text{mm}$

FIGURE 7.2 The domain and boundary conditions (dimensions are not to scale).

TABLE 7.1 Model Dimensions, Properties, and Flow Variables

Variable	Symbol	Value
Base dimensions		
Base length	L_b	20 mm
Base width	W_b	20 mm
Base height	H_b	1 mm
Fins dimensions		
Fins thickness	t_f	1 mm
Fins width	W_f	20 mm
Fins Height	H_f	20 mm
Fluid domain dimensions		
Domain length	L_d	220 mm
Domain width	W_d	480 mm
Domain height	H_d	80 mm
Fluid Properties - (Air)		
Density	ρ	1.225 kg/m^3
Specific Heat	c_p	1006.43 J/kg.K
Thermal Conductivity	k	0.0242 W/m.K
Viscosity	μ	0.01789 mPa.s
Solid Properties - (Aluminum)		
Density	ρ	2719 kg/m^3
Specific Heat	c_p	871 J/kg.K
Thermal Conductivity	k	202.4 W/m.K
Heat Source		
Area	A	4 mm^2
Heat Flux	q	20,000 W/m^2

The heat sink is made of aluminum. The thermal energy from the CPU is conducted through the fins of the heat sink and is then transferred to the surrounding air by convection.

All parameters and dimensions of the heat sink and of the surrounding air are provided in Table 7.1.

7.3 Modeling using Fluent

This section will guide the reader step by step on how to build a model of the heat sink problem using Fluent, with the assumption that the reader has completed the tutorials in the previous chapters. Instructions covered in the previous chapters will not be repeated.

We will import a CAD model of the heat sink to demonstrate how to import geometry into ANSYS. We will also build the geometry of the heat sink in Design Modeler.

Launch ANSYS Workbench. Double-click **Fluid Flow (Fluent)** to create a new **Fluid Flow (Fluent)** analysis into the **Project Schematic** and rename it Chapter7.

Save the workbench project in your working directory and name it Heat Sink.

7.3.1   Import Geometry

The option to import CAD models, and other geometries, into ANSYS is available to the reader. A geometry of the heat sink was created in Solidworks and can be downloaded using the instructions provided in this section. Note that files created in Solidworks should be saved in *igs* format for a successful translation to ANSYS. We will demonstrate in this section how to import the geometry into Fluent.

In **Project Schematic**, right-click **Geometry** and select **Import Geometry Browse....** In the **Open** dialog box, browse and navigate to where the geometry created in CAD was saved. Click **Open**. A green check mark next to geometry indicates a successful geometry import. Right-click **Geometry** and select **Edit Geometry in DesignModeler....** Once DesignModeler is launched, right-click **Import1** in **Tree Outline** view and select **Generate** to generate the geometry and view it in the **Graphics** window.

The geometry can also be imported in DesignModeler by right-clicking **Geometry** and selecting **New DesignModeler Geometry....** Once DesignModeler is launched, click **File** in the menu toolbar and select **Import External Geometry File....** Click **Generate** in the menu toolbar or right-click **Import1** in **Tree Outline** view and select **Generate**.

Once the geometry is imported, expand **2 Parts, 2 Bodies** and right-click the upper **Solid** and rename it `my-heat-sink`. Rename the second solid `my-fluid`.

Control-select **heat-sink** and **fluid-body**. Right-click and select **Form New Part**.

7.3.2   Create Geometry in Design Modeler

In this section, the geometry of the model will be created using ANSYS **DesignModeler**.

Instructions to build the geometry using SpaceClaim are included in Appendix D.

If a geometry was imported in the previous section, reset the **Geometry** application by right-clicking **Geometry** in Workbench **Project Schematic**. Select **Reset** from the drop-down menu. A message will appear warning the user that the component's local and generated data will be deleted. Click **OK**.

Figure 7.3 shows the final geometry that will be created in this section. The computational domain is made of the solid heat sink surrounded by a rectangular fluid body.

Figure 7.4 is a flowchart for the process of creating the geometry using **DesignModeler**.

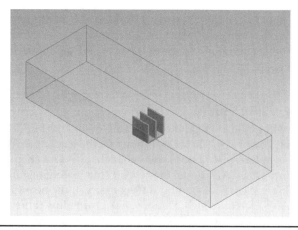

Figure 7.3   Heat sink model.

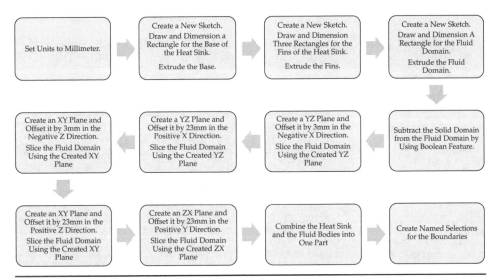

FIGURE 7.4 Flowchart of DesignModeler.

Launch DesignModeler by right-clicking **Geometry** and selecting **New DesignModeler Geometry...**

1. **Set the Units to Millimeter**
   Click **Units** in the menu toolbar and select **Millimeter**.

2. **Create the Base of the Heat Sink**
   Click **ZXPlane** under the **Tree Outline**.

   Click **Look at Face/Plane/Sketch** to display the ZX plane on the **Graphics** window.

   Click **Sketching**. In the **Sketching Toolboxes**, click **Draw** → Click **Rectangle** by two points. A new sketch is automatically being created since there are no existing sketches in the **ZXPlane**. Click the origin to select it as the first point. Make sure the pointer of the mouse displays the letter **P** before selecting the origin. Drag the cursor and click anywhere in the first quadrant to place the opposite corner of the rectangle.

   Click **Dimensions** → Click **General** from **Sketching Toolboxes** and select the upper horizontal edge of the rectangle, then drag the cursor to place the dimension anywhere above the upper edge. Repeat the process to dimension the right vertical side of the rectangle but drag the cursor to the right. Adjust the dimensions of the rectangle in **Details View** by changing both values of **H1** and **V2** to 20.

   Click **Modeling** button and expand **ZXPlane** in **Tree Outline** view. **Sketch1** is the rectangle created. We will extrude the rectangle by clicking **Extrude** in the 3D Feature toolbar. **Extrude1** appears in the **Tree Outline** view with its details in **Details View**.

   Select **Sketch1** in **Tree Outline** view and click **Apply** next to **Geometry**. Enter 1 next to **FD1, Depth(>0)** as the depth of the extrusion. Click **Generate**.

   Click the cyan **ISO** ball in the triad. Right-click anywhere in the **Graphics** window and select **Zoom to Fit** to view the three-dimensional base of the heat sink. The screen should look like Figure 7.5.

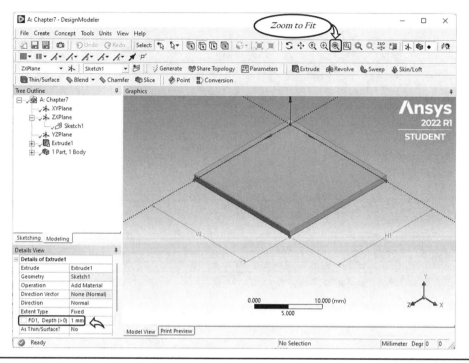

FIGURE 7.5   The base of the heat sink.

3. **Create the Fins**

We will create a new sketch made of three rectangles, as shown in Figure 7.6. We will extrude the three rectangles in the Y direction to create the fins.

Click **ZXPlane** under the **Tree Outline**. Click **Look at Face/Plane/Sketch** to display the ZX plane on the **Graphics** window. Click **New Sketch**. **Sketch2** is created in the **ZXPlane**.

Note that it is important to create a new sketch; otherwise, the fins will be part of **Sketch1** and cannot be extruded separately.

Click **Sketching** to change to the sketching mode. Click **Draw** → Click **Rectangle** by two points.

Make sure the pointer of the mouse changes to the letter **P** before selecting the origin as the first point. Drag the cursor to place the opposite corner anywhere on the upper horizontal edge of the base. Make sure the pointer of the mouse changes to the letter **C** before selecting the second corner, indicating the base of the first fin is aligned with the base of the heat sink.

Repeat and create the middle rectangle making sure the pointer of the mouse changes to the letter **C** when selecting the first point on the lower horizontal edge of the base and the second point on the upper horizontal edge of the base as shown in Figure 7.6.

Repeat and create the third rectangle. Make sure the pointer of the mouse changes to the letter **C** when selecting the first point at the lower edge of the base and to the letter **P** at the right corner of the upper edge of the base. See Figure 7.6.

Click **Dimensions** → Click **Horizontal** from **Sketching Toolboxes** and select the two long vertical edges of the first fin, then select a position for the dimension

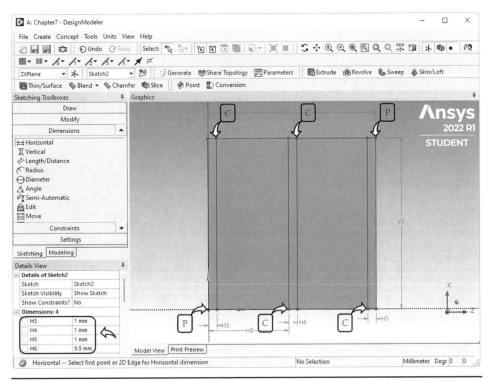

FIGURE 7.6 Create three rectangles to be extruded as three fins.

text below the base. Repeat by selecting the two long edges of the middle fin. Repeat and select the two long edges of the third fin.

Set the thickness of the fins by adjusting the horizontal dimensions created. Change all values of **H3**, **H4**, and **H5** in **Details View** to 1.

We need to add one more horizontal dimension to place the middle fin at the center of the base. While the **Horizontal** dimension is still activated, select the vertical X axis, select the left long edge of the middle fin, then select a location below the base to place the dimension. Change the value of **H6** to 9.5. The screen should look like Figure 7.6. Note in this figure, the ruler is hidden.

Click **Modeling** button and make sure **ZXPlane** in **Tree Outline** view is expanded. **Sketch2** is the base of the three fins to be extruded. Click **Extrude** in the 3D Feature toolbar. Select **Sketch2** in **Tree Outline** view and click **Apply** next to **Geometry** in **Details View**. Enter 20 next to **FD1, Depth (>0)** as the depth of the extrusion and click **Generate**. Click the cyan **ISO** ball in the triad and **Zoom to Fit** to display an isometric view of the heat sink.

Expand **1 Part, 1 Body** in the **Tree Outline** to view the created body named by default **Solid**. Rename it my-heat-sink. In **Details View**, the **Fluid/Solid** option is set to **Solid** by default. Hide **Sketch1** and **Sketch2** in the **Graphics** window by deactivating the icon **Display Plane** as shown in Figure 7.7.

4. **Create the Fluid Domain**

We will build the fluid domain surrounding the heat sink by creating a rectangle in the ZX plane around the heat sink. We will then extrude the rectangle to create

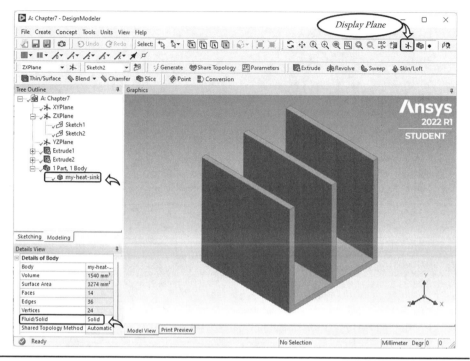

FIGURE 7.7   Heat sink geometry.

a rectangular box surrounding the heat sink. Finally, we will subtract the heat sink from the box to create the fluid domain.

Click the **ZXPlane** in **Tree Outline** view. Click **Look at Face/Plane/Sketch** to display the ZX plane in the **Graphics** window. Click **New Sketch** to create a new sketch in the ZX plane.

Click **Sketching**. Click **Draw** → Click **Rectangle** by two points to create a rectangle around the heat sink. Click anywhere in the third quadrant to select the first point for the rectangle and anywhere in the first quadrant to select the second point such that the rectangle is around the heat sink.

Click **Dimensions** → Click **General** from **Sketching Toolboxes** and select the upper horizontal edge of the rectangle, then drag the cursor to place the dimension anywhere above the upper edge. Repeat the process to dimension the vertical right edge of the rectangle but drag the cursor to the right.

Adjust the dimensions of the rectangle created in **Details View** by changing the values of **H7** to 220 and **V8** to 480.

Right-click anywhere in the **Graphics** window and select **Zoom to Fit**.

We will center the heat sink inside the rectangle along the X axis by setting the distance between the right edges of the heat sink and the rectangle equal to the distance between the left edges of the heat sink and the rectangle.

Click **Constraints** → Click **Equal Distance** in the **Sketching Toolboxes**, as shown in Figure 7.8.

The message displayed at the bottom left of the screen will guide the user through the selection of the edges. The **Box Zoom** icon and scrolling the wheel of the mouse to zoom in and zoom out can make it easier while selecting the edges.

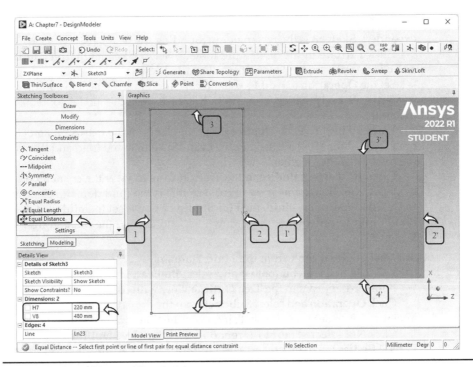

FIGURE 7.8 Equal Distance Constraints.

The first pair of lines to be selected are named 1 and 1' in Figure 7.8. The second pair of lines are named 2 and 2'.

Select the left edge of the fluid rectangle (1) as the first line of first pair for **Equal Distance** constraint. Select the left edge of the heat sink (1') as the second line of first pair for **Equal Distance** constraint.

Select the right edge of the fluid rectangle (2) as the first line of second pair for **Equal Distance** constraint. Select the right edge of the heat sink (2') as the second line of second pair for **Equal Distance** constraint.

The fluid domain will be adjusted in the **Graphics** window such that the heat sink appears centered inside the fluid domain around the X axis.

Repeat the process to center the heat sink inside the fluid rectangle around the Z axis by setting the distance between the top edges of the heat sink and the rectangle equal to the distance between the bottom edges of the heat sink and the rectangle. The first pair of lines to be selected are named 3 and 3' in Figure 7.8. The second pair of lines are named 4 and 4'.

Select the top edge of the fluid rectangle (3) as the first line of the first pair for **Equal Distance** constraint. Select the top edge of the heat sink (3') as the second line of the first pair for **Equal Distance** constraint.

Select the bottom edge (4) of the fluid rectangle as the first line of second pair and the bottom edge of the heat sink (4') as the second line of second pair for equal distance constraints.

The fluid rectangle is adjusted in the **Graphics** window such that the heat sink appears in its center. When completed, the screen should look like Figure 7.8.

Click **Modeling** button and make sure **ZXPlane** in **Tree Outline** view is expanded. **Sketch3** is the rectangle surrounding the heat sink.

Click **Extrude** in the 3D feature toolbar. Select **Sketch3** in **Tree Outline** view and click **Apply** next to **Geometry** in **Details View**.

Set the **Operation** to **Add Frozen**. Recall **Add Frozen** will allow adding material as a separate body, which is useful in this case since the model contains different materials.

Enter 80 next to **FD1, Depth (>0)** as the depth of the extrusion and click **Generate**. Adjust the view by clicking the cyan **ISO** ball in the triad and **Zoom to Fit** in the rotation modes toolbar.

Expand **2 Parts, 2 Bodies** in the **Tree Outline** to view the created body named by default **Solid**. Rename it my-fluid as this is the fluid domain. Click **my-fluid**. In **Details View** of **my-fluid**, the **Fluid/Solid** option is set to **Solid** by default. Click **Fluid/Solid** to activate the drop-down list and change the domain from **Solid** to **Fluid**.

5. **Subtract the Heat Sink from the Fluid Domain**
Click **Create** in the menu toolbar and select **Boolean** from the drop-down list.

Boolean1 appears in the **Tree Outline** view with its details in **Details View**.

Click **Operation** and select **Subtract** from the drop down list next to it, as shown in Figure 7.9, to subtract the heat sink from the fluid domain surrounding it.

Click **Not selected** next to **Target Bodies**. In the **Tree Outline** view, make sure **2 Parts, 2 Bodies** is expanded and select **my-fluid**. Click **Apply** next to **Target Bodies**.

Click **Not selected** next to **Tool Bodies** then select **my-heat-sink** from the **Tree Outline** view and click **Apply** next to **Tool Bodies**. Set **Preserve Tool Bodies?** to **Yes** to preserve the heat sink. Click **Generate**. See Figure 7.9 for details.

6. **Split the Fluid Domain into Six Bodies**
The mesh in the fluid domain surrounding the heat sink needs to be fine. However, the mesh away from the heat sink can be coarse to minimize the computational effort. In order to control the mesh sizing in the fluid domain, we will split it into six bodies that can be mesh controlled independently.

In this section, we will create two YZ planes to split the fluid domain into three bodies. Then we will create two XY planes to split the middle body surrounding the heat sink into three bodies. Lastly, we will create an XZ plane to split the center body surrounding the heat sink into two bodies.

Select **YZPlane** in the **Tree Outline** view.

Click **Create** in the menu toolbar and select **New Plane** from the drop-down list. **Plane4** appears in the **Tree Outline** view.

In **Details View**, click **Transform 1 (RMB)** and from the drop-down list in the cell next to it, scroll down to select **Offset Global X** as shown in Figure 7.10. Set **FD1, Value 1** to -3 and click **Generate**.

Plane4 is a plane perpendicular to the X axis, at $x = -3$ mm. We will use it to slice **my-fluid** into two bodies.

Click **Slice** from the 3D Feature toolbar.

Make sure **Slice Type** is set to **Slice by Plane** in **Details View**. In **Tree Outline** view, select **Plane4** and click **Apply** next to **Base Plane**. Click **Slice Targets**, and from the drop down list in the cell next to it select **Selected Bodies**. Click **0** next to **Bodies** and select **my-fluid** under **2 Parts, 2 Bodies** in the **Tree Outline** view. Click **Apply** next to **Bodies** and click **Generate**.

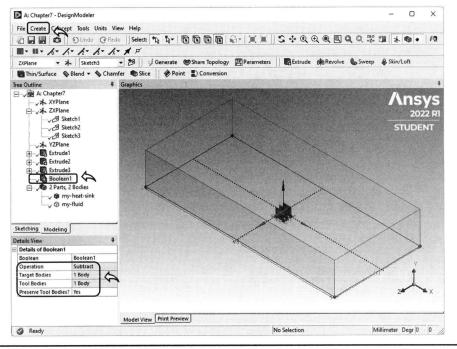

FIGURE 7.9 Subtract the heat sink from the fluid domain.

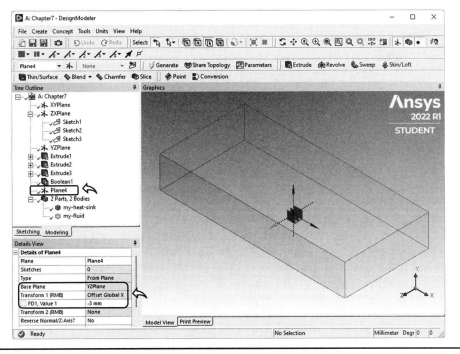

FIGURE 7.10 Details View of Plane4.

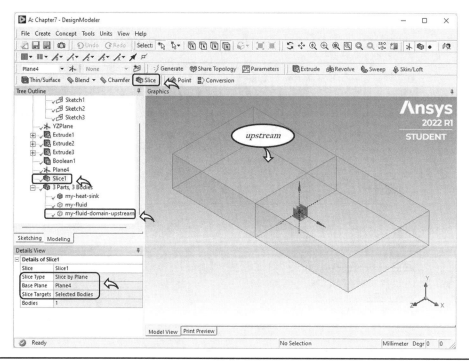

FIGURE 7.11 Slice tool and upstream fluid body.

Plane4 sliced the fluid body, **my-fluid**, into two bodies as shown in Figure 7.11. The two bodies are both named **my-fluid** under **3 Parts, 3 Bodies**. Right-click the third (last) body named **my-fluid** and rename it `my-fluid-domain-upstream`.

We will repeat the above steps to create a second plane parallel to **Plane4** and use it to slice the downstream fluid body, **my-fluid**, into two bodies.

Select **YZPlane** in **Tree Outline** view and click **Create**. Select **New Plane**. **Plane5** appears in the **Tree Outline** view. Click **Transform 1 (RMB)** and from the drop-down list next to it select **Offset Global X**. Set the value of **FD1, Value 1** to 23. Press **Enter** and click **Generate**.

Click **Slice** from the 3D Feature toolbar. In **Details View**, make sure **Slice Type** is set to **Slice by Plane**. In **Tree Outline** view, select **Plane5** and click **Apply** next to **Base Plane**. Click **Slice Targets**, and from the drop down list in the cell next to it select **Selected Bodies**. Click **0** next to **Bodies** and select **my-fluid** under **3 Parts, 3 Bodies** in the **Tree Outline** view. Click **Apply** next to **Bodies** and click **Generate**.

Plane5 sliced **my-fluid** into two bodies as shown in Figure 7.12. The new body is named **my-fluid** and appears under **4 Parts, 4 Bodies**. Select the body **my-fluid** that corresponds to the fluid downstream and rename it `my-fluid-domain-downstream`.

The screen should look like Figure 7.12.

We will create two additional planes perpendicular to Z axis and use them to further divide the middle fluid body.

Select **XYPlane** in the **Tree Outline** view. Click **Create** in the menu toolbar and select **New Plane** from the drop-down list. **Plane6** appears in the **Tree**

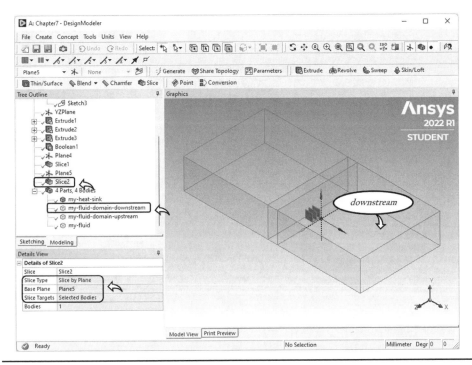

FIGURE 7.12 Downstream fluid body.

Outline. Click **Transform 1 (RMB)**, and from the drop-down list next to it select **Offset Global Z**. Set the value of **FD1, Value 1** to -3. Press **Enter** and click **Generate**.

Click **Slice** from the 3D Feature toolbar. In **Details View**, make sure **Slice Type** is set to **Slice by Plane**. Select **Plane6** and click **Apply** next to **Base Plane**. Click **Slice Targets**, and from the drop down list in the cell next to it select **Selected Bodies**. Click **0** next to **Bodies** and select **my-fluid** under **4 Parts, 4 Bodies** in the **Tree Outline** view. Click **Apply** next to **Bodies** and click **Generate**. A new body named **my-fluid** is created and added to the list under **5 Parts, 5 Bodies**. Right-click the **my-fluid** body that is on the right (negative Z axis) and rename it `my-fluid-domain-right`. See Figure 7.13.

Select **XYPlane** in the **Tree Outline** view. Click **Create** in the menu toolbar and select **New Plane** from the drop-down list. **Plane7** appears in the **Tree Outline**. Click **Transform 1 (RMB)**, and from the drop-down list next to it select **Offset Global Z**. Set the value of **FD1, Value 1** to 23. Press **Enter** and click **Generate**.

Click **Slice** from the 3D Feature toolbar. In **Details View**, make sure **Slice Type** is set to **Slice by Plane**. Select **Plane7** and click **Apply** next to **Base Plane**. Click **Slice Targets**, and from the drop down list in the cell next to it select **Selected Bodies**. Click **0** next to **Bodies** and select **my-fluid** under **5 Parts, 5 Bodies** in the **Tree Outline** view. Click **Apply** next to **Bodies** and click **Generate**. A new body named **my-fluid** is created under **6 Parts, 6 Bodies**. Right-click the body named **my-fluid** that is located on the left (positive Z axis) and rename it `my-fluid-domain-left`. See Figure 7.13.

We will create one last plane perpendicular to the Y axis and use it to split the fluid body around the heat sink into two bodies.

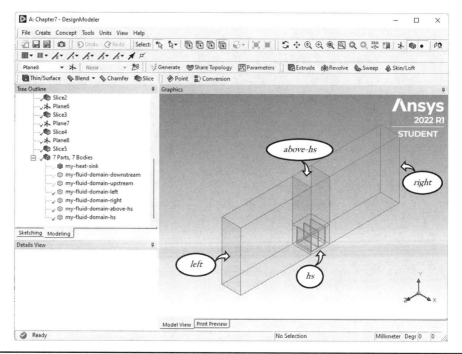

FIGURE 7.13   Right and left fluid domains, fluid domain surrounding the heat sink, and fluid domain above the heat sink.

Select **ZXPlane** in the **Tree Outline** view. Click **Create** in the menu toolbar and select **New Plane** from the drop-down list. **Plane8** appears in the **Tree Outline**. Click **Transform 1 (RMB)**, and from the drop-down list next to it select **Offset Global Y**. Set the value of **FD1, Value 1** to 23. Press **Enter** and click **Generate**.

Click **Slice** from the 3D Feature toolbar. In **Details View**, make sure **Slice Type** is set to **Slice by Plane**. Select **Plane8** and click **Apply** next to **Base Plane**. Click **Slice Targets**, and from the drop down list in the cell next to it select **Selected Bodies**. Click **0** next to **Bodies** and select **my-fluid** under **6 Parts, 6 Bodies** in the **Tree Outline** view. Click **Apply** next to **Bodies** and click **Generate**. A new body named **my-fluid** is created under **7 Parts, 7 Bodies**. Right-click **my-fluid** body located in the upper level (above the heat sink) and rename it `my-fluid-domain-above-hs`. Right-click **my-fluid** body in the lower level (surrounding the heat sink) and rename it `my-fluid-domain-hs`. See Figure 7.13.

7. **Form One Part**

   Hold the **Ctrl** button and select all bodies under **7 Parts, 7 Bodies**. Right-click and select **Form New Part**. Expand **Part** under **1 Part, 7 Bodies** to view the seven bodies. Figure 7.14 displays the six fluid bodies around the heat sink.

8. **Create Named Selections**

   Right-click **my-heat-sink** under **Part** in the **Tree Outline** view and select **Hide Body**. Activate **Selection Filter: Faces**. Rotate the model and select the inlet face shown in Figure 7.14. Right-click anywhere in the **Graphics** window and select **Named Selection** from the drop-down list. Click **Apply** next to **Geometry**. Select **Named-Sel1** and type `inlet`. Press **Enter** and click **Generate**.

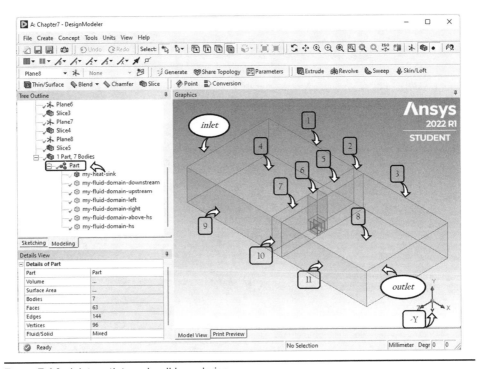

Figure 7.14 Inlet, outlet, and wall boundaries.

Select the outlet face of the domain, shown in Figure 7.14. Right-click anywhere in the **Graphics** window and select **Named Selection**. Click **Apply** next to **Geometry**, type `outlet` and click **Generate**.

Hold the **Ctrl** button and select all the faces on the right side, the left side, and the top of the rectangular fluid domain, for a total of 11 faces. The faces are numbered 1 through 11 in Figure 7.14, but the order of selection of the faces is random. Right-click anywhere in the **Graphics** window and select **Named Selection**. Click **Apply** next to **Geometry**. Type `wall` and click **Generate**.

Rotate the model or click **-Y** axis in the triad as shown in Figure 7.14. Hold the **Ctrl** button and select all the faces on the bottom of the rectangular domain, for a total of 5 Faces, as shown in Figure 7.15. Right-click anywhere in the **Graphics** window and select **Named Selection**. Click **Apply** next to **Geometry** and type `ground` next to **Named Selection**. Click **Generate**.

Click the cyan **ISO** ball in the triad and **Zoom to Fit** to adjust the view.

Right-click **my-heat-sink** in **Tree Outline** view and select **Hide All Other Bodies**. Make sure **Selection Filter: Faces** is activated. Right-click anywhere in the **Graphics** window and select **Select All**. Click **-Y** axis in the triad. Hold the **Ctrl** button and click the bottom face of the heat sink to deselect it. Right-click anywhere in the **Graphics** window and select **Named Selection**. Click **Apply** next to **Geometry** and type `fins` next to **Named Selection**. Click **Generate**.

In the same view, select the bottom face of the heat sink. Right-click anywhere in the **Graphics** window and select **Named Selection**. Click **Apply** next to **Geometry** and type `heat-source` next to **Named Selection**. Click **Generate**.

Right-click anywhere in the **Graphics** window and select **Show All Bodies**.

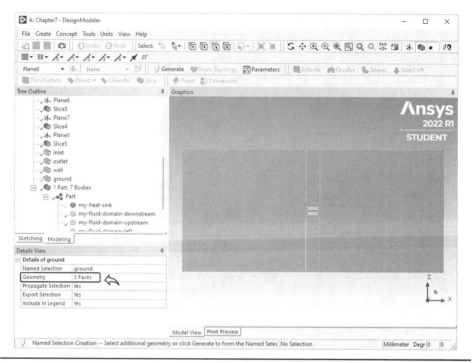

FIGURE 7.15 Total of 5 Faces selected for Named Selection.

9. **Save the Project**
   Click **File** and select **Save Project**. Close the ANSYS DesignModeler application by clicking the **Close** button **X** located in the upper right corner or by clicking **File** and selecting **Close DesignModeler**. The **Geometry** in **Project Schematic** Workbench has now a green check symbol next to it indicating the application is up to date. We are ready to start meshing the model.

Geometry Summary

* Click **Units**. Select **Millimeter**.
* Click **ZXPlane**. Click **Look At Face/Plane/Sketch**.
* Click **Sketching**. Click **Draw** → Click **Rectangle**. Draw a rectangle by two points. Select the origin as the first point. Drag the cursor and place the opposite corner in the first quadrant.
* Click **Dimensions** → Click **General**. Select the upper edge of the rectangle. Drag the cursor and click anywhere to place the dimension. Repeat for the right edge. Adjust the values of **H1** and **V2** to 20. Press **Enter**.
* Click **Extrude**. Expand **ZXPlane**, select **Sketch1** and click **Apply** next to **Geometry**. Enter 1 next to **FD1, Depth (>0)**. Click **Generate**.
* Click **ZXPlane**. Click **Look At Face/Plane/Sketch**. Click **New Sketch**.

* Click **Sketching**. Click **Draw** → Click **Rectangle**. Draw a rectangle by two points as the base of the first fin. Select the origin as the first point. Drag the cursor and place the opposite corner anywhere on the upper edge of the base.

* To create the second fin, select any point on the lower edge of the base as the first point and any point on the upper edge of the base as the second point.

* To create the third fin, select any point on the lower edge of the base as the first point and upper right corner of the base as the second point.

* Click **Dimensions** → Click **Horizontal**. Select the vertical edges of the first fin, then place the dimensions anywhere on the screen. Repeat for the second and third fins. Select the left edge of the middle fin and the X axis and place the dimension anywhere on the screen. Set the values of **H3**, **H4**, and **H5** to 1 and the value of **H6** to 9.5.

* Click **Extrude**. Select **Sketch2** for **Geometry**. Enter 20 next to **FD1, Depth (>0)**. Click **Generate**.

* Expand **1 Part, 1 Body** and rename the created body my-heat-sink.

* Click **ZXPlane**. Click **Look At Face/Plane/Sketch**. Click **New Sketch**.

* Click **Sketching**. Click **Draw** → Click **Rectangle**. Draw a rectangle that surrounds the heat sink.

* Click **Dimensions** → Click **General**. Select the upper edge of the rectangle. Drag the cursor and click anywhere to place the dimension. Repeat for the right edge of the rectangle. Change **H7** to 220 and **V8** to 480.

* Click **Constraints** → Click **Equal Distance**. Select the left edges of the fluid rectangle and the heat sink for the first pair. Select the right edges of the fluid rectangle and the heat sink for the second pair. Select the top edges of the fluid rectangle and the heat sink for the first pair. Select the bottom edges of the fluid rectangle and the heat sink for the second pair.

* Click **Extrude**. Select **Sketch3** for **Geometry**. Set the **Operation** to **Add Frozen**. Enter 80 next to **FD1, Depth (>0)** and click **Generate**.

* Under **2 Parts, 2 Bodies**, rename the created body my-fluid. Select **Fluid** next to **Fluid/Solid**.

* Click **Create** in the menu toolbar and select **Boolean**. Set the **Operation** to **Subtract**.

* Under **2 Parts, 2 Bodies**, select **my-fluid** as the **Target Bodies** and select **my-heat-sink** as the **Tool Bodies**. Set **Preserve Tool Bodies?** to **Yes**. Click **Generate**.

* Select **YZPlane**. Click **Create**. Select **New Plane**. Click **Transform** in **Details View** of **Plane4**. Select **Offset Global X**. Enter -3 next to **FD1, Value 1** and click **Generate**.

* Click **Slice**. Ensure **Slice Type** is set to **Slice by Plane**. Select **Plane4** and click **Apply** next to **Base Plane**. Change **Slice Targets** to **Selected Bodies**.

Select **my-fluid** in the **Tree Outline** view and click **Apply** next to **Bodies**. Click **Generate**.

* Under **3 Parts, 3 Bodies**, rename the upstream body created by slicing the fluid domain `my-fluid-domain-upstream`.

* Select **YZPlane**. Click **Create**. Select **New Plane**. Select **Offset Global X** next to **Transform** and type 23 next to **FD1, Depth (>0)**. Click **Generate**. Click **Slice** and select **Plane5** for **Base Plane**. Change **Slice Targets** to **Selected Bodies**. Select **my-fluid** for **Bodies** and click **Generate**. Rename the downstream body created under **4 Parts, 4 Bodies**, `my-fluid-domain-downstream`.

* Select **XYPlane**. Click **Create**. Select **New Plane**. Select **Offset Global Z** next to **Transform** and type -3 next to **FD1, Value 1**. Click **Generate**. Click **Slice** and select **Plane6** for **Base Plane**. Change **Slice Targets** to **Selected Bodies**. Select **my-fluid** for **Bodies** and click **Generate**. Rename the right body created under **5 Parts, 5 Bodies**, `my-fluid-domain-right`.

* Select **XYPlane**. Click **Create**. Select **New Plane**. Select **Offset Global Z** next to **Transform** and type 23 next to **FD1, Depth (>0)**. Click **Generate**. Click **Slice**, select **Plane7** for **Base Plane**. Change **Slice Targets** to **Selected Bodies**. Select **my-fluid** for **Bodies** and click **Generate**. Rename the left body created under **6 Parts, 6 Bodies**, `my-fluid-domain-left`.

* Select **ZXPlane**. Click **Create**. Select **New Plane**. Select **Offset Global Y** next to **Transform** and type 23 next to **FD1, Value 1**. Click **Generate**. Click **Slice**. Select **Plane8** for **Base Plane**. Change **Slice Targets** to **Selected Bodies**. Select **my-fluid** for **Bodies**. Click **Generate**. Rename the top body created by under **7 Parts, 7 Bodies**, `my-fluid-domain-above-hs`.

* Rename the body surrounding the heat sink `my-fluid-domain-hs`.

* Control-select all bodies under **7 Parts, 7 Bodies**. Right-click and select **Form New Part**.

* Expand **Part** under **1 Part, 7 Bodies**. Right-click **my-heat-sink** and select **Hide Body**. Activate **Selection Filter: Faces**. Select the inlet face. Right-click and select **Named Selection**. Click **Apply** next to **Geometry** and type `inlet` next to **Named Selection**. Click **Generate**.

* Repeat and create a **Named Selection** on the outlet face. Name it `outlet`.

* Control-select all the faces on the right side, the left side, and the top of the rectangular domain. Create a **Named Selection** for the 11 selected faces. Name it `wall`.

* Click **-Y** axis in the triad. Control-select all the faces. Create a **Named Selection** for the five selected faces. Name it `ground`.

* Right-click **my-heat-sink** in **Tree Outline** view and select **Hide All Other Bodies**. Right-click in the **Graphics** window and select **Select All**. Click **-Y** axis in the triad. Hold the **Ctrl** button and click the bottom face of the heat sink to deselect it. Create a **Named Selection** for the 13 selected faces. Name it `fins`.

* Create a **Named Selection** for the bottom face of the heat sink and name it `heat-source`.
* Click **File** and select **Save Project**.
* Click **File** and select **Close DesignModeler**.

7.3.3 Mesh

Open the Mesh application by double-clicking **Mesh** in ANSYS Workbench **Project schematic**.

Figure 7.16 is a flowchart of the steps needed to mesh the model.

1. **Set the Units to Millimeter**
   Click **Metric** in the status bar shown in Figure 7.17. Select the third option which is **Metric (mm, kg, N, s, mV, mA)**.

2. **Mesh the Heat Sink**
   In **Outline** view, expand **Geometry** and then expand **Part**. Right-click **my-heat-sink** and select **Suppress All Other Bodies**.

   Right-click **Mesh** and select **Insert ⇒ Method**.

   Make sure **Body** selection filter is activated. In **Details of "Automatic Method"-Method**, click **No Selection** next to **Geometry** then select the heat sink in the geometry window. Click **Apply** next to **Geometry**.

   Click **Method** and from the drop-down list next to it, select **Tetrahedrons**. Make sure the **Algorithm** is set to **Patch Conforming**. A green check mark appears next to **Patch Conforming Method** under **Mesh** in **Outline** view. See Figure 7.17.

   Right-click **Mesh** in **Outline** view and select **Insert ⇒ Sizing** to set the element size to 0.5 mm. Activate the **Body** selection filter and select the heat sink in the geometry window. Click **Apply** next to **Geometry** in **Details of "Sizing" - Sizing**. Change the **Element Size** from its default value to `0.5` and the **Growth Rate** to `1.1`. Right-click **Mesh** in **Outline** view and select **Generate Mesh**.

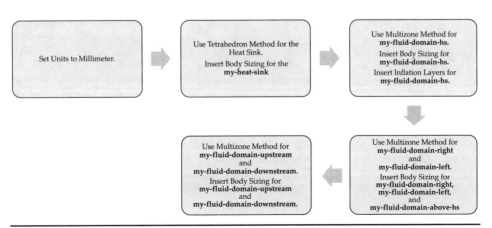

FIGURE 7.16 Flowchart of Mesh.

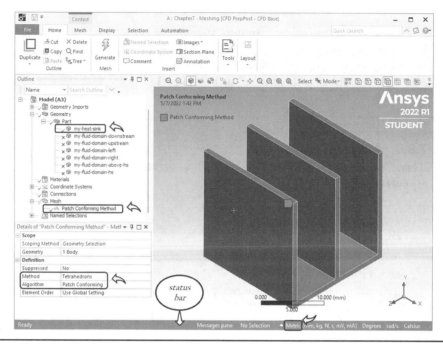

FIGURE 7.17   Metric and Patch Conforming Method Mesh.

3. **Mesh my-fluid-domain-hs**

We will mesh the fluid body surrounding the heat sink first. Under **Geometry** → **Part** in **Outline** view, right-click **my-fluid-domain-hs** and select **Unsuppress Body**. Right-click **my-fluid-domain-hs** and select **Hide All Other Bodies**.

Right-click **Mesh** in **Outline** view and select **Insert** ⇒ **Method**. Make sure **Body** selection filter is activated and select the fluid domain in the geometry window. Click **Apply** next to **Geometry** in **Details of "Automatic Method"-Method**. Click **Method**, and from the drop-down list next to it select **MultiZone**. A green check mark appears next to **MultiZone** under **Mesh** in **Outline** view.

Right-click **Mesh** in **Outline** view and select **Insert** ⇒ **Sizing** to set the element size to 0.6 mm. Make sure **Body** selection filter is activated and select the fluid domain in the geometry window. Click **Apply** next to **Geometry** in **Details of "Sizing"-Sizing**. Set the value of the **Element Size** to 0.6 and the **Growth Rate** to 1.1.

We will add inflation layers since the flow is turbulent. Right-click **Mesh** in **Outline** view and select **Insert** ⇒ **Inflation**. Make sure **Body** selection filter is activated and select the fluid domain in the geometry window. Click **Apply** next to **Geometry** in **Details of "Inflation"-Inflation**. Expand **Named Selections** in **Outline** view. Right-click **fins** and select **Select Items in Group**. Click **Inflation** under **Mesh**, and click **No Selection** next to **Boundary** in **Details of "Inflation"-Inflation**. Click **Apply**. The cell next to **Boundary** displays **13 Faces**. Click **Inflation Option** and from the drop-down list next to it select **First Layer Thickness**. Click **Please Define** next to **First Layer Height** and type 0.1 to set the thickness of the first layer to 0.1 mm. Set **Maximum Layers** to 8 and the **Growth Rate** to 1.13. Right-click **Mesh** in **Outline** view and select **Generate Mesh**.

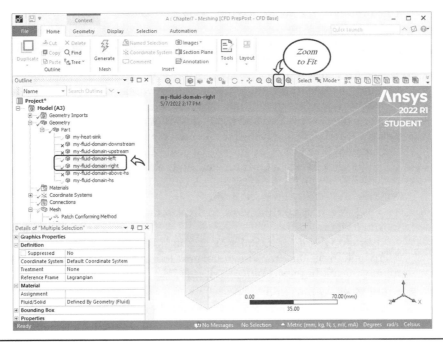

FIGURE 7.18 Right and left fluid bodies.

4. **Mesh my-fluid-domain-left and my-fluid-domain-right**
   Hold the **Ctrl** button and select **my-fluid-domain-left** and **my-fluid-domain-right** under **Geometry** → **Part** in **Outline** view. Right-click and select **Unsuppress Body**. While both bodies are still selected, right-click again and select **Hide All Other Bodies**.
      Click **Zoom To Fit** in the graphics toolbar. The screen should look like Figure 7.18.
      Right-click **Mesh** in **Outline** view and select **Insert** ⇒ **Method**. Make sure **Body** selection filter is activated. Right-click anywhere in the geometry window and select **Select All**. Click **Apply** next to **Geometry** in **Details of "Automatic Method"-Method**. **Apply** will change into **2 Bodies**. Click **Method** and from the drop-down list next to it select **MultiZone**.
      A green check mark appears next to **Multizone 2** under **Mesh** in **Outline** view.

5. **Insert Element Size on my-fluid-domain-above-hs, my-fluid-domain-left and my-fluid-domain-right.**
   Right-click **my-fluid-domain-above-hs** under **Geometry** → **Part** in **Outline** view and select **Unsuppress Body**. Right-click **Mesh** in **Outline** view and select **Insert** ⇒ **Sizing** to change the element size to 6 mm. Make sure **Body** selection is activated. Right-click anywhere in the geometry window and select **Select All**. Click **Apply** next to **Geometry** in **Details of "Sizing"-Sizing**. **Apply** will change into **3 Bodies**. Change the **Element Size** to 6. Change the **Growth Rate** to 1.1. Right-click **Mesh** in **Outline** view and select **Generate Mesh**.

6. **Mesh my-fluid-domain-downstream and my-fluid-domain-upstream.**
   Hold the **Ctrl** button and select **my-fluid-domain-downstream** and **my-fluid-domain-upstream** under **Geometry** → **Part** in **Outline** view. Right-click and select

Unsuppress Body. While both bodies are still selected, right-click again and select **Hide All Other Bodies**.

Right-click **Mesh** in **Outline** view and select **Insert** ⇒ **Method**. Make sure **Body** selection filter is activated and right-click anywhere in the geometry window. Select **Select All.** Click **Apply** next to **Geometry** in **Details of "Automatic Method"-Method**. Click **Method** and from the drop-down list next to it select **MultiZone**.

Right-click **Mesh** in **Outline** view and select **Insert** ⇒ **Sizing**. Make sure **Body** selection filter is activated and right-click anywhere in the geometry window. Select **Select All** and click **Apply** next to **Geometry** in **Details of "Sizing" -Sizing**. Change the **Element Size** to 12 mm and the **Growth Rate** to 1.1.

Right-click **Mesh** in **Outline** view and select **Update**.

Click **Zoom To Fit** in the graphics toolbar. Right-click anywhere in the geometry window and select **Show All Bodies**.

Click **Mesh** in **Outline** view to display the mesh on the screen which should look like Figure 7.19.

Expand **Quality** in **Details of "Mesh"**, as shown in Figure 7.19. Click **Mesh Metric** and from the drop-down list next to it select **Aspect Ratio**. Note the maximum value of the aspect ratio, **Max**, is **24.74**.

7. **Close Mesh Application**

Click **File** and select **Save Project** from the drop-down list.

Close or minimize ANSYS Mesh application and return to Workbench. A green check symbol next to **Mesh** indicates the model is ready for the ANSYS **Setup** Application. Right-click **Mesh** and select **Update** if needed.

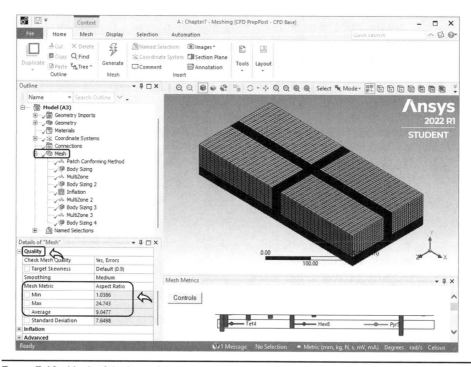

FIGURE 7.19 Mesh of the heat sink.

Mesh Summary

* Click **Metric** in the status bar and select **Metric (mm, kg, N, s, V, A)**.

* Expand **Geometry** → Expand **Part**. Right-click **my-heat-sink** and select **Suppress All Other Bodies**.

* Right-click **Mesh** and select **Insert ⇒ Method**. Activate **Body** selection filter and select the heat sink in the geometry window for **Geometry**. Select **Tetrahedrons** next to **Method**.

* Right-click **Mesh** and select **Insert ⇒ Sizing**. Make sure **Body** selection filter is activated and select the heat sink in the geometry window for **Geometry**. Set the **Element Size** to 0.5 and the **Growth Rate** to 1.1. Right-click **Mesh** and select **Generate Mesh**.

* Right-click **my-fluid-domain-hs** under **Geometry** → **Part** and select **Unsuppress Body**. Right-click **my-fluid-domain-hs** and select **Hide All Other Bodies**.

* Right-click **Mesh** and select **Insert ⇒ Method**. Activate **Body** selection filter and select the fluid domain in the geometry window for **Geometry**. Select **MultiZone** next to **Method**.

* Right-click **Mesh** and select **Insert ⇒ Sizing**. Make sure **Body** selection filter is activated and select the fluid domain in the geometry window for **Geometry**. Set the **Element Size** to 0.6 and the **Growth Rate** to 1.1.

* Right-click **Mesh** and select **Insert ⇒ Inflation**. Make sure **Body** selection filter is activated and select the fluid domain in the geometry window for **Geometry**. Expand **Named Selections**, right-click **fins** and select **Select Items in Group**. Click **Inflation**, click **No Selection** next to **Boundary** and click **Apply**. Select **First Layer Thickness** next to **Inflation Option**. Set the **First Layer Height** to 0.1, the **Maximum Layers** to 8 and the **Growth Rate** to 1.13. Right-click **Mesh** and select **Generate Mesh**.

* Control-select **my-fluid-domain-left** and **my-fluid-domain-right** under **Geometry** → **Part**. Right-click and select **Unsuppress Body**. While both bodies are still selected, right-click and select **Hide All Other Bodies**.

* Right-click **Mesh** and select **Insert ⇒ Method**. Make sure **Body** selection filter is activated and right-click anywhere in the geometry window. Select **Select All** and click **Apply**. Select **MultiZone** next to **Method**.

* Right-click **my-fluid-domain-above-hs** under **Geometry** → **Part** and select **Unsuppress Body**.

* Right-click **Mesh** and select **Insert ⇒ Sizing**. Make sure **Body** selection filter is activated and right-click anywhere in the geometry window. Select **Select All** and click **Apply**. Set the **Element Size** to 6 and the **Growth Rate** to 1.1. Right-click **Mesh** and select **Generate Mesh**.

* Control-select **my-fluid-domain-upstream** and **my-fluid-domain-downstream** under **Geometry** → **Part**. Right-click and select **Unsuppress**

Body. While both bodies are still selected, right-click and select **Hide All Other Bodies**.

* Right-click **Mesh** and select **Insert** ⇒ **Method**. Make sure **Body** selection filter is activated and right-click anywhere in the geometry window. Select **Select All** and click **Apply**. Select **MultiZone** next to **Method**.

* Right-click **Mesh** and select **Insert** ⇒ **Sizing**. Make sure **Body** selection filter is activated and right-click anywhere in the geometry window. Select **Select All** and click **Apply**. Set the **Element Size** to 12 and the **Growth Rate** to 1.1. Right-click **Mesh** and select **Update**.

* Click **Mesh**. Expand **Quality** and select **Aspect Ratio** next to **Mesh Metric**.

* Click **File** and select **Save Project**.

* Click **File** and select **Close Meshing**.

7.3.4 Setup

Figure 7.20 is a flowchart of the steps needed to set up the model in Fluent.

Launch Fluent in ANSYS Workbench by double-clicking **Setup** in **Project Schematic**. Enable **Double Precision** and increase the number of **Solver Processes** depending on the available processors for usage. Click **Start**.

1. **General**
   Expand **Setup** → Double-click **General**. Keep the default options of **Pressure-Based** for **Type** solver in the **Task Page** view. Keep the **Absolute** for **Velocity Formulation**, and **Steady** for **Time** options.

2. **Models**
   Expand **Setup** → Double-click **Models**. Double-click **Energy-Off** in the **Task Page** view. Enable the **Energy Equation** and click **OK**.

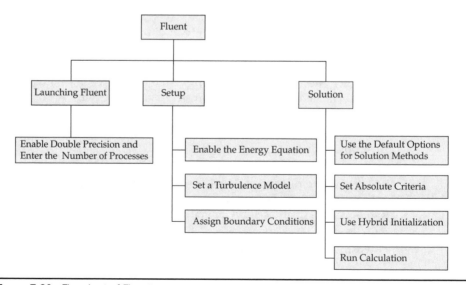

FIGURE 7.20 Flowchart of Fluent.

We will use the k-omega turbulence model. Keep the default option of **SST k-omega** for **Viscous**.

3. **Boundary Conditions**

   Expand **Setup** → Double-click **Boundary Conditions** to assign boundary conditions to the listed zones in the **Task Page**.

   If the boundaries are not grouped together in the **Task Page** view, click **Toggle Tree View** and select **Zone Type** under **Group By** from the list of options to make it easier to assign boundary conditions.

   Select **inlet** and make sure its **Type** is **velocity-inlet**. Click **Edit...** and choose **Magnitude, Normal to Boundary** as the **Velocity Specification Method**. Set the **Velocity Magnitude [m/s]** value to 4 and click **Apply**. Click the **Thermal** tab and set the inlet **Temperature [K]** value to 280. Click **Apply** and click **Close** to close the **Velocity Inlet** dialog box.

   Select **outlet**. Make sure its **Type** is **pressure-outlet**. Click **Edit...** and confirm the value of **Gauge Pressure [Pa]** at the outlet is 0. Click **Close**.

   Select **fins** which is made of the faces interfacing the solid and the fluid bodies in the model. Make sure its **Type** is **wall**. Click **Edit...** and confirm the **Thermal** conditions for the **fins** are set to **Coupled**. Click **Close**.

   Click **heat-source** and make sure its **Type** is **wall**. Click **Edit....** Select **Heat Flux** under **Thermal Conditions** and set the value of **Heat Flux [W/m^2]** to 20000. Click **Apply** and click **Close**.

4. **Methods**

   Expand **Solution** → Double-click **Methods**. Keep the default options for the **Solution Methods**:

   > **Pressure-Velocity Coupling** → **Scheme** → **Coupled**.

   > **Pressure-Velocity Coupling** → **Flux Type** → **Rhie-Chow: distance based**.

   > **Spatial Discretization** → **Gradient** → **Least Squares Cell Based**.

   > **Spatial Discretization** → **Pressure** → **Second Order**.

   > **Spatial Discretization** → **Momentum** → **Second Order Upwind**.

   > **Spatial Discretization** → **Turbulent Kinetic Energy** → **Second Order Upwind**.

   > **Spatial Discretization** → **Specific Dissipation Rate** → **Second Order Upwind**.

   > **Spatial Discretization** → **Energy** → **Second Order Upwind**.

   Scroll down if needed and make sure **Pseudo Time Method** is enabled.

5. **Residuals**

   Expand **Solution** → Expand **Monitors** → Double-click **Residual**.

   In addition to the residuals for the continuity, x-velocity, y-velocity, z-velocity, and energy, we have two residuals corresponding to the **k-omega** turbulence model for a total of seven residuals.

   Change the **Absolute Criteria** to 1e-14 for all seven equations. Click **OK** to close the **Residual Monitors** dialog box.

6. **Initialization**

   Expand **Solution** → Double-click **Initialization**. Select **Hybrid Initialization** under **Initialization Methods** in the **Task Page** view and click **Initialize**. The **Console** will display a message when **Hybrid Initialization** is done.

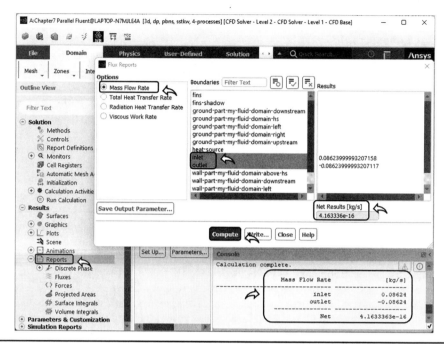

FIGURE 7.21 Mass Flow Rate Flux Reports.

7. **Run Calculation**
Expand **Solution** → Double-click **Run Calculation**. Set the **Number of Iterations** to
1000 in the **Task Page** view. Click **Calculate** button under **Solution Advancement**.
Click **OK** when the **Calculation complete** message appears.

8. **Flux Reports**
Expand **Results** → Double-click **Reports** to check the total mass flux through the
inlet and outlet of the model.
Double-click **Fluxes** in the **Task Page** view.
Select **Mass Flow Rate** for **Options**, then select **inlet** and **outlet** from the list
of **Boundaries** in the **Flux Reports** dialog box to check the mass balance for the
model. Click **Compute**. The **Net Results** is displayed in the dialog box, and also it is
displayed in the **Console** window as shown in Figure 7.21. The net mass imbalance
is almost zero and is an indication of convergence.
While the **Flux Reports** dialog box is still open, select **Total Heat Transfer Rate**
under **Options**. Click **Deselect All Shown**, shown in Figure 7.22, then select **fins**.
Click **Compute**.
The value of the heat flux at the boundary named **heat-source** is set by the user
as 20,000 W/m^2. Given than the **heat-source** face has a total surface area of 4 mm^2,
the total heat released by the heat source is 8 W. This value is in agreement with
the heat released by the fins.
Click **Close** to close the **Flux Reports** dialog box.

9. **Save the Project and Close Fluent**
Save the project by clicking **File** and selecting **Save Project** from the drop-down list.
Close the Fluent application by clicking **File** and selecting **Close Fluent** and return
to Workbench.

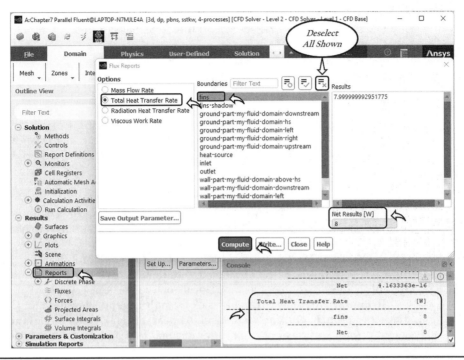

FIGURE 7.22 Total Heat Transfer Rate Flux Reports.

Setup Summary

* Expand **Setup** → Double-click **General**. Make sure **Pressure-Based** is selected for **Type, Absolute** is selected for **Velocity Formulation** and **Steady** is selected for **Time**.

* Expand **Setup** → Double-click **Models**. Double-click **Energy - Off**. Enable **Energy Equation**. Click **OK**. Keep the default **Viscous - SST k-omega** for the viscous model.

* Expand **Setup** → Double-click **Boundary Conditions**. Double-click **inlet**. Set the **Velocity Magnitude [m/s]** to 4 and click **Apply**. Click **Thermal** tab. Change the **Temperature [K]** to 280. Click **Apply** and click **Close**. Double-click **heat-source** and set **Heat Flux [W/m²]** to 20000. Click **Close**.

* Expand **Solution** → **Methods**. Use the default options for **Solution Methods**.

* Expand **Solution** → Expand **Monitors** → Double-click **Residual**. Set the **Absolute Criteria** for all equations to 1e-14 Click **OK**.

* Expand **Solution** → Double-click **Initialization**. Click **Initialize**.

* Expand **Solution** → Double-click **Run Calculation**. Set the **Number of Iterations** to 1000. Click **Calculate**.

* Expand **Results** → Expand **Reports**.

∗ Double-click **Fluxes**. Select **Mass Flow Rate**. Select **inlet** and **outlet**. Click **Compute**.

∗ Select **Total Heat Transfer Rate**. Select **fins** and click **Compute**.

7.3.5 Solution

Double-click **Results** in ANSYS Workbench **Project Schematic** to launch CFD-Post where the ANSYS Fluent results are automatically loaded. Note that **DesignModeler**, **Mesh**, and **Fluent** applications should be closed in order to open **Results** application or an error might occur.

1. **Heat Sink Temperature Distribution**

   Click **Contour**, shown in Figure 7.23, to create the temperature contours. You can also access **Contour** from the **Insert** menu. Type `My Temp HS` next to **Name**. Click **OK**.

   In **Details of My Temp HS** view, select **fins** next to **Locations**. Select **Temperature** next to **Variable**. Select **Local** next to **Range**. Increase the # **of Contours** to `30` to capture the variation of the temperature on the fins. Click **Apply**.

   To improve the legend, double-click **Default Legend View 1** under **User Locations and Plots** in **Outline** view. Click **Appearance** in **Details of Default Legend View 1**. Set the number of digits after decimal to `0` and the format type to **Fixed**

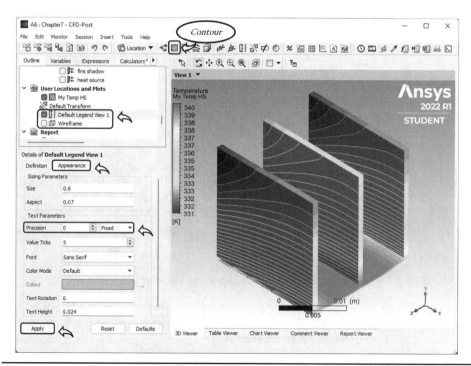

FIGURE 7.23 Heat sink temperature distribution.

next to **Precision** under **Text Parameters**. Click **Apply**. Disable **Wireframe** in **Outline** view and then click **Fit View** in the 3D viewer toolbar. Figure 7.23 displays the temperature distribution on the surfaces of the heat sink.

2. **Temperature Contours**

We will create two planes inside the model in order to display the temperature contours on a surface containing the heat sink and surrounding air.

We will create the first plane using a point whose coordinates are (0.02,0.018, 0.01) and a normal whose direction vector is (0,1,0).

Click **Location** and select **Plane** from the list of options. Type My Top Plane for **Name** and click **OK**.

Select **Point and Normal** next to **Method** in **Details of My Top Plane** view. Enter the coordinates of the points: (0.02,0.018 0.01), and the direction vector of the normal (0,1,0). Select **Rectangular** next to **Type**. Type 0.08 next to **X Size** and 0.08 next to **Y Size**. Click **Apply**.

A bounded plane cutting through the domain is created as shown in Figure 7.24. **My Top Plane** appears under **User Locations and Plots** in the **Outline** view.

We will create the second plane by duplicating **My Top Plane** and changing the coordinates of the **Point** used to create the plane.

Right-click **My Top Plane** under **User Locations and Plots** in **Outline** view. Select **Duplicate**. Type My Bottom Plane next to **Name** and click **OK**. Double-click **My Bottom Plane** in **Out-line** view. In **Details of My Bottom Plane**, set the

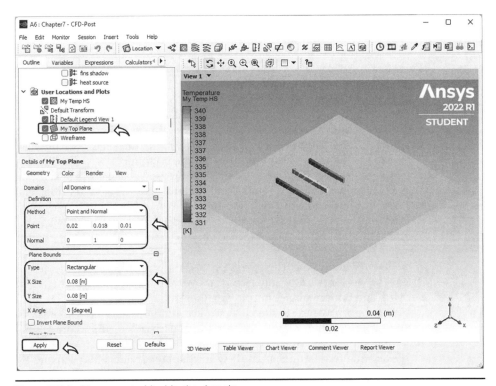

FIGURE 7.24 Plane created inside the domain.

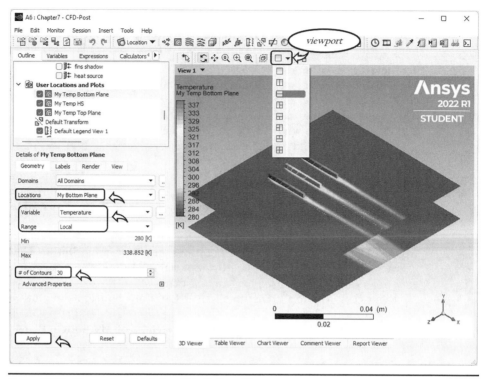

FIGURE 7.25 Viewport and details of My Bottom Plane.

Point to (0.02, 0.002, 0.01), and the normal to (0, 1, 0). Make sure that **X Size** and **Y Size** are set to **0.08 [m]**. Click **Apply**.

Click the **Contour** icon. Type My Temp Top Plane next to **Name**. Click **OK**. Select **My Top Plane** next to **Locations** in **Details of My Temp Top Plane** view. Select **Temperature** next to **Variable**. Select **Local** next to **Range**. Increase the # of **Contours** to 30 to capture the variation of the temperature. Click **Apply**.

To create the temperature contours on the bottom plane, right-click **My Temp Top Plane** and select **Duplicate**. Type My Temp Bottom Plane next to **Name** and click **OK**. Double-click **My Temp Bottom Plane** and select **My Bottom Plane** next to **Locations** in **Details of My Temp Bottom Plane**. Click **Apply**. Adjust the view by clicking the cyan **ISO** ball in the triad and **Fit View** in the 3D viewer toolbar. Figure 7.25 displays the temperature contours on the horizontal planes in isometric view.

CFD-Post allows the user to postprocess multiple files simultaneously. We will divide the viewer window into two windows and plot the temperature contours in both top and bottom planes to compare them side by side.

Click the arrow next to the viewport layout in the 3D viewer toolbar shown in Figure 7.25. Select the two windows on top of each other, which corresponds to the third option.

Deactivate **Synchronize visibility in displayed views** and activate **Synchronize camera in displayed views** in the 3D viewer toolbar. The screen should look like Figure 7.26.

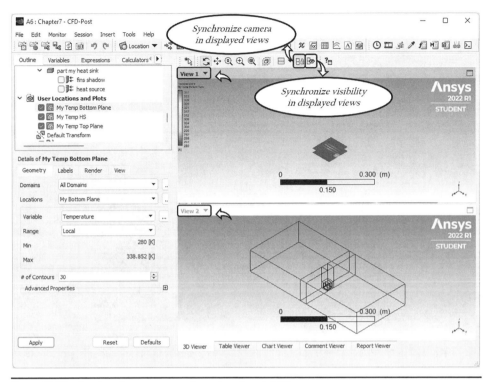

FIGURE 7.26 Synchronize camera and synchronize visibility.

By activating **Sychronized camera in displayed views** shown in Figure 7.26, the camera orientation and zoom level are identical for the selected viewports.

By activating **Sychronized visibility in displayed views**, the visibility of all objects under **User Locations and Plots** is identical for the selected viewports.

Click **View 1**, shown in Figure 7.26, to display the temperature contours of the bottom plane in the upper viewport.

Disable **My Temp HS**, **My Temp Top Plane**, **My Bottom Plane** and **My Top Plane** under **User Locations and Plots** in **Outline** view.

Double-click **Default Legend View 1** and click **Appearance** in **Details of Default Legend View 1**. Set the **Size** equal to 0.9. Make sure the number of digits after decimal is **0** and the format type is **Fixed** next to **Precision**. Set the **Text Height** to 0.042. Click **Apply**.

Click **Y** axis in the triad of the upper viewport to display a top view of the temperature contours. Notice that the orientation and zoom level in the lower viewport are identical to those in the upper viewport.

Click **View 2**, shown in Figure 7.26, to display the temperature contours of the upper plane in the lower viewport.

Disable **Wireframe**, and enable **My Temp Top Plane** under **User Locations and Plots** in **Outline** view.

Double-click **Default Legend View 2** and click **Appearance** in **Details of Default Legend View 2**. Set the **Size** to 0.9 under **Details of Default Legend View 2**. Set

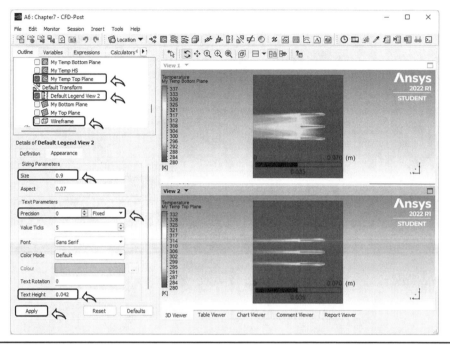

FIGURE 7.27   Temperature contours of the heat sink and surrounding air at two different elevations along the Z axis.

the number of digits after decimal to 0 and format type to **Fixed** next to **Precision**. Set the **Text Height** 0.042. Click **Apply**.

The screen should look like Figure 7.27.

7.4   Verification

In this section, we will report the metrics of the grid independent study performed using the Richardson Extrapolation scheme described in Section 2.9. The mesh statistics for this study are summarized in Table 7.2.

The parameter investigated is the average temperature along the side of one of the fins. The convergence metrics are reported at 11 points on the left fin. The points were selected at different locations along the Y axis.

Table 7.3 displays the convergence metrics calculated using the Richardson Extrapolation scheme. We observe that the local order of accuracy ranges between 1.5 and 1.65 with a global average of 1.56. This value is close to the formal order of accuracy of the

TABLE 7.2   Mesh Statistics

Mesh # j	Quality	N_j	h_j(mm)	$r_{j,j-1}$	$max(y^+)$
1	Fine	3,856,857	0.0013		1.65
2	Medium	1,338,780	0.0018	1.42	2.04
3	Coarse	485,892	0.0026	1.4	2.61

TABLE 7.3 Convergence Metrics Using Richardson Extrapolation for the Temperature

i	y_i (mm)	ϕ_{l_1} (K)	ϕ_{l_2} (K)	ϕ_{l_3} (K)	s_i	$(e_a^{21})_i$ %	$(e_a^{32})_i$ %	p_i	$(GCI^{21})_i$ ×100	EB_i
1	0	337.86	337.96	338.12	1	0.029	0.049	1.63	0.049	0.17
2	2	336.90	337.01	337.18	1	0.031	0.052	1.66	0.052	0.18
3	4	335.88	335.98	336.14	1	0.029	0.048	1.60	0.049	0.17
4	6	334.99	335.08	335.23	1	0.028	0.045	1.56	0.047	0.16
5	8	334.21	334.30	334.44	1	0.026	0.042	1.51	0.045	0.15
6	10	333.55	333.63	333.77	1	0.025	0.041	1.52	0.043	0.14
7	12	332.99	333.07	333.20	1	0.024	0.039	1.51	0.041	0.14
8	14	332.54	332.62	332.74	1	0.023	0.038	1.51	0.040	0.13
9	16	332.20	332.27	332.40	1	0.023	0.037	1.54	0.039	0.13
10	18	331.96	332.04	332.16	1	0.022	0.036	1.56	0.038	0.13
11	20	331.86	331.93	332.05	1	0.022	0.036	1.60	0.037	0.12

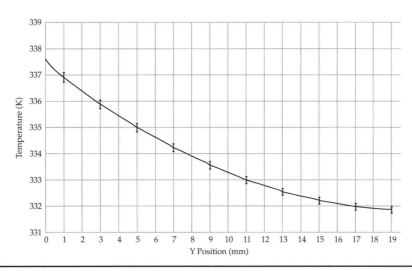

FIGURE 7.28 Error bars for the temperature of the fin.

numerical schemes employed. Figure 7.28 displays the variation of the average temperature along the Y axis, where the hottest point (at $y = 0$) corresponds to the location where the fin intersects the base of the heat sink. The maximum value of the error bar is ± 0.18 K which corresponds to a discretization error of 0.1%.

References

[1]   John R Howell, M Pinar Mengüç, Kyle Daun, and Robert Siegel. *Thermal Radiation Heat Transfer*. CRC Press, 2020.

[2]   James A Fay. *Introduction to Fluid Mechanics*. MIT Press, 1994.

[3]   TJ Chung. *Computational Fluid Dynamics*. Cambridge University Press, 2010.

[4]   Dale Anderson, John C Tannehill, Richard H Pletcher, Ramakanth Munipalli, and Vijaya Shankar. *Computational Fluid Mechanics and Heat Transfer*. CRC Press, 2020.

[5]   Peter D Lax and Robert D Richtmyer. "Survey of the stability of linear finite difference equations." *Communications on Pure and Applied Mathematics* 9.2 (1956), pp. 267–293.

[6]   Milton Abramowitz, Irene A Stegun, and Robert H Romer. *Handbook of Mathematical Functions with Formulas, Graphs, and Mathematical Tables*. American Journal of Physics, 56.10 (1988), pp. 958–958

[7]   SV Patankar. "Numerical heat transfer and fluid flow." McGraw-Hill Book Company (1980).

[8]   Richard Courant, Eugene Isaacson, and Mina Rees. "On the solution of nonlinear hyperbolic differential equations by finite differences." *Communications on Pure and Applied Mathematics* 5.3 (1952), pp. 243–255.

[9]   Richard Courant, Kurt Friedrichs, and Hans Lewy. "Über die partiellen Differenzengleichungen der mathematischen Physik." *Mathematische Annalen* 100.1 (1928), pp. 32–74.

[10] Gene H. Golub. "Matrix computations." Baltimore, MD (1983).

[11] Jiri Blazek. *Computational Fluid Dynamics: Principles and Applications*. Butterworth-Heinemann, 2015.

[12] Yousef Saad. *Iterative Methods for Sparse Linear Systems*. SIAM, 2003.

[13] Gene H Golub and Charles F Van Loan. *Matrix Computations*. JHU Press, 2013.

[14] Henk Kaarle Versteeg and Weeratunge Malalasekera. *An Introduction to Computational Fluid Dynamics: The Finite Volume Method*. Pearson Education, 2007.

[15] S.V Patankar and D.B Spalding. "A calculation procedure for heat, mass and momentum transfer in three-dimensional parabolic flows." *International Journal of Heat and Mass Transfer* 15.10 (1972), pp. 1787–1806.

[16] Suhas V Patankar. "A calculation procedure for two-dimensional elliptic situations." *Numerical Heat Transfer* 4.4 (1981), pp. 409–425.

[17] Jeffrey P Van Doormaal and George D Raithby. "Enhancements of the SIMPLE method for predicting incompressible fluid flows." *Numerical Heat Transfer* 7.2 (1984), pp. 147–163.

[18] R Issa. *Solution of the Implicit Discretized Fluid Flow Equations by Operator Splitting Mechanical Engineering Rep.* 1982.

285

[19] Tyrone S Phillips and Christopher J Roy. "Richardson extrapolation-based discretization uncertainty estimation for computational fluid dynamics." *Journal of Fluids Engineering* 136.12 (2014).

[20] Ismail Celik and Wei-Ming Zhang. "Calculation of numerical uncertainty using richardson extrapolation: application to some simple turbulent flow calculations." *J. Fluids Eng.* Sep 1995, 117(3): 439–445.

[21] Patrick J Roache and Patrick M Knupp. "Completed Richardson extrapolation." *Communications in Numerical Methods in Engineering* 9.5 (1993), pp. 365–374.

[22] Wei Shyy, Marc Garbey, A Appukuttan, and J Wu. "Evaluation of Richardson extrapolation in computational fluid dynamics." *Numerical Heat Transfer: Part B: Fundamentals* 41.2 (2002), pp. 139–164.

[23] Ishmail B Celik, Urmila Ghia, Patrick J Roache, and Christopher J Freitas. "Procedure for estimation and reporting of uncertainty due to discretization in CFD applications." *Journal of Fluids Engineering-Transactions of the ASME* 130.7 (2008).

[24] Ismail Celik and Ozgur Karatekin. "Numerical experiments on application of richardson extrapolation with nonuniform grids." *Journal of Fluids Engineering* 119.3 (1997), pp. 584–590.

[25] P. J. Roache. "Perspective: A method for uniform reporting of grid refinement studies." *Journal of Fluids Engineering* 116.3 (1994), pp. 405–413.

[26] Patrick Roache. "Error Bars for CFD." *41st Aerospace Sciences Meeting and Exhibit*, 06 January 2003–09 January 2003, Reno, Nevada.

[27] Ismail Celik, Jun Li, Gusheng Hu, and Christian Shaffer. "Limitations of Richardson extrapolation and some possible remedies." *J. Fluids Eng.* Jul 2005, 127(4): 795–805.

[28] Frank M White. *Fluid Mechanics*. Tata McGraw-Hill Education, 1979.

[29] Glenn O Brown. "The history of the Darcy-Weisbach equation for pipe flow resistance." *Environmental and Water Resources History*. 2003, pp. 34–43.

[30] Hermann Schlichting. "Boundary-Layer theory." (1987).

[31] Frank M White and Joseph Majdalani. *Viscous Fluid Flow*. Vol. 3. McGraw-Hill New York, 2006.

[32] Patrick Knupp. *Remarks on Mesh Quality*. Tech. rep. Sandia National Lab. (SNL-NM), Albuquerque, NM (United States), 2007.

[33] Hugh Thornburg. "Overview of the PETTT workshop on mesh quality/resolution, practice, current research, and future directions." *50th AIAA Aerospace Sciences Meeting including the New Horizons Forum and Aerospace Exposition*. 2012, p. 606.

[34] Ins ANSYS. "ANSYS FLUENT user's guide." *Canonsburg, PA* 15317 (2011).

[35] Yunus A Cengel. *Fluid Mechanics*. Tata McGraw-Hill Education, 2010.

[36] William Morrow Kays. *Convective Heat and Mass Transfer*. Tata McGraw-Hill Education, 2011.

[37] Frank P Incropera, David P DeWitt, Theodore L Bergman, Adrienne S Lavine, et al. *Fundamentals of Heat and Mass Transfer*. Vol. 6. Wiley New York, 1996.

[38] BP Leonard. "A survey of finite differences of opinion on numerical muddling of the incomprehensible defective confusion equation." *Finite Element Methods for Convection Dominated Flows* 34 (1979), pp. 1–10.

Appendix A
Upwind Schemes to Evaluate the Advection Term

The first order upwind scheme (or upwind differencing scheme [UDS]) for approximating ϕ at the centroid of face $\Bbbk$ was presented in Section (2.4.6). We present below, two of the higher order upwind schemes, the second order upwind scheme and the third order QUICK scheme.

A.1 Second Order Upwind Scheme (LUDS)

The second order upwind scheme or linear upwind differencing scheme (LUDS) approximates $\phi_\Bbbk$ by linear extrapolation of the linear profile made up of the values of ϕ at the center of the upwind cell and the following cell in the upwind direction. Figure A.1 shows the cells involved in evaluating ϕ at the face $\mathbb{e}$ for the following two cases: (a) the

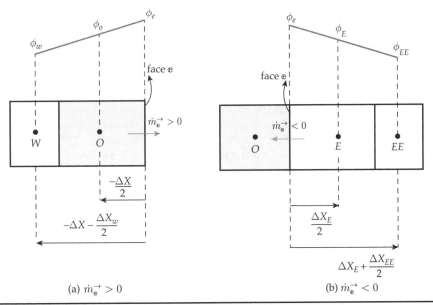

(a) $\dot{m}_\mathrm{e}^{\rightarrow} > 0$

(b) $\dot{m}_\mathrm{e}^{\rightarrow} < 0$

FIGURE A.1 Schematic of the cells employed in approximating ϕ at the center of face $\mathbb{e}$ using a second order upwind scheme for (a) the mass flow rate across face $\mathbb{e}$ is out of cell $\mathbb{O}$ and (b) the mass flow rate across face $\mathbb{e}$ is into cell $\mathbb{O}$.

TABLE A.1 Expressions for the Coefficients in Eq. (A.1)

$c_\mathbb{O}$	$1 - c_\mathbb{L}$
$c_\mathbb{L}$	$-\dfrac{\Delta_\mathbb{O}}{\Delta_\mathbb{O} + \Delta_\mathbb{L}}$
$c_\mathbb{K}$	$1 - c_\mathbb{KK}$
$c_\mathbb{KK}$	$-\dfrac{\Delta_\mathbb{K}}{\Delta_\mathbb{K} + \Delta_\mathbb{KK}}$

TABLE A.2 Terms Needed in the Second Order Upwind Scheme

$\mathbb{k}$	$\mathbb{K}$	$\mathbb{L}$	$\mathbb{KK}$	$\Delta_\mathbb{O}$	$\Delta_\mathbb{L}$	$\Delta_\mathbb{K}$	$\Delta_\mathbb{KK}$
$\mathbb{e}$	$\mathbb{E}$	$\mathbb{W}$	$\mathbb{EE}$	Δx	$\Delta x_\mathbb{W}$	$\Delta x_\mathbb{E}$	$\Delta x_\mathbb{EE}$
$\mathbb{w}$	$\mathbb{W}$	$\mathbb{E}$	$\mathbb{WW}$	Δx	$\Delta x_\mathbb{E}$	$\Delta x_\mathbb{W}$	$\Delta x_\mathbb{WW}$
$\mathbb{n}$	$\mathbb{N}$	$\mathbb{S}$	$\mathbb{NN}$	Δy	$\Delta y_\mathbb{N}$	$\Delta y_\mathbb{S}$	$\Delta y_\mathbb{NN}$
$\mathbb{s}$	$\mathbb{S}$	$\mathbb{N}$	$\mathbb{SS}$	Δy	$\Delta y_\mathbb{S}$	$\Delta y_\mathbb{N}$	$\Delta y_\mathbb{SS}$

mass flow rate across face $\mathbb{e}$ is out of cell $\mathbb{O}$, that is, $\dot{m}_\mathbb{e}^\rightarrow > 0$ and (b) the mass flow rate across face $\mathbb{e}$ is into cell $\mathbb{O}$, that is, $\dot{m}_\mathbb{e}^\rightarrow < 0$.

Linear extrapolation yields the following expression for ϕ at the faces of cell (i, j):

$$\phi_\mathbb{k} \simeq \begin{cases} c_\mathbb{O}\phi_\mathbb{O} + c_\mathbb{L}\phi_\mathbb{L} & : \dot{m}_\mathbb{k}^\rightarrow > 0 \\ c_\mathbb{K}\phi_\mathbb{K} + c_\mathbb{KK}\phi_\mathbb{KK} & : \dot{m}_\mathbb{k}^\rightarrow < 0 \end{cases} \tag{A.1}$$

Expressions for $c_\mathbb{O}, c_\mathbb{L}, c_\mathbb{K}$, and $c_\mathbb{KK}$ are listed in Table A.1. For each value of $\mathbb{k} \in \{\mathbb{e}, \mathbb{w}, \mathbb{n}, \mathbb{s}\}$, the corresponding values of $\mathbb{K}, \mathbb{L}, \mathbb{KK}, \Delta_\mathbb{O}, \Delta_\mathbb{L}, \Delta_\mathbb{K}$, and $\Delta_\mathbb{KK}$ are listed in Table A.2.

A.2 Third Order Upwind Scheme (QUICK)

The third order upwind scheme approximates $\phi_\mathbb{k}$ by interpolation using a parabolic fit of the values of ϕ at the centroids of three cells. The Quadratic upwind interpolation for convective kinematics (QUICK) scheme, proposed by Leonard [38], employs an additional cell to those employed in the second order upwind scheme presented in Figure A.1. Figure A.2 shows the cells involved in evaluating ϕ at the face $\mathbb{e}$ for (a) the mass flow rate across face $\mathbb{e}$ is out of cell $\mathbb{O}$, that is, $\dot{m}_\mathbb{e}^\rightarrow > 0$ and (b) the mass flow rate across face $\mathbb{e}$ is into cell $\mathbb{O}$, that is, $\dot{m}_\mathbb{e}^\rightarrow < 0$. The QUICK scheme yields the following expression for ϕ at the faces of cell $\mathbb{O}$:

$$\phi_\mathbb{k} \simeq \begin{cases} c_{\mathbb{O}1}\phi_\mathbb{O} + c_\mathbb{L}\phi_\mathbb{L} + c_{\mathbb{K}1}\phi_\mathbb{K} & : \dot{m}_\mathbb{k}^\rightarrow > 0 \\ c_{\mathbb{O}2}\phi_\mathbb{O} + c_{\mathbb{K}2}\phi_\mathbb{K} + c_{\mathbb{KK}}\phi_\mathbb{KK} & : \dot{m}_\mathbb{k}^\rightarrow < 0 \end{cases} \tag{A.2}$$

For each value of $\mathbb{k} \in \{\mathbb{e}, \mathbb{w}, \mathbb{n}, \mathbb{s}\}$, the corresponding values of $\mathbb{K}, \mathbb{L}, \mathbb{KK}, \Delta_\mathbb{O}, \Delta_\mathbb{L}, \Delta_\mathbb{K}$, and $\Delta_\mathbb{KK}$ are listed in Table A.2. The corresponding values of $c_{\mathbb{O}1}, c_\mathbb{L}$, and $c_{\mathbb{K}1}$ are listed in Table A.3 and those of $c_{\mathbb{O}2}, c_{\mathbb{K}2}$, and $c_\mathbb{KK}$ are listed in Table A.4.

Because the QUICK scheme is based on interpolation from a parabolic fit, it is third order accurate, which makes it a popular choice. Recall that expressing the flux terms in

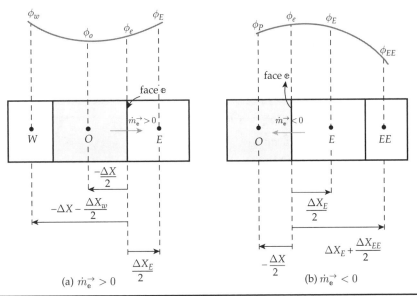

FIGURE A.2 Schematic of the cells employed in approximating ϕ at the center of face e using the third order QUICK upwind scheme for (a) the mass flow rate across face e is out of cell O and (b) the mass flow rate across face e is into cell O.

TABLE A.3 Terms Needed in the QUICK Upwind Scheme for $\dot{m}_k^{\rightarrow} > 0$

k	C_{O1}	C_{L}	C_{K1}
e	$\dfrac{(2\Delta x+\Delta x_W)\Delta x_E}{4\delta x_W \delta x_E}$	$\dfrac{\Delta x\Delta x_E}{4\delta x_W(\delta x_E+\delta x_W)}$	$\dfrac{(\Delta x+2\delta x_W)\Delta x}{4(\delta x_E+\delta x_W)\delta x_E}$
w	$\dfrac{(2\Delta x+\Delta x_E)\Delta x_W}{4\delta x_W \delta x_E}$	$\dfrac{\Delta x\Delta x_W}{4\delta x_E(\delta x_E+\delta x_W)}$	$\dfrac{(\Delta x+2\delta x_E)\Delta x}{4(\delta x_E+\delta x_W)\delta x_W}$
n	$\dfrac{(2\Delta y+\Delta y_S)\Delta y_N}{4\delta y_S \delta y_N}$	$\dfrac{\Delta y\Delta y_N}{4\delta y_S(\delta y_N+\delta y_S)}$	$\dfrac{(\Delta y+2\delta y_S)\Delta y}{4(\delta y_N+\delta y_S)\delta y_N}$
s	$\dfrac{(2\Delta y+\Delta y_N)\Delta y_S}{4\delta y_S \delta y_N}$	$\dfrac{\Delta y\Delta y_S}{4\delta y_N(\delta y_N+\delta y_S)}$	$\dfrac{(\Delta y+2\delta y_N)\Delta y}{4(\delta y_N+\delta y_S)\delta y_S}$

TABLE A.4 Terms Needed in the QUICK Upwind Scheme for $\dot{m}_k^{\rightarrow} < 0$

k	C_{O2}	C_{K2}	C_{KK}
e	$\dfrac{(2\Delta x_E+\Delta x_{EE})\Delta x_E}{4(\delta x_E+\delta x_{EE})\delta x_E}$	$\dfrac{(2\Delta x_E+\Delta x_{EE})\Delta x}{4\delta x_E \delta x_{EE}}$	$\dfrac{\Delta x\Delta x_E}{4\delta x_{EE}(\delta x_E+\delta x_{EE})}$
w	$\dfrac{(2\Delta x_W+\Delta x_{WW})\Delta x_W}{4(\delta x_W+\delta x_{WW})\delta x_W}$	$\dfrac{(2\Delta x_W+\Delta x_{WW})\Delta x}{4\delta x_W \delta x_{WW}}$	$\dfrac{\Delta x\Delta x_W}{4\delta x_{WW}(\delta x_W+\delta x_{WW})}$
n	$\dfrac{(2\Delta y_N+\Delta y_{NN})\Delta y_N}{4(\delta y_N+\delta y_{NN})\delta y_N}$	$\dfrac{(2\Delta y_N+\Delta y_{NN})\Delta y}{4\delta y_N \delta y_{NN}}$	$\dfrac{\Delta y\Delta y_N}{4\delta y_{NN}(\delta y_N+\delta y_{NN})}$
s	$\dfrac{(2\Delta y_S+\Delta y_{SS})\Delta y_S}{4(\delta y_S+\delta y_{SS})\delta y_S}$	$\dfrac{(2\Delta y_S+\Delta y_{SS})\Delta y}{4\delta y_S \delta y_{SS}}$	$\dfrac{\Delta y\Delta y_S}{4\delta y_{SS}(\delta y_S+\delta y_{SS})}$

terms of nodal quantities requires two approximations, one for approximating the surface integrals in terms of quadrature points distributed on the faces, and one for approximating ϕ at these points in terms of its values at the nodes, which, for a collocated grid, consist of the cell centroids. Using the face centroid as the only quadrature point makes the first approximation second order. The overall accuracy of approximating the flux term is limited by the smaller order of these two approximations.

TABLE A.5 Coefficients $c_{\mathbb{O},a}$, $c_{\mathbb{K},a}$, $c_{\mathbb{L},a}$, and $c_{\mathbb{KK},a}$ for the different schemes. $\mathcal{H} \equiv \mathcal{H}(\dot{m}_{\mathbb{k}}^{\rightarrow})$ and $\mathcal{H}^- \equiv \mathcal{H}(-\dot{m}_{\mathbb{k}}^{\rightarrow})$. Expressions for the terms within each scheme are available in the corresponding equations/tables entry

Scheme	Equations/Tables	$c_{\mathbb{O},a}$	$c_{\mathbb{K},a}$	$c_{\mathbb{L},a}$	$c_{\mathbb{KK},a}$
CDS	Equations (2.32–2.35)	$\alpha_{\mathbb{K}}$	$1 - \alpha_{\mathbb{K}}$	0	0
UDS		$\mathcal{H}$	$\mathcal{H}^-$	0	0
LUDS	Tables A.1 and A.2	$(1 - c_{\mathbb{L}})\,\mathcal{H}$	$(1 - c_{\mathbb{KK}})\,\mathcal{H}^-$	$c_{\mathbb{L}}\mathcal{H}$	$c_{\mathbb{KK}}\mathcal{H}^-$
QUICK	Tables A.3 and A.4	$c_{\mathbb{O}1}\mathcal{H} + c_{\mathbb{O}2}\mathcal{H}^-$	$c_{\mathbb{K}1}\mathcal{H} + c_{\mathbb{K}2}\mathcal{H}^-$	$c_{\mathbb{L}}\mathcal{H}$	$c_{\mathbb{KK}}\mathcal{H}^-$

Appendix B
Time Integration Schemes

Another class of numerical methods for solving the initial value problem is the *Multipoint* methods, which are based on fitting, using Lagrange polynomials, of f at a finite set of the discrete times $b\Delta t$. Multipoint methods are listed in Table B.1. Challenges include starting because of the unavailability of needed past solutions and the possibility of producing nonphysical solutions due to instability.

The Adams-Bashforth explicit scheme and the Adams-Moulton semi-implict scheme can be combined within a predictor-corrector scheme to yield a highly accurate solution. In this case, the Adams-Bashforth explicit scheme is used for the predictor to yield $\phi^*((\ell+1)\Delta t)$. The predicted solution, $\phi^*((\ell+1)\Delta t)$, is then used in the Adams-Moulton semi-implict scheme to produce the corrected or improved solution.

In contrast to multipoint methods, *Runge-Kutta* methods employ solutions at intermediate times between $t_1 = \ell\Delta t$ and $t_2 = (\ell+1)\Delta t$, which makes them more expensive due to the associated cost of computing f at these intermediate times. They are more accurate and stable than multipoint methods of the same order.

It is obvious that any of the schemes presented above can be used to numerically integrate Eq. (2.60) in time. The choice of the scheme that is most appropriate to solving the general transport equation (including the Navier-Stokes equations) is a trade off between accuracy, complexity, and cost. Explicit schemes such as the the first order Euler explicit method is the simplest to implement as it does not require matrix inversion since, for node i, the solution at time step $\ell + 1$ is a function of the value of f at the previous time step, that is,

$$\phi_i^{(\ell+1)} = \phi_i^{(\ell)} + \frac{1}{\rho_0 \mathcal{V}_i} f_i^{(\ell)} \Delta t \tag{B.1}$$

where the density is assumed to be constant. Although simple to implement and requires little memory storage, this method is conditionally stable. For the one-dimensional advection diffusion problem discretized using central difference for the advection and diffusion fluxes, the stability of the scheme requires the time step and grid cell size to satisfy

TABLE B.1 Multipoint Methods

Method	Type	$a_{\ell-2}$	$a_{\ell-1}$	a_ℓ	$a_{\ell+1}$	Order	Stability
Adams-Bashforth	explicit	0	$-\frac{1}{2}$	$\frac{3}{2}$	0	2	
Adams-Bashforth	explicit	$\frac{5}{12}$	$-\frac{16}{12}$	$\frac{23}{12}$	0	3	
Adams-Moulton	semi-implicit	0	0	$\frac{1}{2}$	$\frac{1}{2}$	2	
Adams-Moulton	semi-implicit	0	$-\frac{1}{12}$	$\frac{8}{12}$	$\frac{5}{12}$	3	

the conditions:

$$\Delta t < \frac{\rho \, \Delta x_i^2}{2\Gamma} \tag{B.2}$$

$$\Delta x_i < \frac{2\Gamma}{\rho u_i} \tag{B.3}$$

where Γ is the diffusion coefficient. The stability can be improved upon using the first order upwind scheme proposed by Courant, Friedrichs, and Lewy [8] requiring the time step to satisfy:

$$\Delta t < \frac{1}{\frac{2\Gamma}{\rho \, \Delta x_i^2} + \frac{u_i}{\Delta x_i}} \tag{B.4}$$

For convection dominated flows, $\left| \frac{u_i}{\Delta x_i} \right| >> \frac{2\Gamma}{\rho \Delta x_i^2}$ so that condition (B.4) simplifies to

$$CFL_{conv} \le 1, \text{ where } CFL_{conv} = \frac{|u_i| \, \Delta t}{\Delta x_i} \tag{B.5}$$

which requires the time step to be smaller than or equal to the advection time scale $\Delta x_i / |u_i|$. For diffusion dominated flows, condition (B.4) simplifies to

$$CFL_{diff} \le 1, \text{ where } CFL_{diff} = \frac{2\Gamma \Delta t}{\rho \Delta x_i^2} \tag{B.6}$$

which requires the time step to be smaller than or equal to the diffusion time scale $\rho \Delta x_i^2 / (2\Gamma)$. Another explicit scheme is the three-level Leapfrog method which approximates the time derivative using the second order central differentiation formula:

$$\frac{\partial \phi}{\partial t} \simeq \frac{\phi(t + \Delta t) - \phi(t - \Delta t)}{2\Delta t} \tag{B.7}$$

The method effectively uses the midpoint rule to integrate in time

$$\phi_i^{(\ell+1)} = \phi_i^{(\ell-1)} + \frac{1}{\rho_0 \mathcal{V}_i} f_i^{(\ell)} \, 2\Delta t \tag{B.8}$$

Although this scheme is second order accurate in time, it is unconditionally unstable when using Central Differencing Schemes for the advection and diffusion terms inside f.

Stability can be improved using the DuFort-Frankel method which replaces $\phi_i^{(\ell)}$ inside $f_i^{(\ell)}$ with $\frac{1}{2} \left(\phi_i^{(\ell-1)} + \phi_i^{(\ell+1)} \right)$. If the advection term is alternatively discretized using the upwind scheme, the stability condition $CFL_{conv} \le 2$ is realized by the Crank-Nicolson Scheme, which is less restricting than the first order scheme. An unconditionally stable explicit three-level scheme is the Adams-Moulton scheme which approximates the time derivative using the second order backward differentiation formula

$$\frac{\partial \phi}{\partial t} \simeq \frac{3\phi(t) - 4\phi(t - \Delta t) + \phi(t - 2\Delta t)}{2\Delta t} \Rightarrow \phi_i^{(\ell+1)} = \frac{4}{3}\phi_i^{(\ell-1)} - \frac{1}{3}\phi_i^{(\ell-2)} + \frac{1}{\rho_0 \mathcal{V}_i} f_i^{(\ell)} \left(\frac{2}{3}\Delta t \right) \tag{B.9}$$

Appendix C
Instructions to
Download ANSYS

The following steps will guide the student on the download.

1. Click **START HERE** next to **FREE STUDENT SOFTWARE**.

2. Click **DOWNLOAD NOW** under **Ansys Student**.

3. Click **DOWNLOAD ANSYS STUDENT 2022 R1**. Note that the listed number of the version will be the latest released version.

4. Right-click **ANSYSACADEMICSSTUDENT_2022R1_WINX64** in the **Downloads** directory, and select **Extract Files** to extract the compressed product files to the directory of choice. Click **Extract**.

5. Double-click the folder created **ANSYSACADEMICSTUDENT_2022R1.1_WINX64** to open it. Scroll down and right-click **setup.exe**. Select **Run as administrator**. The following message may be displayed in a **User Account Control** dialog window: **"Do you want to allow this app to make changes to your device?"**. Select **Yes**.

6. Select **I AGREE** when the license agreement **2022 R1 Product Installation - "Windows x64"** dialog box appears. Click **Next>**.

7. Use the default directory or change the directory by clicking **Browse** and selecting a directory where to install the software. Click **Next>** to start the process of installing the different systems in ANSYS.

8. A status bar at the bottom of the installation window displays the progress of the installation.

    When installation is complete, the dialog box displays **Installation Complete. Please review the above information**, the **Estimated remaining time** displays **0** and the status bar displays **100%**. Click **Exit**.

9. To launch ANSYS, go to **Start Menu** → **ANSYS 2022 R1**. A list of all the files that were installed is displayed. Scroll down to the bottom and select **Workbench 2022 R1**.

    We can alternatively launch Workbench by going to the **Start Menu** and typing `Workbench 2022 R1`.

10. Pin workbench to the taskbar for easy access by right-clicking the Workbench logo in the taskbar and selecting **Pin to taskbar**.

    Check www.ANSYS.com/academic for additional detailed instructions on the download.

Appendix D
SpaceClaim Tutorials

In this appendix, the geometry of all the Fluent models will be created using Space-Claim application. Note that the instructions provided are based on version 2022 R1 which is slightly different from previous versions of SpaceClaim.

Chapter 3: Two-Dimensional Steady State Laminar Incompressible Fluid Flow

Right-click **Geometry** and from the drop-down menu, select **New SpaceClaim Geometry...** to launch SpaceClaim. When SpaceClaim is launched, the screen looks like Figure D.1. The window where the model is built is called the design window. The design window contains the sketch grid in the 2D working plane. The mini toolbar at the bottom of the design window contains useful options. **Design1** is the default title of the design and is shown at the bottom left of the design window.

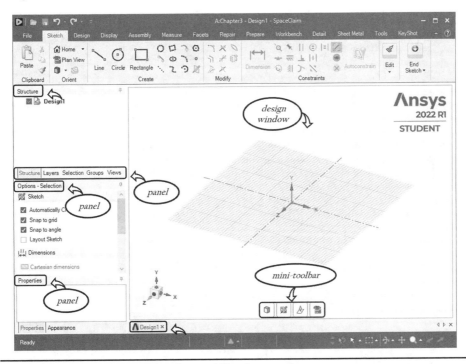

FIGURE D.1 Design window, mini-toolbar, and Structure panel.

The **Structure** panel to the left of the design window will populate the objects created by the user in the current design titled **Design1**. The user can rename the design by right-clicking **Design1** in the **Structure** tree and selecting **Rename**.

The user can display a different panel by clicking one of the four choices located below the **Structure** panel. Grouping objects by line style, color, visibility, and materials can be done in the **Layers** panel. The **Selection** panel is for selecting objects related to the one currently selected. The **Groups** panel allows the user to group different bodies, or faces. It also allows the user to create named selections on the domain's boundaries. Standard views are stored in the **Views** panel where the user can adjust the view by double-clicking one of the standard available views.

The **Options** panel is located in the middle left of the screen and allows the user to modify the functions of the SpaceClaim tools. The **Properties** panel at the bottom left displays details of the selected object where the user can make changes to its properties.

The user can create a new design by clicking **File** and selecting **New** ⇒ **Design**. The new design will appear on a tab in the design window. A design can be saved by clicking **File** and selecting **Save**. The user can close a design by clicking the **Close** button × on the design window tab bar at the bottom of the design window. A design can also be closed by clicking **File** and selecting **Close**. Make sure the design window to be closed is active before closing it.

To restore the same window layout as when SpaceClaim application is launched, click **File** and select **SpaceClaim Options**. Click **Appearance** from the **SpaceClaim Options** dialog box and click **Reset Docking Layout** as shown in Figure D.2. This may

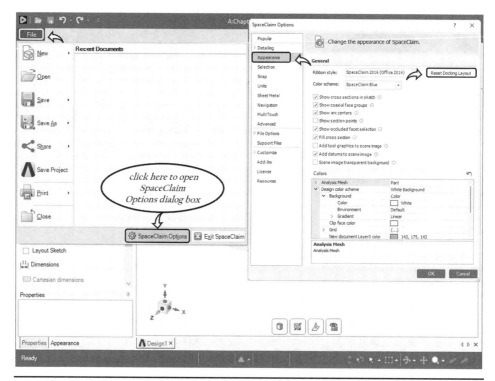

FIGURE D.2 SpaceClaim Options and Reset Docking Layout.

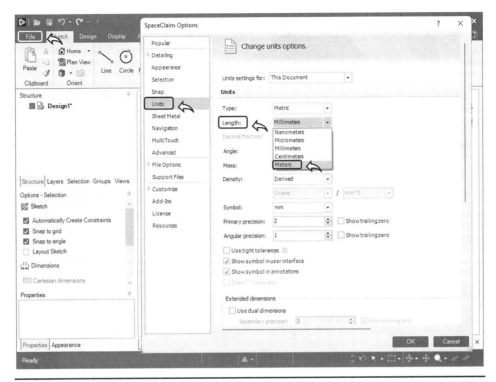

FIGURE D.3 Change units options.

be useful if the layout is accidentally modified by the user. Click **OK** to close the **Space-Claim Options** dialog box.

1. **Set the Units**

   Click **File** and select **SpaceClaim Options**. Click **Units** in the **SpaceClaim Options** dialog box as shown in Figure D.3. From the drop-down list next to **Length:**, select **Meters**.

   Click **OK** to close the **SpaceClaim Options** dialog box. We note here that when the dialog box closes, the name of the window changes from **Design1** to **Design1***. A * displayed next to **Design1** indicates the user has not saved the latest changes to the model.

2. **Create a Rectangle**

   Right-click anywhere in the design window and select **View ⇒ Front**, as shown in Figure D.4, to display the XY plane. Clicking the **Z** axis in the triad will also display the XY plane. Click **Select New Sketch Plane** in the mini-toolbar of the design window, shown in Figure D.4. Click anywhere in the design window to select the XY plane as the sketching plane.

   Click **Rectangle** in the **Create** group on the **Sketch** tab, shown in Figure D.4.

   We will select the origin as the first point of the rectangle. The pointer of the mouse will display a green circle • when we hover over the origin indicating a point is detected at the location. Notice that if we hover over the **X** axis and **Y** axis, the pointer of the mouse will display a green square ■ indicating

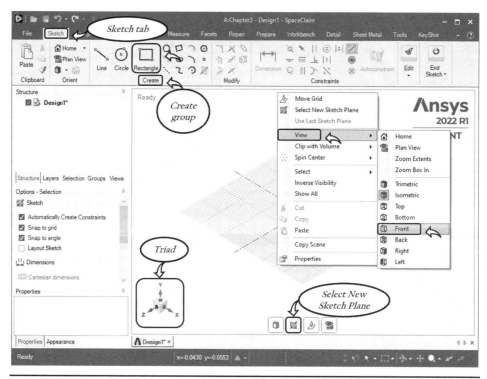

FIGURE D.4   Display Front View, Select New Sketch Plane, and create Rectangle.

the detection of a curve. Make sure the pointer displays a green circle before selecting the origin as the first point.

Drag the mouse over the sketch grid to preview the rectangle being created. Click anywhere in the first quadrant to create the rectangle. *Make sure not to draw a square as this will add constraints to the dimensions and set all sides to be equal.*

Expand **Sketching Plane1** in the **Structure** tree. **Sketch1** is the rectangle created. Expand **Sketch1** to view the four lines that make the rectangle as shown in Figure D.5.

3. **Dimension the Rectangle**

Click **Dimension** in the **Constraints** group on the **Sketch** tab, shown in Figure D.5.

Select the upper horizontal edge of the rectangle. Drag the mouse in the positive Y direction and click anywhere to place the dimension. Select the right vertical edge of the rectangle. Drag the mouse in the positive X direction and click anywhere to place the dimension. Select the numerical value of the horizontal dimension to adjust it. Type 10 and press **Enter**. Right-click anywhere in the design window and select **View** ⇒ **Zoom Extents** to display the full model. Select the numerical value of the vertical dimension and type 0.1. Press **Enter**. **Zoom Extents** is very useful and can also be used to view a specific component in our model by right-clicking any component and selecting **View** ⇒ **Zoom Extents** to zoom into the extents of the selected component.

Press **Esc** to deactivate **Dimension**.

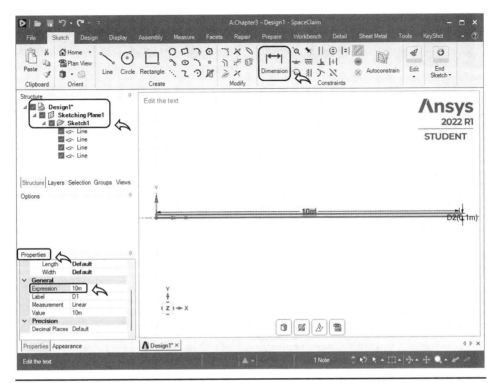

FIGURE D.5 Expand Sketching Plane1, Expand Sketch1, and Dimension.

4. **Create the Surface Inside the Rectangle**
   Click **Design** tab. Click the **3D Mode** in the **Mode** group on the **Design** tab, shown in Figure D.6, to switch from 2D sketching to 3D modeling. Once the user switches to **3D Mode**, SpaceClaim will transform all closed curves in the sketch into surfaces.

   The four lines in the rectangle are transformed into a surface named **Surface** which appears under **Design1*** in the **Structure** tree. Right-click **Surface** and select **Rename**. Type `Surface Body` and press **Enter**. Click anywhere in the design window. The screen should look like Figure D.6.

5. **Change the Material of Surface Body**
   The material of **Surface Body** is solid by default. Click **Surface Body** under **Design1*** in the **Structure** tree. See Figure D.7. Select **Unknown Material** in the **Properties** panel and type `My Fluid`. Press **Enter**. The **Fluid** cell below **Material Name** becomes active. Click **False** to activate the drop-down list. Click the arrow and from the drop-down list select **True**.

6. **Save the Project**
   Click **File** and select **Save Project** to save the Workbench project. To save the geometry as a separate SpaceClaim file, the user can click **File** and select **Save As**. Notice that in SpaceClaim application the name of the design window changes from **Design1*** to **FFF** after saving it. **FFF** stands for Fluid Flow Fluent differentiating the design that will be used by the current workspace from other created designs in SpaceClaim. Close the ANSYS SpaceClaim application

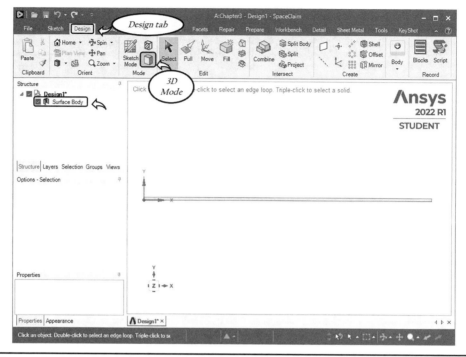

FIGURE D.6   3D Mode and renaming the surface created.

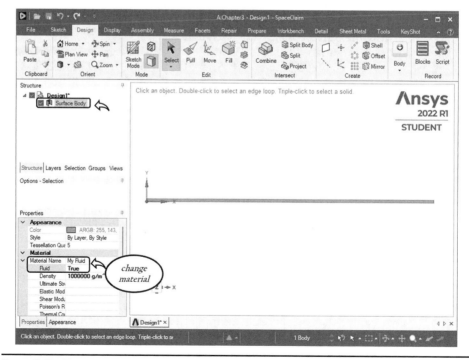

FIGURE D.7   Change the material from solid to fluid.

by clicking **File** and selecting **Exit SpaceClaim** or by clicking the **Close** button ✕ at the top right corner of the Spaceclaim window.

Chapter 4: Three-Dimensional Steady State Turbulent Incompressible Fluid Flow

Right-click **Geometry** in **Project Schematic** and from the drop-down menu, select **New SpaceClaim Geometry...** to launch SpaceClaim application.

1. **Set the Units**
   Click **File** and select **SpaceClaim Options**. Click **Units** in the **SpaceClaim Options** dialog box, and select **Meters** from the drop-down list next to **Length:**. Click **OK** to close the **SpaceClaim Options** dialog box.

2. **Create the Cylinder**
   Click **Select New Sketch Plane** in the mini-toolbar. Hover over the axes in the design window to display the sketching planes available. When we hover the mouse over an axis, a sketching plane that is normal to the axis is displayed. Click the **X** axis in the design window (*and not in the triad*), as shown in Figure D.8, to set the YZ plane as the sketching plane. The sketch grid in the YZ plane is available to start the sketch.

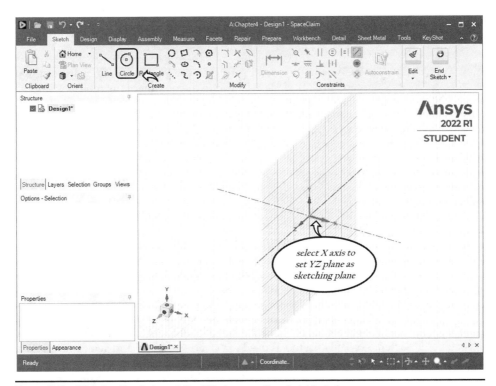

FIGURE D.8   Click the **X** axis to set the YZ plane as the sketching plane.

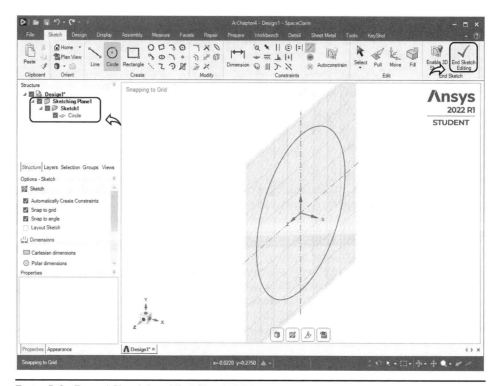

Figure D.9   Expand Sketch1 and End Sketch Editing.

Click **Circle** in the **Create** group on the **Sketch** tab. Click the origin to select it as the center of the circle. Make sure the pointer of the mouse displays a green circle before selecting the origin. Drag the mouse over the sketch grid to preview the circle being created *but don't click any point in the design window.* We will dimension the circle while sketching it. The dimension of the circle is highlighted while dragging the mouse. Type 0.2 and press **Enter** to set the diameter of the circle to 0.2 m.

Adjust the view by right-clicking anywhere in the design window and selecting **View** ⇒ **Zoom Extents**. Expand **Sketching Plane1** → **Sketch1** in the **Structure** tree to view **Circle** which is the circle created. See Figure D.9.

Hover the mouse in the design window and press **Esc** to deactivate **Circle** or click **End Sketch Editing** in the **End Sketch** group on the **Sketch** tab shown in Figure D.9. Exiting the sketch by pressing **Esc** will deactivate the operation but will allow the user to make modifications to the sketch such as changing dimensions, adding components, etc. The **End Sketch Editing** completes the sketch editing and will not allow the user to make future modifications to the sketch. Furthermore, it automatically activates the **Pull** tool in the **Design** tab.

Click **Design** tab. Click **Pull** in the **Edit** group on the **Design** tab if it is not activated. The **Pull** tool is used to extrude, revolve, and sweep faces. We will use **Pull** to extrude the circle and create a cylinder.

Notice that when **Pull** tool is activated, the circle is transformed into a surface. A message at the top of the screen will guide the user in completing the

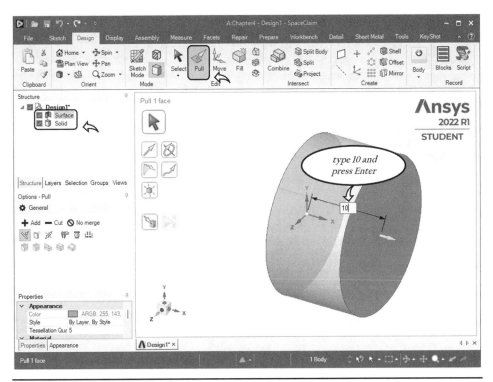

FIGURE D.10 Pull Command.

Pull command. Select the surface in the design window, or we can alternatively select **Surface** under **Design1*** in the **Structure** tree. The selected surface changes color. In the design window, press and hold the left mouse button and drag the mouse in the positive X direction to create the cylinder. It is easier to move the cursor away from the pull arrow while pulling the surface. As we are pulling the surface, **Solid** is being added to the list of **Design1*** as shown in Figure D.10. Release the cursor. Type 10 and press **Enter** to dimension the length of the cylinder.

Press **Esc** to release the surface selected for pulling. Press **Esc** one more time to deactivate the **Pull** operation.

We note here that the surface has transformed into a cylinder, and **Solid** replaced **Surface** under **Design1*** in the **Structure** tree. Right-click **Solid** and select **Rename**. Type my-domain and press **Enter**. The screen should look like Figure D.11.

Adjust the view by right-clicking anywhere in the design window and selecting **View** ⇒ **Zoom Extents**. **Zoom Extents** can also be accessed through **Orient** group. Click the arrow next to **Zoom**, shown in Figure D.11 and select **Zoom Extents**.

3. **Change the Material of My Domain**
   Click **my-domain** under **Design1*** in the **Structure** tree. Select **Unknown Material** in the **Properties** panel and type My Fluid. Press **Enter**. The **Fluid**

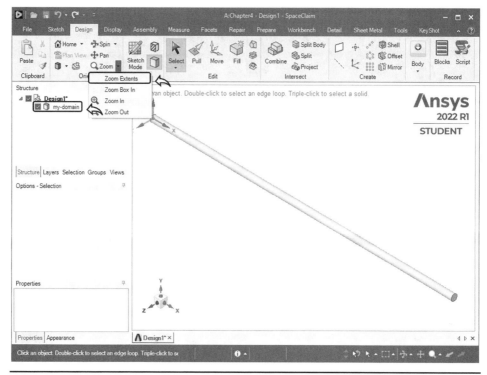

FIGURE D.11 Rename the solid created and Zoom under Orient group.

cell below **Material Name** becomes active. Click **Fluid** and select **True** from the drop-down list next to it.

4. **Create Named Selections**

Creating named selections in SpaceClaim will make it easier to mesh and apply boundary conditions to the model. Click **Groups** panel shown in Figure D.12 to activate the **Groups** panel. We will create groups and then use these groups as **Named Selections**.

Rotate the model by clicking **Spin** shown in Figure D.12 and zoom in to make it easy to select the left face (inlet) of the cylinder. Deactivate **Spin** by clicking it. Select the left face of the cylinder which will change color. Click **Create NS** shown in Figure D.12 and type `inlet` to name the group created under **Named Selections**. Press **Enter**. A message will appear when **Create NS** is clicked informing the user of the creation of a group. Adjust the view by right-clicking anywhere in the design window and selecting **View ⇒ Isometric**.

Select the curved face of the cylinder. Click **Create NS** in the **Groups** panel. Type `wall` and press **Enter**.

Select the right face (outlet) of the cylinder. Click **Create NS** in the **Groups** panel. Type `outlet` and press **Enter**.

5. **Save the Project**

Click **File** and select **Save Project** to save the Workbench project. Close the Space-Claim application by clicking **File** and selecting **Exit SpaceClaim**.

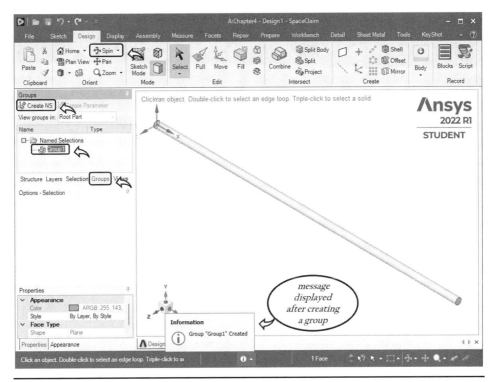

FIGURE D.12   Groups panel, Spin, and Create NS.

Chapter 5: Convection Heat Transfer for Two-Dimensional Steady State Incompressible Flow

Right-click **Geometry** in **Project Schematic** and from the drop-down menu, select **New SpaceClaim Geometry...** to launch SpaceClaim application.

1. **Set the Units**
    Click **File** and select **SpaceClaim Options**. Click **Units** in the **SpaceClaim Options** dialog box, and select **Meters** from the drop-down list next to **Length:**. Click **OK** to close the **SpaceClaim Options** dialog box.

2. **Create the Fluid Domain**
    Click the **Z** axis in the triad to display the XY plane. We can also right-click anywhere in the design window and select **View** ⇒ **Front**. Click **Select New Sketch Plane** in the mini-toolbar of the design window. Click anywhere in the design window to select the XY plane as the sketching plane.

    The **Ctrl+Z**, or the **Undo** button, shown in Figure D.13, and the **Ctrl+Y**, or the **Redo** button, also shown in Figure D.13, are available in SpaceClaim and can be very useful to undo actions performed and to correct mistakes. SpaceClaim stores all the actions performed by the user from the moment SpaceClaim application is opened. Every action is recorded and can be undone and redone.

    Click **Rectangle** in the **Create** group on the **Sketch** tab. Click the origin to select it as the first corner. Make sure the pointer of the mouse displays a green

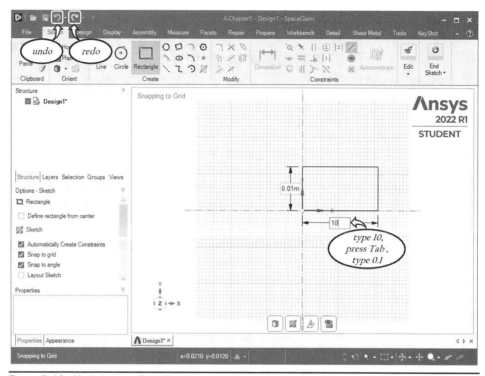

Figure D.13   Undo button, Redo button, and dimension the rectangle while sketching it.

circle before selecting the origin. Drag the mouse in the positive X and Y directions over the sketch grid to preview the rectangle being created. We will dimension the rectangle while sketching it, so *don't click to select the opposite corner in the design window*. Notice that the dimensions of the rectangle are displayed as we drag the mouse. The horizontal dimension is highlighted indicating it can be modified. Type 10 and then press the **Tab** button on the keyboard. See Figure D.13. The vertical dimension becomes highlighted. Type 0.1 and then press **Enter**.

The rectangle created is dimensioned and will be used for the fluid domain.

3. **Create the Solid Domain**

Rectangle is still activated. Let's sketch the rectangle for the solid domain. Click the top left corner of the existing rectangle as the first point. Make sure the pointer of the mouse displays a green circle before selecting the corner. Zooming in by scrolling the wheel of the mouse is helpful in selecting the top left corner of the existing rectangle.

Drag the mouse in the positive X and Y directions above the first rectangle, to preview the rectangle being created. We will dimension the rectangle while sketching it, so *don't click to select the opposite corner in the design window yet*. The dimensions of the rectangle are displayed as we drag the mouse. The horizontal dimension is highlighted. Type 10 and then press the **Tab** button on the keyboard. The vertical dimension becomes highlighted. Type 0.1 and then press **Enter**. Press **Esc** to deactivate **Rectangle**.

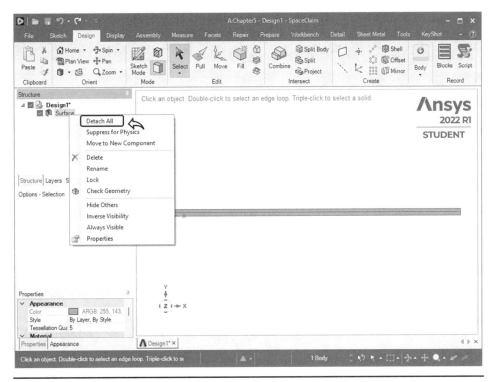

Figure D.14   Surface: Detach All.

Zoom out by scrolling the mouse wheel or right-clicking anywhere in the design window and selecting **View ⇒ Zoom Extents**.

Expand **Sketching Plane1 → Sketch1** in the **Structure** tree to view the eight lines created.

Click **Design** tab. Click **3D Mode** in the **Mode** group on the **Design** tab to switch from 2D sketching to 3D modeling. When switching to 3D Mode, Space-Claim will transform all closed curves in the sketch into surfaces. The eight lines of the rectangles are transformed into a surface named **Surface** which appears under **Design1*** in the **Structure** tree. **Surface** is made of the two rectangles combined into one surface body.

Right-click **Surface** under **Design1*** and select **Detach All**, as shown in Figure D.14, to split it into two separate surfaces based on the two rectangles created.

Two **Surface** bodies are now displayed under **Design1*** in the **Structure** tree. The first **Surface** corresponds to the fluid body and the second **Surface** corresponds to the solid body.

Right-click the first **Surface** under **Design1***. Select **Rename** and type my-fluid. Press **Enter**. Right-click **Surface**. Select **Rename** and type my-solid. Press **Enter**.

4. **Change the Material of My Fluid**
   The material of a surface body is set to solid by default in SpaceClaim; therefore, we only need to change the material of the fluid surface (**my-fluid**).

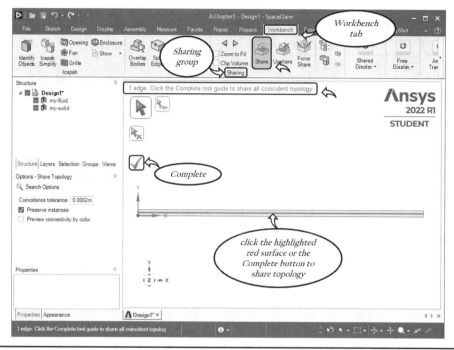

FIGURE D.15 Workbench tab and Sharing group.

Click **my-fluid** under **Design1*** in the **Structure** tree. Select **Unknown Material** in the **Properties** panel and type My Fluid. Press **Enter** The **Fluid** cell below **Material Name** becomes active. Click **Fluid** and select **True** from the drop-down list next to it.

Adjust the view by right-clicking anywhere in the design window and selecting **View** ⇒ **Zoom Extents**.

5. **Share Topolgy**

To share the interface between **my-solid** and **my-fluid**, click **Workbench** tab shown in Figure D.15. Click **Share** in the **Sharing** group on the **Workbench** tab.

A message indicating one edge is detected appears at the top of the design window. Click the **Complete** button to fix the selected problem area. When sharing is complete, a message at the bottom of the design window displays **Shared 1 edge**. Press **Esc** to deactivate **Share**.

6. **Create Named Selections**

Click **Groups** panel. Zoom in and out by using the wheel of the mouse to make it easier to select the small edges.

Select the left edge of the fluid rectangle. Click **Create NS** in the **Groups** panel and type inlet. Press **Enter**.

Select the right edge of the fluid rectangle, and click **Create NS** in the **Groups** panel. Type outlet and press **Enter**.

Repeat the process and create a named selection on each of the three sides of the solid rectangle of the model according to their locations: top-wall, right-wall, and left-wall.

Create a named selection at the bottom edge of the fluid rectangle and name it `centerline`.

7. **Save the Project**
Click **File** and select **Save Project** to save the Workbench project. Close the Space-Claim application by clicking **File** and selecting **Exit SpaceClaim**.

Chapter 6: Three-Dimensional Fluid Flow and Heat Transfer Modeling in a Heat Exchanger

Right-click **Geometry** in **Project Schematic** and from the drop-down menu, select **New SpaceClaim Geometry...** to launch SpaceClaim application.

1. **Set the Units**
Click **File** and select **SpaceClaim Options**. Click **Units** in the **SpaceClaim Options** dialog box and select **Millimeters** from the drop-down list next to **Length:**. Click **OK** to close the **SpaceClaim Options** dialog box.

2. **Create the Outer Pipe Cylinder**
Click **Select New Sketch Plane** in the mini-toolbar. Click the **Z** axis in the design window and *not in the triad*, to set the XY plane as the sketching plane. The sketch grid in the XY plane is now available to start the sketch.

Click **Circle** in the **Create** group on the **Sketch** tab. Click the origin to select it as the center of the circle. Make sure the pointer of the mouse displays a green circle before selecting the origin. Drag the mouse over the sketch grid to pre-view the circle. Type 62 and press **Enter** to set the diameter of the circle to 62 mm. Press **Esc** to deactivate **Circle**. Expand **Sketching Plane1** → **Sketch1** in the **Structure** tree to view **Circle**.

Click **Design** tab. Click **Pull** in the **Edit** group on the **Design** tab. Select **Surface** under **Design1*** in the **Structure** tree. Activate **+Add** and **Pull Both Sides** in the **Options** panel as shown in Figure D.16.

In the design window, press and hold the left mouse button while dragging the mouse to pull the surface in both directions. Release the mouse once the cylinder is created and type 602 to set the length of the cylinder. Press **Enter**.

Press **Esc** twice to deactivate the **Pull** operation.

The cylinder created is the outer pipe of the heat exchanger. It is named **Solid** and is listed under **Design1*** in the **Structure** tree. We will create the inlet and outlet of the pipe by sketching concentric circles in the XZ plane and pulling the surfaces inside the circles in opposite directions.

Click **Sketch Mode** in the **Mode** group on the **Design** tab. Click the **Y** axis in the design window and *not in the triad*, to set the XZ plane as the sketching plane. The sketching plane in the design window should look like Figure D.17.

Click **Circle** in the **Create** group on the **Sketch** tab. Click any point on the **Z** axis in the positive direction as the center of the circle. Make sure the mouse displays a green square before selecting the point. The green square means the point detected is on a curve, which is the **Z** axis in this case. Drag the mouse, type 20 and press **Enter**. Click any point on the **Z** axis in the negative direction as the center of another circle. Make sure the mouse of the circle displays a green

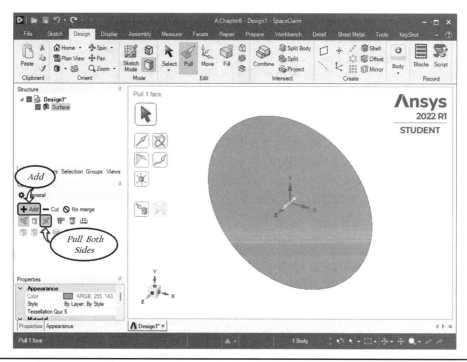

FIGURE D.16 Add and Pull Both Sides options.

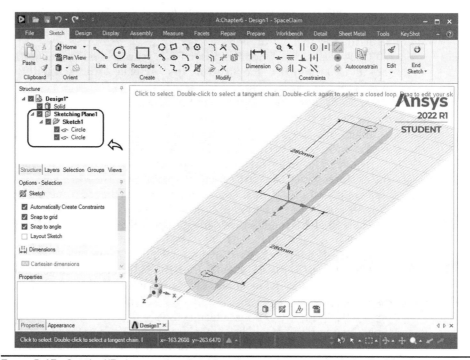

FIGURE D.17 Set the XZ plane as the sketching plane and create two circles in this plane.

square before selecting the center of the circle. Drag the mouse, type 20 and press **Enter**.

Click **Dimension** in the **Constraints** group on the **Sketch** tab.

Select the center of the first circle and then select the **X** axis. Drag the mouse and click anywhere to place the dimension. Select the numerical value of the horizontal dimension, type 280 and press **Enter**.

Repeat the process to place the center of the second circle at a distance of 280 mm from the **X** axis in the negative **Z** direction.

Press **Esc** to deactivate **Dimension**. Expand **Sketching Plane1** → **Sketch1**, in the **Structure** tree, to view the circles created as shown in Figure D.17.

Click **Design** tab. Click the **3D Mode** in the **Mode** group on the **Design** tab to switch from 2D sketching to 3D modeling. SpaceClaim will transform the circles created in the sketch into a surface named **Surface** which appears under **Design1*** in the **Structure** tree.

Right-click **Surface** under **Design1*** in the **Structure** tree and select **Detach All** to break it into two surfaces in order to select and pull the different surfaces in different directions.

Click **Pull** in the **Edit** group on the **Design** tab. Select the first **Surface** created under **Design1*** in the **Structure** tree. An arrow ready to pull the surface is displayed as shown in Figure D.18.

Activate **- Cut** in the **Options** panel shown in Figure D.18. In the design window, press and hold the left mouse button while dragging the cursor upwards in the positive Y direction to cut through the material in the outer pipe. Release

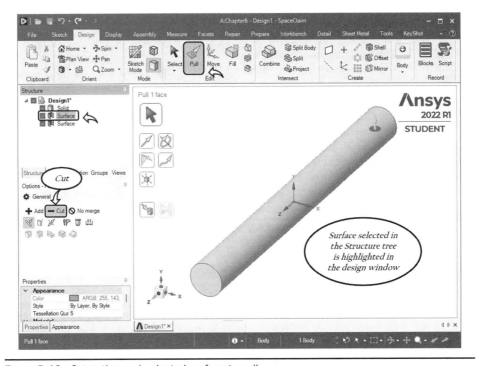

FIGURE D.18 Cut option and selected surface to pull.

the mouse once the cut is visible or when the dimension displayed while pulling exceeds **30**.

The number of **Surface** bodies listed under **Design1*** in the **Structure** tree is reduced to one.

Select **Surface** under **Design1*** in the **Structure** tree. The surface becomes highlighted in the design window. Activate **- Cut** in the **Options** panel. In the design window, press and hold the left mouse button while dragging the mouse downwards in the negative Y direction to cut through the material in the outer pipe. Release the mouse when the dimension displayed while pulling exceeds **30**.

Press **Esc** twice to deactivate the **Pull** operation.

Click **Sketch Mode** in the **Mode** group on the **Design** tab. Click the **Y** axis, in the design window and *not in the triad*, to set the XZ plane as the sketching plane.

Click **Circle** in the **Create** group on the **Sketch** tab. Click the center of the circle in the positive direction as the center of the new circle. Make sure the mouse displays a green circle before selecting the point. Drag the mouse, type 22 and press **Enter**. Create a concentric circle with a smaller diameter. Click the center of the first circle created in the XZ plane to select it as the center of the smaller circle. Make sure the pointer of the mouse displays a green circle before selecting the center. Drag the mouse, type 20 and press **Enter**.

Repeat the same process by creating two concentric circles whose centers are located on the **Z** axis in the negative direction and whose dimensions are identical to the first pair of circles created. *Note that the center of the second pair of circles is coincident with the center of the circle created in the negative Z direction.*

Click **Design** tab. Click **3D Mode** in the **Mode** group on the **Design** tab to switch from 2D sketching to 3D modeling. SpaceClaim will transform the rings created in the sketch into a surface named **Surface** which appears under **Design1*** in the **Structure** tree.

Right-click **Surface** under **Design1*** and select **Detach All** to break it into two surfaces in order to select and pull the different surfaces in different directions.

Click **Pull** in the **Edit** group on the **Design** tab.

Select the second **Surface** under **Design1*** in the **Structure** tree. This is the ring between the two concentric circles on the negative side of the **Z** axis. The surface becomes highlighted in the design window once selected. An arrow ready to pull the selected surface is also displayed. Activate **+ Add** in the **Options** panel. In the design window, press and hold the left mouse button while dragging the cursor in the positive Y direction. Release the mouse once the cylinder is created and type 50. Press **Enter**.

Adjust the view by right-clicking anywhere in the design window and selecting **View** ⇒ **Isometric**. The screen should look like Figure D.19. The inlet for the shell is now created. The number of surfaces is reduced to one under **Design1*** in the **Structure** tree.

Select **Surface** under **Design1*** in the **Structure** tree and activate **+ Add** in the **Options** panel. In the design window, press and hold the left mouse button while dragging the cursor downwards in the negative Y direction. Release the mouse once the cylinder is created. Type 50 and press **Enter**.

Press **Esc** twice to deactivate the **Pull** operation.

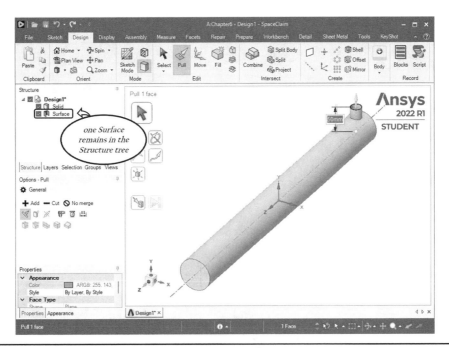

FIGURE D.19 Isometric view with created inlet for the shell.

To create a cavity in the outer pipe, we will sketch a circle of 60 mm diameter in the XY plane centered at the origin. Pulling the circle in both directions while activating **- Cut** feature will convert the outer pipe from a solid cylinder to a hollow cylinder that is 1 mm thick.

Click **Sketch Mode** in the **Mode** group on the **Design** tab. Click the **Z** axis in the design window to set the XY plane as the sketching plane. Click **Circle** in the **Create** group on the **Sketch** tab. Click the origin to select it as the center of the circle. Make sure the pointer of the mouse displays a green circle before selecting the origin. Drag the mouse. Type 60 and press **Enter**. Press **Esc** to deactivate **Circle**. Click **Design** tab. Click **Pull** in the **Edit** group on the **Design** tab. Select **Surface** in the **Structure** tree. Activate **- Cut** and **Pull Both Sides** in the **Options** panel.

In the design window, press and hold the left mouse button while dragging the mouse to pull the surface in both directions. Release the mouse at any point that belongs to the interior of the outer pipe cylinder (*the dimension displayed while dragging the mouse should not exceed **602 mm***). Type 600 and press **Enter**. Press **Esc** once to release the surface selected for pulling.

Let's blend the intersections of the inlet with the outer pipe and the intersection of the outlet with the outer pipe. The intersections to blend are a total of four edges.

Pull is still activated. Select the edge shown in Figure D.20, that is the intersection of the inlet with the outer pipe. An arrow appears next to the selected edge. Click **Selection** panel, also shown in Figure D.20. Expand **Surface holes equal to or smaller** under **Same Size**. Hold the **Ctrl** button and select the first three **Surface hole** as shown in Figure D.20. A total of four edges selected are

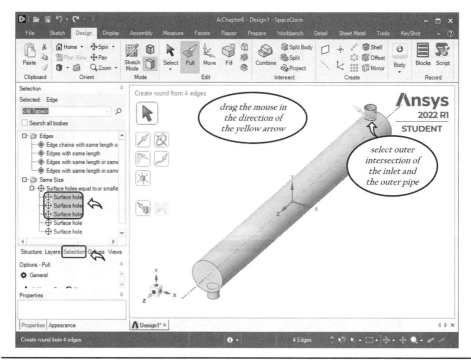

FIGURE D.20 Blend the intersections of the inlet and outlet with the pipe.

highlighted on the design window and represent the intersections to blend. In the design window, press and hold the left mouse button. Drag the mouse in the direction of the yellow arrow. Note that nothing will happen in the design window if the mouse is dragged in the opposite direction. Release the mouse, type 5 and press **Enter**.

Press **Esc** twice to deactivate the **Pull** operation. Click **Structure** panel to display the objects created. Right-click **Solid** under **Design1*** in the **Structure** tree and select **Rename**. Type my-hx and press **Enter**.

3. **Create the Inner Pipe**
   Click **Sketch Mode** in the **Mode** group on the **Design** tab. Click the **Z** axis in the design window (*not the Z axis in the triad*), to set the XY plane as the sketching plane. Click **Circle** in the **Create** group on the **Sketch** tab. Click the origin to select it as the center of the circle. Make sure the mouse displays a green circle before selecting the origin. Drag the mouse, type 28, and press **Enter**.

   Press **Esc** to deactivate **Circle**. Click **Design** tab. Click **Pull** in the **Edit** group on the **Design** tab and select the **Surface** under **Design1*** in the **Structure** tree. Activate **- Cut** and **Pull Both Sides** in the **Options** panel.

   In the design window, press and hold the left mouse button while dragging the mouse. Release the mouse once the dimension displayed on the screen exceeds **602 mm** or when the cut is visible. Press **Esc** twice to deactivate the **Pull** operation.

   Click **Sketch Mode** in the **Mode** group on the **Design** tab. Click the **Z** axis in the design window (*not the Z axis in the triad*), to set the XY plane as the sketching

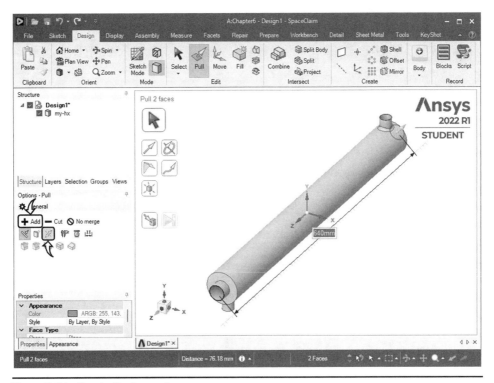

FIGURE D.21 Create the inner pipe.

plane. Click **Circle** in the **Create** group on the **Sketch** tab. Click the origin to select it as the center of the circle. Make sure the mouse displays a green circle before selecting the origin. Drag the mouse, type 30 and press **Enter**.

We will create a concentric circle with a smaller diameter while **Circle** is still activated. Click the origin and make sure the pointer of the mouse displays a green circle before selecting it. Drag the mouse, type 28, and press **Enter**.

Press **Esc** to deactivate **Circle**. Click **Design** tab.

Click **Pull** in the **Edit** group on the **Design** tab and select the **Surface** under **Design1*** in the **Structure** tree. This is the ring between the two concentric circles. Activate **+Add** and **Pull Both Sides** in the **Options** panel. In the design window, press and hold the left mouse button. Drag the mouse in any direction and release it. Type 640 and press **Enter**. The screen should look like Figure D.21.

Press **Esc** twice to deactivate the **Pull** operation.

The heat exchanger is now completed and is made of a 1 mm thick solid body with a void inside its inner pipe and a void between the inner pipe and the outer pipe. The next step is to create the geometry for the hot fluid inside the inner pipe and the geometry of the cold fluid inside the outer pipe.

4. **Create the Hot and Cold Fluids**
   To create the geometry for the hot fluid inside the inner pipe and the geometry of the cold fluid inside the outer pipe we will use the **Volume Extract** tool. The **Volume Extract** tool creates a body based on an enclosed region within a part.

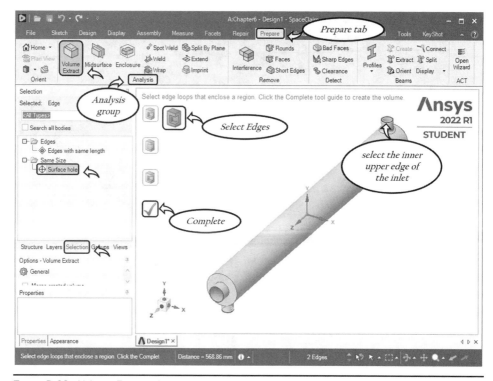

Figure D.22   Volume Extract, Select Edges, and Selection panel.

Click **Prepare** tab. Click **Volume Extract** in the **Analysis** group on the **Prepare** tab. Click **Select Edges**, shown in Figure D.22, to select the edges that loop the enclosed region.

Select the inner edge of the inlet highlighted in Figure D.22. Click **Selection** panel and select **Surface hole** under **Same Size** as shown in Figure D.22. Click **Complete** in the design window.

Alternatively, the user can select the inner edge of the inlet, rotate the model, hold the **Ctrl** button, select the inner edge of the outlet, and click **Complete**.

Click **Structure** panel and expand **Volume** under **Design1*** in the **Structure** tree. A body named **Volume** is created under **Volume** in the **Structure** tree. Right-click the body **Volume** and select **Rename**. Type my-cold-fluid and press **Enter**.

Volume Extract is still activated. Click **Select Edges** from the list of options in the design window. Hold the **Ctrl** button and select the inner edges of the inlet and the outlet of the inner pipe. Click **Complete**.

Press **Esc** to deactivate **Volume Extract**. Expand the second **Volume** under **Design1*** in the **Structure** tree. A body named **Volume** is created under the second volume. Right-click the body **Volume** and select **Rename**. Type my-hot-fluid and press **Enter**. See Figure D.23.

5. **Change Material of my-cold-fluid and my-hot-fluid**
   Click **my-cold-fluid** under **Design1*** in the **Structure** tree. Select **Unknown Material** in the **Properties** panel and type My Fluid. Press **Enter**. The **Fluid**

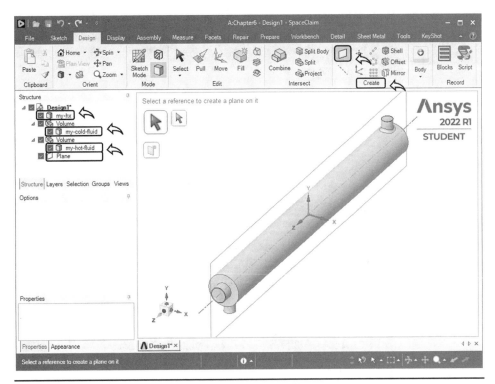

FIGURE D.23   Hot and Cold fluid bodies and Split Body by Plane.

cell below **Material Name** becomes active. Click **Fluid** and select **True** from the drop-down list next to it. Click **my-hot-fluid** under **Design1*** in the **Structure** tree. Click **Unknown Material** in the **Properties** panel and click the arrow. Select **My Fluid** from the list displayed and click **OK**. Click anywhere in the design window.

6. **Split Body**

The model exhibits symmetry around the YZ plane, and therefore we will be modeling half of the created geometry.

To split the geometry we will use the **Split Body** tool used to split a body by its faces, edges or a plane.

Click **Design** tab. Click **Plane** in the **Create** group on the **Design** tab as shown in Figure D.23. Click the **X** axis in the design window (*not the X axis in the triad*). A plane that goes through the origin and perpendicular to the **X** axis is created as shown in Figure D.23. We will use this plane as a tool to split the model.

Press **Esc** to deactivate **Plane**.

Click **Split Body** in the **Intersect** group on the **Design** tab.

Select Target is activated. Hold the **Ctrl** button and select **my-hx, my-cold-fluid** and **my-hot-fluid** under **Design1*** in the **Structure** tree, as shown in Figure D.24.

Click **Select Cutter** from the list of options in the design window, as shown in Figure D.24. Select **Plane** in the **Structure** tree.

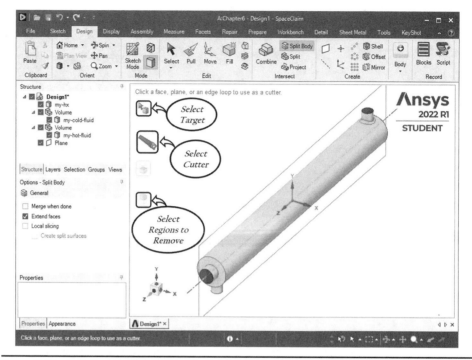

Figure D.24 Split Body options.

Each body is split into two bodies in the **Structure** tree and in the design window.

Click **Select Regions to Remove** from the list of options. Select **my-hx1, my-cold-fluid1** and **my-hot-fluid1**, under **Design1*** in the **Structure** tree. The bodies will be removed from the model as they are being selected.

Press **Esc** to deactivate **Split Body**. Disable **Plane** under **Design1*** in the **Structure** tree to hide it in the design window.

7. **Create Named Selections**

We will name all the boundaries of the heat exchanger. We will start with the inlets and outlets shown in Figure D.25.

Adjust the view by right-clicking anywhere in the design window and selecting **View ⇒ Isometric**.

Click **Groups** panel. Select the inlet of the cold fluid and click **Create NS** in the **Groups** panel. Type `inlet-cold` and press **Enter**.

Select the outlet of the hot fluid and click **Create NS** in the **Groups** panel. Type `outlet-hot` and press **Enter**.

Rotate the model to access the outlet of the cold fluid face. Create a **Named Selection** on the outlet of the cold fluid and name it `outlet-cold`.

Rotate the model to access the inlet of the hot fluid face. Create a **Named Selection** on the inlet of the hot fluid and name it `inlet-hot`.

The last named selection is the symmetry plane shown in Figure D.26. There are five faces in this plane. Hold the **Ctrl** button and select the five faces. Click **Create NS** in the **Groups** panel and type `symmetry`.

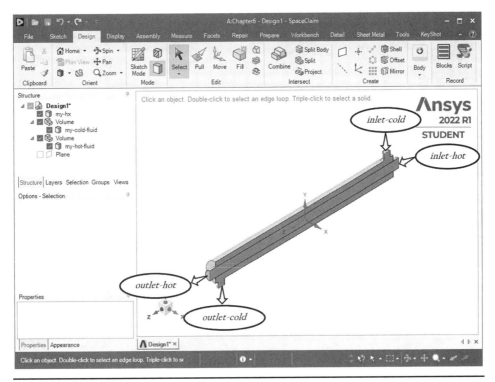

FIGURE D.25   Inlets and outlets of the heat exchanger.

Another way to select the five faces is provided in this paragraph in case the user encounters difficulty selecting them. Click the **X** axis in the triad. Click **Selection Filter** in the **Status Bar** and select **Visible** under **Box Selection** as shown in Figure D.26. Click the arrow under **Select** in the **Edit** group on the **Design** tab and select **Using Box** from the drop-down list. In the design window, press and hold the left mouse button at a point above the heat exchanger. Drag the mouse downwards and to the left. Release the mouse at a point below the heat exchanger as shown in Figure D.27. It is very important to drag the cursor in the negative horizontal direction, otherwise no selections will be made. **Visible** is selected in the selection filter, therefore only visible faces are selected. Click **Create NS** in the **Groups** panel, type `symmetry` and press **Enter**.

8. **Share Topology**
    Click **Workbench** tab. Click **Share** in the **Sharing** group on the **Workbench** tab to share the interfaces between **my-hx**, **my-cold-fluid**, and **my-hot-fluid**. A message indicating **9 faces, 34 edges. Click the Complete guide to share all coincident topology** appears on the screen. Click the **Complete** button to fix the selected problem areas. When sharing is complete, a message appears at the bottom of the design window and confirms sharing 9 faces, 34 edges. Press **Esc** to deactivate **Share**.

9. **Save the Project**
    Click **File** and select **Save Project** to save the Workbench project. Close the Space-Claim application by clicking **File** and selecting **Exit SpaceClaim**.

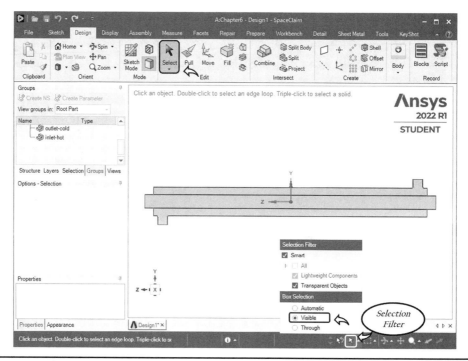

Figure D.26 Symmetry plane and Selection Filter.

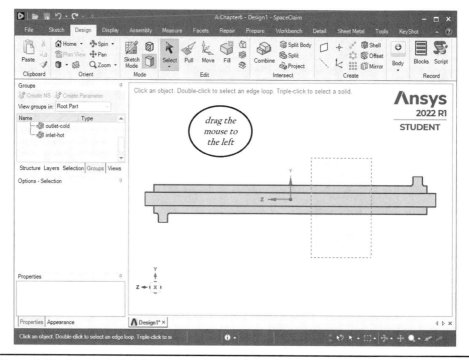

Figure D.27 Faces selected for symmetry Named Selection.

Chapter 7: Three-Dimensional Fluid Flow and Heat Transfer Modeling in a Heat Sink

Figure D.28 shows the geometry of the heat sink model. The computational domain is made of the solid heat sink surrounded by the fluid region. The base of the heat sink will be created first, followed by creating the three fins. Last, the fluid domain will be created and will be made of six regions in order to control the mesh sizing around the heat sink.

Right-click **Geometry** in **Project Schematic** and from the drop-down menu, select **New SpaceClaim Geometry...** to launch SpaceClaim application.

1. **Set the Units**
   Click **File** and select **SpaceClaim Options**. Click **Units** in the **SpaceClaim Options** dialog box, and select **Millimeters** from the drop-down list next to **Length:**. Click **OK** to close the **SpaceClaim Options** dialog box.

2. **Create the Base of the Heat Sink**
   Click **Select New Sketch Plane** in the mini-toolbar. Click the **Y** axis, in the design window *not the Y axis in the triad*, to set the ZX plane as the sketching plane. The sketch grid in the ZX plane is now available to start the sketch. Click **Y** axis in the triad to adjust the view to the ZX plane. Note that the positive Z direction is downwards.

   Click **Rectangle** in the **Create** group on the **Sketch** tab. Click the origin to select it as the first corner of the rectangle. Make sure the pointer displays a green circle before selecting the origin. Drag the mouse in the positive X direction (to the right) and the positive Z direction (downward) over the sketch grid to preview the rectangle being created. Type 20, press **Tab** to highlight the second dimension. Type 20 and press **Enter**. Press **Esc** to deactivate **Rectangle**.

   Click **Design** tab. Click **Pull** in the **Edit** group on the **Design** tab. Right-click anywhere in the design window and select **View ⇒ Isometric**. Select **Surface** under **Design1*** in the **Structure** tree. In the design window, press and hold the left-mouse button while dragging the mouse in the positive **Y** direction. Release the mouse and type 1. Press **Enter** and press **Esc** twice to deactivate **Pull**. The screen should look like Figure D.29.

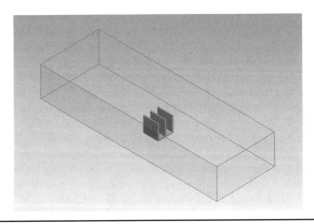

FIGURE D.28 Heat sink model.

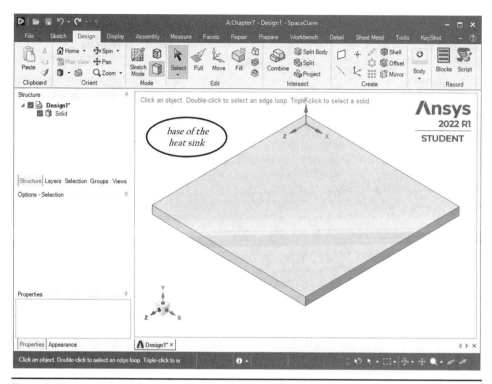

FIGURE D.29 The base of the heat sink.

3. **Create the Fins of the Heat Sink**

The fins are created by sketching three rectangles that are 1 mm wide and 20 mm long on the face of the base. The rectangles are then pulled to create the fins.

Click **Sketch Mode** in the **Mode** group on the **Design** tab. Click the upper horizontal surface of the heat sink base, shown in Figure D.30, to select it as the sketching plane.

The default origin of the sketching plane is the center of the surface selected. Click the **Y** axis in the triad to display the ZX plane. Right-click anywhere in the design window and select **View** ⇒ **Zoom Extents**.

Click **Rectangle** in the **Create** group on the **Sketch** tab. Click the lower left corner of the base to select it as the first corner of one of the three fins. Drag the mouse upwards and to the right. Type 20, press **Tab**, and type 1. Press **Enter**. Click the upper right corner of the square to select it as the first corner of another fin. Drag the mouse downwards and to the left. Type 20 and press **Tab**. Type 1 and press **Enter**. Enable **Define rectangle from center** in the **Options** panel and click the center of the square to select it as the center of the rectangle. Drag the mouse and type 20. Press **Tab**, type 1, and press **Enter**. The screen should look like Figure D.31. Press **Esc** to deactivate **Rectangle**.

Right-click anywhere in the design window and select **View** ⇒ **Isometric**. Disable **Solid** under **Design1*** in the **Structure** tree to hide it in the design window. Click **Design** tab. Click **Pull** in the **Edit** group on the **Design** tab. Select **Surface** under **Design1*** in the **Structure** tree. In the design window, press and hold the left mouse button while dragging the mouse in the positive Y direction.

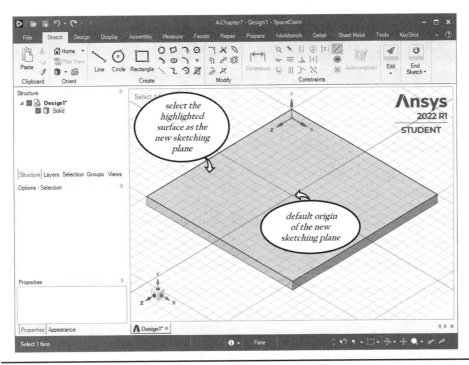

Figure D.30 Select upper surface of the heat sink as the sketching plane.

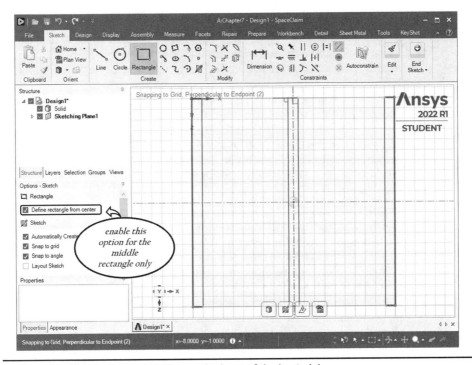

Figure D.31 Sketch three rectangles on the base of the heat sink.

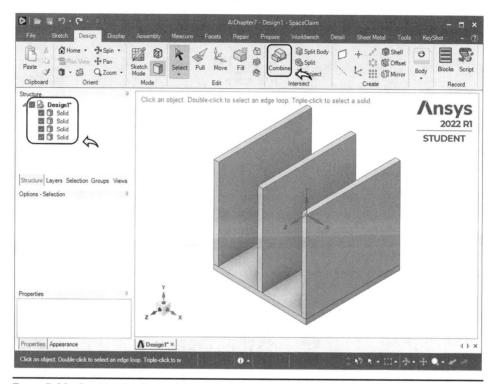

FIGURE D.32 Combine the four solids created.

Release the mouse and type 19. Press **Enter**. Press **Esc** twice to deactivate **Pull**. Right-click anywhere in the design window and select **Show All**. Notice that four **Solid** bodies are created under **Design1*** in the **Structure** tree as shown in Figure D.32.

We will combine all bodies created in the design window into one body by using the **Combine** tool. Hold the **Ctrl** button and select the four **Solid** bodies created under **Design1*** in the **Structure** tree. Click **Combine** in the **Intersect** group on the **Design** tab as shown in Figure D.32. The four bodies are now united into one body. Press **Esc** to deactivate **Combine**. Right-click **Solid** under **Design1*** in the **Structure** tree and select **Rename**. Type my-heat-sink and press **Enter**.

4. **Create Fluid Region**

The mesh in the fluid region surrounding the heat sink needs to be fine, however, the mesh away from the heat sink can be coarse to minimize the computational effort. In order to control the mesh sizing in the domain, we will divide the fluid domain into six regions that can be mesh controlled independently.

Click **Sketch Mode** in the **Mode** group on the **Design** tab. Click the **Y** axis in the design window (*not the Y axis in the triad*) to set ZX plane as the sketching plane. Recall the base of the heat sink was created in this plane and therefore should be displayed on it. Click the **Y** axis in the triad to display the ZX plane.

Click **Rectangle** in the **Create** group on the **Sketch** tab. Click any point in the quadrant located at the top left of the design window to set it as the first corner of the rectangle and any point in the quadrant located at the bottom right of the design window to set it as the second corner of the rectangle. Make sure not to draw a square and the rectangle is surrounding the heat sink.

Click **Dimension** in the **Constraints** group on the **Sketch** tab. Select the upper horizontal edge of the fluid rectangle. Drag the mouse in the negative Z direction (upwards) and click anywhere to place the dimension. Select the numerical value of the horizontal dimension, type 480, and press **Enter**. Select the left vertical edge of the fluid rectangle. Drag the mouse in the negative X direction (to the left) and click anywhere to place the dimension. Select the numerical value of the vertical dimension, type 220, and press **Enter**. Right-click anywhere in the design window and select **View** ⇒ **Zoom Extents**.

We will center the heat sink inside the rectangle around the **Z** axis by setting the distance between the right edge of the heat sink and the right side of rectangle equal to the distance between the left edge of the heat sink and the left side of the rectangle.

Click **Equal Distance Constraint** in the **Constraints** group on the **Sketch** tab. The message displayed at the top left of the design window guides the user through the selection of the edges. Zooming in and out can make it easier while selecting the edges.

The first pair of lines are named 1 and 1' as shown in Figure D.33. The second pair of lines are named 2 and 2'. Select the left edge of the fluid rectangle (1) as the first line of the first pair of **Equal Distance Constraint**. Select the **Z** axis (1') as the second line of the first pair of **Equal Distance Constraint**.

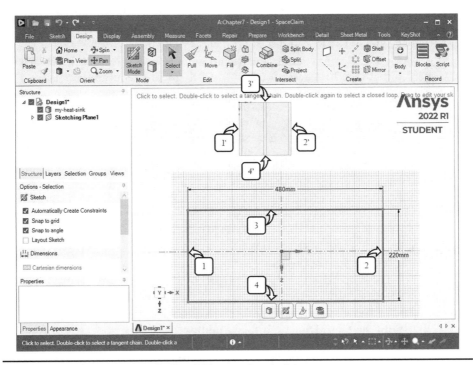

FIGURE D.33   Pairing the edges for Equal Distance Constraint.

Select the right edge of the fluid rectangle (2) as the first line of the second pair of the **Equal Distance Constraint**. Select the right edge (2′) of the heat sink as the second line of the second pair of the **Equal Distance Constraint**.

The fluid domain is adjusted in the design window such that the heat sink appears centered inside the fluid domain around the **Z** axis. We will now center the heat sink around the **X** axis by setting the distance between the top edge of the heat sink and the top side of the rectangle equal to the distance between the bottom edge of the heat sink and the bottom side of the rectangle. The first pair of lines to be selected are named 3 and 3′ as shown in Figure D.33. The second pair of lines are named 4 and 4′.

Select the top edge of the fluid rectangle (3) as the first line of the first pair for **Equal Distance Constraint**. Select the top edge of the heat sink (3′) as the second line for **Equal Distance Constraint**.

Select the bottom edge of the fluid rectangle (4) as the first line of the second pair and the **X** axis as the second line of the second pair for equal distance constraints.

The fluid rectangle is adjusted in the design window such that the heat sink is at its center. When completed, the screen should look like Figure D.33.

Disable **my-heat-sink** under **Design1*** in the **Structure** tree to hide it. Right-click anywhere in the design window and select **View ⇒ Isometric**.

Click **Design** tab. Click **Pull** in the **Edit** group on the **Design** tab. Select **Surface** under **Design1*** in the **Structure** tree. Click **No merge** in the **Options** panel. In the design window, press and hold the left-mouse button while dragging the mouse in the positive **Y** direction. Release the mouse, type 80 and press **Enter**.

Press **Esc** twice to deactivate **Pull**. Right-click **Solid** under **Design1*** in the **Structure** tree and select **Rename**. Type my-fluid and press **Enter**.

5. **Subtract the Heat Sink from the Fluid Body**
Right-click anywhere in the design window and select **Show All**. Click **Combine** in the **Intersect** group on the **Design** tab. **Combine** tool is used to combine or split objects. We will use it to split the fluid domain using the heat sink as a cutter. Activate **Select Target** in the list of options in the design window, shown in Figure D.34, if it is not activated. Select **my-fluid** under **Design1*** in the **Structure** tree as the target object to split. Activate **Select Cutter** in the list of options in the design window if it is not activated and select **my-heat-sink** under **Design1*** in the **Structure** tree. A new body named **Solid** appears under **Design1*** in the **Structure** tree, and this body is a duplicate of the heat sink. Activate **Select Regions to Remove** in the list of options in the design window and select **Solid** under **Design1*** in the **Structure** tree. Hover the mouse over the design window and Press **Esc** to deactivate the **Combine** tool.

6. **Change the Material of my-fluid**
Click **my-fluid** under **Design1*** in the **Structure** tree. Select **Unknown Material** in the **Properties** panel and type My Fluid. Press **Enter**. The **Fluid** cell below **Material Name** becomes active. Click **Fluid** and select **True** from the drop-down list next to it.

7. **Split the Fluid Domain into Six Bodies**
The mesh in the fluid domain surrounding the heat sink needs to be fine. However, the mesh away from the heat sink can be coarse to minimize the

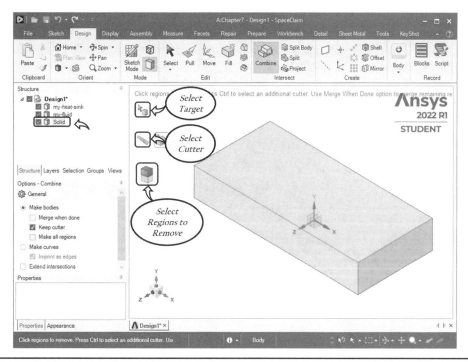

FIGURE D.34 Combine tool options.

computational effort. In order to control the mesh sizing in the fluid domain, we will split it into six bodies that can be mesh controlled independently. In this section, we will create two YZ planes to split the fluid domain into three bodies. Then, we will create two XY planes to split the middle body surrounding the heat sink into three bodies. Lastly, we will create an XZ plane to split the center body surrounding the heat sink into two bodies.

Click **Plane** in the **Create** group on the **Design** tab. Click the **X** axis in the design window and *not the triad* to create a YZ plane. Press **Esc** to deactivate **Plane**.

Click **Move** in the **Edit** group on the **Design** tab. Select **Plane** under **Design1*** in the **Structure** tree. A handle appears on the selected plane. Click the arrow that is normal to the selected plane (blue arrow) as shown in Figure D.35 to move the plane in the direction of the arrow. In the design window, press and hold the left mouse button while dragging the mouse in the negative X direction (to the left). Release the mouse and type 3. Press **Enter**. Press **Esc** twice to deactivate the **Move** operation.

Plane is a plane perpendicular to the **X** axis located at x = −3 mm. We will use it to slice **my-fluid** into two bodies.

Click **Split Body** in the **Intersect** group on the **Design** tab. Activate **Select Target** in the list of options if it is not activated and select **my-fluid** under **Design1*** in the **Structure** tree. Activate **Select Cutter** in the list of options if it is not activated and select **Plane** under **Design1*** in the **Structure** tree. Hover the mouse in the design window and press **Esc** to deactivate **Split Body**.

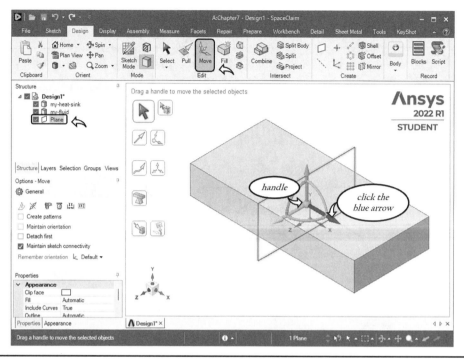

FIGURE D.35 Move tool options.

Right-click **my-fluid** under **Design1*** in the **Structure** tree and select **Rename**. Type `my-fluid-domain-upstream` and press **Enter**. Right-click **my-fluid1** under **Design1*** in the **Structure** tree and select **Rename**. Type `my-fluid` and press **Enter**.

Right-click **Plane** under **Design1*** in the **Structure** tree and select **Delete**.

We will repeat the process to split the downstream fluid body into two bodies. Click **Plane** in the **Create** group on the **Design** tab. Click the **X** axis in the design window and *not the triad* to create a YZ plane. Press **Esc** to deactivate **Plane**.

Click **Move** in the **Edit** group on the **Design** tab. Select **Plane** under **Design1*** in the **Structure** tree. A handle appears on the selected plane. Click the arrow that is normal to the selected plane (blue arrow) to move the plane in the direction of the arrow. In the design window, press and hold the left mouse button while dragging the mouse in the positive X direction. Release the mouse and type `23`. Press **Enter** and press **Esc** twice to deactivate the **Move** operation.

Click **Split Body** in the **Intersect** group on the **Design** tab. Activate **Select Target** in the list of options if it is not activated and select **my-fluid** under **Design1*** in the **Structure** tree. Activate **Select Cutter** in the list of options if it is not activated and select **Plane** under **Design1*** in the **Structure** tree. Hover the mouse over the design window and press **Esc** to deactivate **Split Body**.

Right-click **my-fluid1** under **Design1*** in the **Structure** tree and select **Rename**. Type `my-fluid-domain-downstream` and press **Enter**.

Right-click **Plane** under **Design1*** in the **Structure** tree and select **Delete**.

We will now split the middle fluid region into three bodies. Click **Plane** in the **Create** group on the **Design** tab. Click the **Z** axis in the design window and *not the triad* to create an XY plane. Press **Esc** to deactivate **Plane**.

Click **Move** in the **Edit** group on the **Design** tab. Select **Plane** under **Design1*** in the **Structure** tree. A handle appears on the selected plane. Click the arrow that is normal to the selected plane (blue arrow) to move the plane in the direction of the arrow. In the design window, press and hold the left mouse button while dragging the mouse in the negative Z direction. Release the mouse and type 3. Press **Enter** and press **Esc** twice to deactivate the **Move** operation.

Click **Split Body** in the **Intersect** group on the **Design** tab. Activate **Select Target** in the list of options if it is not ativated and select **my-fluid** under **Design1*** in the **Structure** tree. Activate **Select Cutter** in the list of options if it is not activated and select **Plane** under **Design1*** in the **Structure** tree. Hover the mouse over the design window and press **Esc** to deactivate **Split Body**.

Right-click **my-fluid** under **Design1*** in the **Structure** tree and select **Rename**. Type my-fluid-domain-right and press **Enter**. Right-click **my-fluid1** under **Design1*** in the **Structure** tree and select **Rename**. Type my-fluid and press **Enter**.

Right-click **Plane** under **Design1*** in the **Structure** tree and select **Delete**.

We will repeat the process to split the left middle fluid body into two bodies. Click **Plane** in the **Create** group on the **Design** tab. Click the **Z** axis in the design window and *not the triad* to create an XY plane. Press **Esc** to deactivate **Plane**.

Click **Move** in the **Edit** group on the **Design** tab. Select **Plane** under **Design1*** in the **Structure** tree. A handle appears on the selected plane. Click the arrow that is normal to the selected plane (blue arrow) to move the plane in the direction of the arrow. In the design window, press and hold the left mouse button while dragging the mouse in the positive Z direction. Release the mouse and type 23. Press **Enter** and press **Esc** twice to deactivate the **Move** operation.

Click **Split Body** in the **Intersect** group on the **Design** tab. Activate **Select Target** in the list of options if it is not activated and select **my-fluid** under **Design1*** in the **Structure** tree. Activate **Select Cutter** in the list of options if it is not activated and select **Plane** under **Design1*** in the **Structure** tree. Hover the mouse in the design window and press **Esc** to deactivate **Split Body**.

Right-click **my-fluid1** under **Design1*** in the **Structure** tree and select **Rename**. Type my-fluid-domain-left and press **Enter**.

Right-click **Plane** under **Design1*** in the **Structure** tree and select **Delete**.

Click **Plane** in the **Create** group on the **Design** tab. Click the **Y** axis in the design window and *not the triad* to create a ZX plane. Press **Esc** to deactivate **Plane**.

Click **Move** in the **Edit** group on the **Design** tab. Select **Plane** under **Design1*** in the **Structure** tree. A handle appears on the selected plane. Click the arrow that is normal to the selected plane (blue arrow) to move the plane in the direction of the arrow. In the design window, press and hold the left mouse button while dragging the mouse in the positive Y direction. Release the mouse and type 23. Press **Enter** and press **Esc** twice to deactivate the **Move** operation.

Click **Split Body** in the **Intersect** group on the **Design** tab. Activate **Select Target** in the list of options if it is not activated and select **my-fluid** under **Design1*** in the **Structure** tree. Activate **Select Cutter** in the list of options if it is not activated and select **Plane** under **Design1*** in the **Structure** tree. Hover the mouse in the design window and press **Esc** to deactivate **Split Body**.

Right-click **my-fluid1** under **Design1*** in the **Structure** tree and select **Rename**. Type my-fluid-domain-above-hs and press **Enter**. Right-click **my-fluid** under **Design1*** in the **Structure** tree and select **Rename**. Type my-fluid-domain-hs and press **Enter**.

Right-click **Plane** under **Design1*** in the **Structure** tree and select **Delete**.

8. **Share Topology**

   Click **Workbench** tab. Click **Share** in the **Sharing** group on the **Workbench** tab to share the interfaces between the different bodies created. A message indicating **26 faces, 68 edges** appears on the screen. Click the Complete tool guide to share all coincident topology. Click the **Complete** button to fix the selected problem areas. When sharing is complete, a message appears at the bottom of the design window to confirm sharing 26 faces, 68 edges. Press **Esc** to deactivate **Share**.

9. **Create Named Selections**

   Disable my-heat-sink under **Design1*** in the **Structure** tree to hide it. Click **Groups** panel. Rotate the model and select the inlet face of the fluid domain as shown in Figure D.36. Click **Create NS** in the **Groups** panel. Type inlet and press **Enter**.

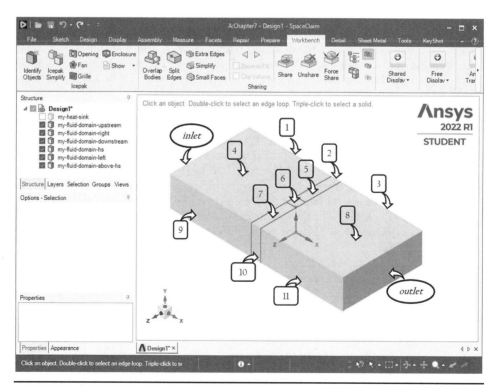

FIGURE D.36 Inlet, outlet, and wall named selection.

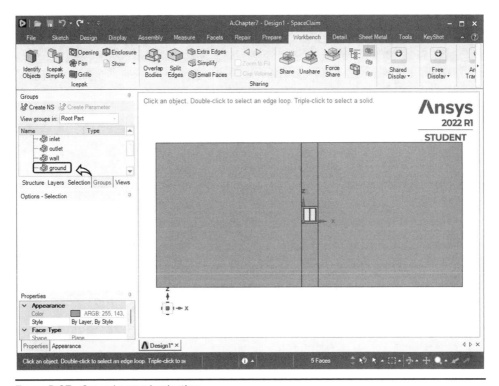

Figure D.37 Ground named selection.

Adjust the view and select the outlet face of the fluid domain shown in Figure D.36. Click **Create NS** in the **Groups** panel. Type `outlet` and press **Enter**.

Hold the **Ctrl** button and select all the faces on the right side, the left side and the top of the rectangular fluid domain for a total of 11 faces. The faces are numbered 1 through 11 in Figure D.36, but the order of selecting them is random. Click **Create NS** in the **Groups** panel. Type `wall` and press **Enter**.

Right-click anywhere in the design window and select **View** ⇒ **Bottom**. Hold the **Ctrl** button and select all the faces on the bottom of the rectangular domain for a total of 5 faces as shown in Figure D.37. Click **Create NS** in the **Groups** panel. Type `ground` and press **Enter**.

Right-click anywhere in the design window and select **View** ⇒ **Isometric**. Right-click anywhere in the design window and select **Show All**. Click **Structure** panel. Right-click **my-heat-sink** under **Design1*** in the **Structure** tree and select **Hide Others**.

Click **Groups** panel. Right-click anywhere in the design window and select **View** ⇒ **Bottom**. Right-click anywhere in the design window and **Select** ⇒ **Select All**. Hold the **Ctrl** button and click the bottom surface of the heat sink to deselect it as shown in Figure D.38. Click **Create NS** in the **Groups** panel. Type `fins` and press **Enter**.

Select the bottom face of the heat sink and click **Create NS** in the **Groups** panel. Type `heat-source` and press **Enter**.

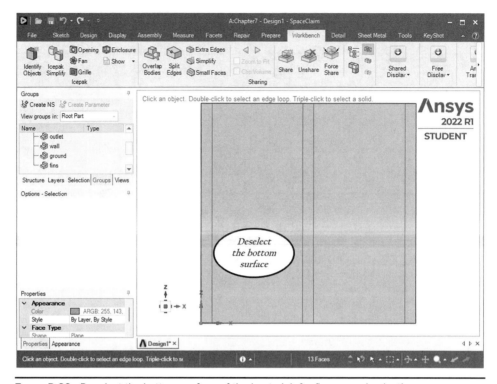

FIGURE D.38 Deselect the bottom surface of the heat sink for fins named selection.

10. **Save the Project**
    Click **File** and select **Save Project** to save the Workbench project. Close the Space-Claim application by clicking **File** and selecting **Exit SpaceClaim**.

Index

Note: Page numbers with *f* or *t* refer to figures or tables.

G